Lecture Notes in Computer Science   11328

Commenced Publication in 1973
Founding and Former Series Editors:
Gerhard Goos, Juris Hartmanis, and Jan van Leeuwen

More information about this series at http://www.springer.com/series/7407

Angelo Sifaleras · Said Salhi ·
Jack Brimberg (Eds.)

Variable Neighborhood Search

6th International Conference, ICVNS 2018
Sithonia, Greece, October 4–7, 2018
Revised Selected Papers

Editors
Angelo Sifaleras
University of Macedonia
Thessaloniki, Greece

Said Salhi
University of Kent
Canterbury, UK

Jack Brimberg
Royal Military College of Canada
Kingston, ON, Canada

ISSN 0302-9743   ISSN 1611-3349   (electronic)
Lecture Notes in Computer Science
ISBN 978-3-030-15842-2   ISBN 978-3-030-15843-9   (eBook)
https://doi.org/10.1007/978-3-030-15843-9

Library of Congress Control Number: 2019934739

LNCS Sublibrary: SL1 – Theoretical Computer Science and General Issues

This Springer imprint is published by the registered company Springer Nature Switzerland AG
The registered company address is: Gewerbestrasse 11, 6330 Cham, Switzerland

Preface

This volume edited by Angelo Sifaleras, Said Salhi, and Jack Brimberg contains peer-reviewed papers from the 6th International Conference on Variable Neighborhood Search (ICVNS 2018) held in Sithonia, Halkidiki, Greece, during October 4–7, 2018.

The conference follows previous successful meetings that were held in Puerto de La Cruz, Tenerife, Spain (2005); Herceg Novi, Montenegro (2012); Djerba, Tunisia (2014); Malaga, Spain (2016); Ouro Preto, Brazil, (2017). This edition was organized by Angelo Sifaleras, from the University of Macedonia (Greece), who was the conference chair, Nenad Mladenović, from the Mathematical Institute, of the Serbian Academy of Sciences and Arts (Serbia), who was the general chair, and Pierre Hansen, from GERAD and HEC Montreal (Canada), who was the honorary chair.

Like its predecessors, the main goal of ICVNS 2018 was to provide a stimulating environment in which researchers coming from various scientific fields can share and discuss their knowledge, expertise, and ideas related to the VNS metaheuristic and its applications. The location of ICVNS 2018 in Porto Carras Meliton Hotel, Sithonia, Greece, allowed us to combine academic presentations and social networking.

The following three plenary lecturers shared their current research directions with the ICVNS 2018 participants:

- Panos M. Pardalos, from the Center for Applied Optimization, Department of Industrial and Systems Engineering, of the University of Florida, USA, "On VNS for Hard Optimization Problems and the Power of Heuristics"
- Abraham Duarte, from the Department of Computer Sciences, of the Universidad Rey Juan Carlos, Spain, "Multi-objective VNS"
- Daniel Aloise, from the GERAD and Department of Computer Engineering, of Polytechnique Montréal, Canada, "Clustering and Variable Neighborhood Search: A Love Story"

Around 50 participants took part in the ICVNS 2018 conference and a total of 37 papers were accepted for oral presentation. A total of 23 long papers were accepted for publication in this LNCS volume after thorough peer reviewing by the members of the ICVNS 2018 Program Committee. These papers describe recent advances in methods and applications of variable neighborhood search.

The editors thank all the participants in the conference for their contributions and for their continuous effort to disseminate VNS and are grateful to the reviewers for preparing excellent reports. The editors wish to acknowledge the Springer LNCS editorial staff for their support during the entire process of making this volume. Finally, we express our gratitude to the organizers and sponsors of the ICVNS 2018 meeting:

- The Research Committee of the University of Macedonia
- The Computational Methodologies and Operations Research (CMOR Lab)
- The EURO Working Group on Metaheuristics (EWG EU/ME)
- The GERAD Group for Research in Decision Analysis

- The DIMOULAS Special Cables S.A.
- The Marathon Data Systems
- The TZIOLA Publications
- The Museum for the Macedonian Struggle, in Greece

Their support is greatly appreciated for making ICVNS 2018 a great scientific event.

February 2019

Angelo Sifaleras
Said Salhi
Jack Brimberg

Organization

Honorary Chair

Pierre Hansen — GERAD and HEC Montreal, Canada

General Chair

Nenad Mladenović — Mathematical Institute, Serbian Academy of Sciences and Arts, Serbia

Conference Chair

Angelo Sifaleras — University of Macedonia, Greece

Program Chairs

Angelo Sifaleras — University of Macedonia, Greece
Said Salhi — University of Kent, UK
Jack Brimberg — Royal Military College of Canada

Program Committee

Daniel Aloise — GERAD, Montréal, Canada
Ada Alvarez — Universidad Autonoma de Nuevo Leon, Mexico
John Beasley — Brunel University, UK
Jack Brimberg — Royal Military College of Canada
Mirjana Cangalović — University of Belgrade, Serbia
Gilles Caporossi — HEC Montréal, Canada
Emilio Carrizosa — University of Seville, Spain
Sergio Consoli — European Commission, Italy
Teodor Gabriel Crainic — CIRRELT, Canada
Dragoš Cvetković — Mathematical Institute, Serbian Academy of Sciences and Arts, Serbia
Tatjana Davidović — Serbian Academy of Sciences and Arts, Serbia
Abraham Duarte — Universidad Rey Juan Carlos, Spain
Karl Dörner — University of Vienna, Austria
Anton Eremeev — Sobolev Institute of Mathematics, Russia
Laureano Escudero — Universidad Rey Juan Carlos, Spain
Michel Gendreau — Polytechnique Montréal, Canada
Andreas C. Georgiou — University of Macedonia, Greece
Said Hanafi — University of Valenciennes, France
Pierre Hansen — GERAD and HEC Montreal, Canada

Richard Hartl	University of Vienna, Austria
Dimitrios Hristu-Varsakelis	University of Macedonia, Greece
Chandra Irawan	Nottingham University Business School, China
Bassem Jarboui	Emirates College of Technology, United Arab Emirates
Konstantinos Kaparis	University of Macedonia, Greece
Yury Kochetov	Sobolev Institute of Mathematics, Novosibirsk, Russia
Ioannis Konstantaras	University of Macedonia, Greece
Vera Kovačević-Vujčić	University of Belgrade, Serbia
Leo Liberti	CNRS LIX Ecole Polytechnique, France
Igor Machado Coelho	Universidade do Estado do Rio de Janeiro, Brazil
Yannis Marinakis	Technical University of Crete, Greece
Belén Melián-Batista	University La Laguna, Spain
Athanasios Migdalas	Lulea University of Technology, Sweden
Nenad Mladenović	Mathematical Institute, Serbian Academy of Sciences and Arts, Serbia
José Andrés Moreno Pérez	Universidad de La Laguna, Spain
Vitor Nazário Coelho	Universidade Federal Fluminense, Brazil
Luiz Satoru Ochi	Fluminense Federal University, Brazil
Joaquín Pacheco	University of Burgos, Spain
Eduardo G. Pardo	Universidad Politécnica de Madrid, Spain
Leonidas Pitsoulis	Aristotle University of Thessaloniki, Greece
Justo Puerto	University of Seville, Spain
Günther Raidl	TU Wien, Austria
Ioannis Refanidis	University of Macedonia, Greece
Celso Ribeiro	Universidade Federal Fluminense, Brazil
Said Salhi	University of Kent, UK
Nikolaos Samaras	University of Macedonia, Greece
Haroldo Santos	Universidade Federal de Ouro Preto, Brazil
Marc Sevaux	Université de Bretagne-Sud, France
Patrick Siarry	Université Paris-Est Creteil, France
Angelo Sifaleras	University of Macedonia, Greece
Marcone Souza	Federal University of Ouro Preto, Brazil
Thomas Stützle	Université libre de Bruxelles (ULB), Belgium
Kenneth Sörensen	University of Antwerp, Belgium
Christos D. Tarantilis	Athens University of Economics, Greece
Raca Todosijević	University of Valenciennes, France
Dragan Urošević	Mathematical Institute, Serbian Academy of Sciences and Arts, Serbia

Contents

Improved Variable Neighbourhood Search Heuristic for Quartet Clustering

Sergio Consoli[1,2]($\boxtimes$), Jan Korst[2], Steffen Pauws[2,3], and Gijs Geleijnse[2,4]

[1] European Commission, Joint Research Centre, Directorate A-Strategy,
Work Programme and Resources, Scientific Development Unit,
Via E. Fermi 2749, 21027 Ispra, VA, Italy
sergio.consoli@ec.europa.eu
[2] Philips Research,
High Tech Campus 34, 5656 AE Eindhoven, The Netherlands
[3] TiCC, Tilburg University,
Warandelaan 2, 5037 AB Tilburg, The Netherlands
[4] Netherlands Comprehensive Cancer Organisation (IKNL),
Zernikestraat 29, 5612 HZ Eindhoven, The Netherlands

Abstract. Given a set of n data objects and their pairwise dissimilarities, the goal of quartet clustering is to construct an optimal tree from the total number of possible combinations of quartet topologies on n, where optimality means that the sum of the dissimilarities of the embedded (or consistent) quartet topologies is minimal. This corresponds to an NP-hard combinatorial optimization problem, also referred to as minimum quartet tree cost (MQTC) problem. We provide details and formulation of this challenging problem, and propose a basic greedy heuristic that is characterized by a very high speed and some interesting implementation details. The solution approach, though simple, substantially improves the performance of a Reduced Variable Neighborhood Search for the MQTC problem. The latter is one of the most popular heuristic algorithms for tackling the MQTC problem.

Keywords: Combinatorial optimization · Quartet trees ·
Hierarchical clustering · Metaheuristics ·
Variable Neighbourhood Search · Graph theory

1 Introduction

Quartet clustering methods are popular in computational biology, where dendrograms (or phylogenies) are ubiquitous. These methods aim at reconstructing a rooted dendrogram from a set of pairwise distant objects (or taxa). Given a set of objects, define Q to be the set of all the possible quartets, and Q_t to be the set of consistent quartets being embedded in a dendrogram t. The problem of recombining the quartet topologies of Q to form an estimate of the correct tree diagram can be naturally formulated as an optimization problem. Steel [19]

© The Author(s) 2019
A. Sifaleras et al. (Eds.): ICVNS 2018, LNCS 11328, pp. 1–12, 2019.
https://doi.org/10.1007/978-3-030-15843-9_1

formulated the *maximum quartet consistency (MQC) problem*, which looks for a dendrogram tree t maximizing the number of consistent quartets Q_t belonging to a subset $P \subseteq Q$ of quartet topologies. This problem has been shown to be NP-hard [19], and Jiang et al. [15] proved that the problem admits a polynomial time approximation scheme by using the technique of smooth integer polynomial programming and by exploiting the natural denseness of the set Q. However, this scheme only guarantees a dendrogram that may deviate from Q by εn^4 quartet topologies for any small constant $\varepsilon > 0$, where n is the number of taxa.

Due to these results, most quartet methods are heuristics which attempt to solve the MQC problem, or some variants of the MQC problem with weaker optimization requirements. Strimmer and von Haeseler [20] formulated the *quartet tree-puzzling problem*, which is a variant of the MQC problem where each quartet is provided with a probability value to be embedded, and for each set of four objects the quartet with the highest probability is selected (at random in case of ties) to form a "maximum-likelihood dendrogram". Felsenstein [11] presented a heuristic which solves the MQC by incrementally growing the tree diagram in random order by stepwise addition of objects in the local optimal way. This procedure is repeated iteratively for different object orders, adding agreement values on the branches of the tree. Both agglomerative approaches are quite fast, but suffer from the usual bottom-up problem: a wrong decision early on cannot be corrected later. Berry et al. [1] reported an interesting result. They presented two "quartet cleaning" algorithms for correcting bounded numbers of quartet errors (i.e. incorrect inferences of simple quartet topologies) for many popular quartet problems.

Cilibrasi and Vitányi [2] proposed for the first time the minimum quartet tree cost (MQTC) problem. Given a set N of $n \geq 4$ objects, the MQTC deals with a *full unrooted binary tree* with n leaves, a special topology dendrogram having all internal nodes connected exactly with three other nodes, the n objects assigned as leaf nodes, and without any distinction between parent and child nodes [12]. A full unrooted binary tree with $n \geq 4$ leaves has exactly $n - 2$ internal nodes, and consequently has a total of $2n - 2$ nodes. Full unrooted binary trees are of primary interest in clustering contexts because, of all tree diagrams with a fixed number of nodes, they have the richest internal structure (most differentiated paths between nodes). They are therefore very suitable for representing the structure of a set of objects [12]. A full unrooted binary tree with exactly $n = 4$ leaves is also referred to as *simple quartet topology*, or just as *quartet* [10,12]. Given a set N of $n \geq 4$ objects, the number of sets of four objects from the set N is given by:

$$\binom{n}{4} = \frac{n!}{4!(n-4)!} = \frac{n(n-1)(n-2)(n-3)}{24}.$$

Given four generic objects $\{a, b, c, d\} \in N$, there exist exactly three different quartets: $ab|cd$, $ac|bd$, $ad|bc$, where the vertical bar divides the two pairs of leaves, with each pair labelled by the corresponding objects and attached to the same internal node. Therefore the total number of possible simple quartet topologies of N is: $3 \cdot \binom{n}{4}$.

A full unrooted binary tree is said to be "consistent" with respect to a simple quartet topology, say $ab|cd$, *if and only if* the path from a to b does not cross the path from c to d. This quartet $ab|cd$ is also said to be "embedded" in the given full unrooted binary tree.

Considering the set N, the MQTC problem accepts as input a *distance matrix*, D, which is a matrix containing the dissimilarities, taken pairwise, among the n objects[1]. To extract a hierarchy of clusters from the distance matrix, the MQTC problem determines a full unrooted binary tree with n leaves that visually represents the symmetric $n \times n$ distance matrix as well as possible according to a cost measure. Consider the set Q of all possible $3 \cdot \binom{n}{4}$ quartets, and let $C : Q \rightarrow \Re^{+}$ be a cost function assigning a real valued cost $C(ab|cd)$ to each quartet topology $ab|cd \in Q$. The cost assigned to each simple quartet topology is the sum of the dissimilarities (taken from D) between each pair of neighbouring leaves [4]. For example, the cost of the quartet $ab|cd$ is $C(ab|cd) = D_{(a,\ b)} + D_{(c,\ d)}$, where $D_{(a,\ b)}$ and $D_{(c,\ d)}$ indicate, respectively, the dissimilarities among (a and b) and (c and d), obtained from the D.

Consider now the set Γ of all full unrooted binary trees with $2n - 2$ nodes (i.e. n leaves and $n - 2$ internal nodes), obtained by placing the n objects to cluster as leaf nodes of the trees. For each $t \in \Gamma$, precisely one of the three possible simple quartet topologies for any set of four leaves is consistent [4]. Thus, there exist precisely $\binom{n}{4}$ consistent quartet topologies (one for each set of four objects) for each $t \in \Gamma$.

The cost associated with a full unrooted binary tree $t \in \Gamma$ is the sum of the costs of its $\binom{n}{4}$ consistent quartet topologies, that is: $C(t) = \sum_{\forall ab|cd \in Q_t} C(ab|cd)$, where Q_t is the set of such $\binom{n}{4}$ quartet topologies embedded in t.

In a hierarchical clustering context, we do not even have a priori knowledge that certain simple quartet topologies are objectively true and must be embedded. Thus, the MQTC problem assigns a cost value to each simple quartet topology, in order to express the relative importance of the simple quartet topologies to be embedded in the full unrooted binary tree having the n objects as leaves. The full unrooted binary tree with the minimum cost balances the importance of embedding different quartet topologies against others, leading to a binary tree that visually represents the symmetric distance matrix $n \times n$ as well as possible. The solution of this problem allows the hierarchical representation of a set of n objects within a full unrooted binary tree [12]. That is, the resulting binary tree will have the n objects assigned as leaves such that objects with short relative dissimilarities will be placed close to each other in the tree. This hierarchical clustering approach coming from the MQTC problem is also referred in the literature to as *quartet method* [4]. Such method is more sensitive and objective than other quartet clustering methods, which are usually too slow when they are exact or global, and too inaccurate or uncertain when they are statistical incremental, like the case of quartet tree-puzzling. In [4] the MQTC problem was shown to be NP-hard, and a Randomized Hill Climbing heuristic was also

---

[1] It is therefore a symmetric $n \times n$ matrix, with $n \geq 4$, containing non-negative reals, normalized between 0 and 1, as entries.

proposed to obtain approximate problem solutions. Other MQTC metaheuristics based on Greedy Randomized Adaptive Search Procedure, Simulated Annealing, and Variable Neighbourhood Search were proposed in [6]. These metaheuristics performed well for the problem, although the best performance was obtained by a Reduced Variable Neighbourhood Search (RVNS) implementation [6].

In this paper we propose some improved metaheuristics for the MQTC problem, to be used to get solutions of higher quality, in terms of reduced costs and computational running times. In particular we first propose a basic greedy heuristic which is characterized by a very high speed and some interesting implementation improvements, which can be used to enhance the MQTC metaheuristics to date in the literature. This greedy algorithm is characterized by its ease of implementation and simplicity, and it takes inspiration from the recently proposed "less is more approach" [9,18], which supports the adoption of non-sophisticated and effective metaheuristics instead of hard-to-reproduce and complex solution approaches. In particular we will show how the performance of the RVNS quartet heuristic is improved, with particular emphasis to computational running time, by adopting our proposed basic greedy heuristic to construct initial solutions for the algorithm.

The rest of the paper is organized as follows. Section 2 presents the related work. Section 3 describes the details of the proposed heuristics, along with their main implementation concepts and pseudo-code formulations. Our computational experience is reported in Sect. 4, and finally the paper ends with conclusions in Sect. 5.

2   Related Work

The MQTC problem was originally proposed in [2]. There the main focus was on compression-based distances, but the authors visually presented the tree reconstruction results by full unrooted binary trees deriving by their MQTC problem formulation. Hence, they developed the quartet method for hierarchical clustering, a new approach aimed at general hierarchical clustering of data from different domains, not necessarily biological phylogenies. Several practical applications of the quartet method have been explored in the literature. In particular, Cilibrasi et al. [5] proposed a robust automatic music classification procedure consisting of two steps. The first step consisted of extracting the "Normalized Compression Distances" [16] among some considered pieces of music. The Normalized Compression Distance is a similarity metric based on string compression which mimics the ideal performance of Kolmogorov complexity [16]. The second step consisted of creating an efficient visualization of the extracted pairwise distances by means of the quartet method of hierarchical clustering. To substantiate the claims of universality and robustness of this automatic classification method, evidence of other successful applications in areas as diverse as genomics, virology, languages, literature, handwriting, astronomy and combinations of objects from completely different domains, were reported in [2]. In addition, Cilibrasi and Vitányi [3] reported an interesting application of this theory, consisting of the automatic extraction of similarities among words and phrases from the WWW using Google page counts. Granados et al. [13] studied the impact of several kinds of information distortion on compression-based text clustering, showing their results as ternary trees by means of the quartet method of hierarchical clustering. In a recent application, a variant of the quartet method based on the Variable Neighborhood Search metaheuristic was used for biomedical literature extraction and clustering [7,8]. The proposed application was able to retrieve relevant references for systematic reviews and meta-analysis from the Medline/PubMed database, and for visualizing the retrieved bibliography through an intuitive graph layout.

In [4], the authors presented the minimum quartet tree cost problem in a more formal way. They showed the main concepts, components, advantages and disadvantages of the quartet method of hierarchical clustering, particularly underlining the similarities and differences with respect to other methods from biological phylogeny. Cilibrasi and Vitányi [4] also showed that the MQTC problem is NP-hard by reduction from the MQC problem, and provided a Randomized Hill Climbing heuristic to obtain approximate problem solutions. Several other efficient metaheuristics based on Greedy Randomized Adaptive Search Procedure, Simulated Annealing, and Variable Neighbourhood Search were proposed and compared for the MQTC problem in [6]. The best reported performance was obtained by an implementation of a Reduced Variable Neighbourhood Search metaheuristic, which we will use as a reference benchmark in our paper and try to overcome its performance.

3   Description of the Solution Algorithms

3.1   Greedy Constructive Heuristic

We first propose a new greedy heuristic for the MQTC problem, used to construct initial solutions of good quality requiring short computational running time [9, 18]. In the metaheuristics to date used for solving the MQTC problem [4,6], the initial solution was usually set either completely at random, or by selecting the corresponding flat structure, and then this solution was iteratively improved towards local optimality using the different heuristic guidelines of the specific metaheuristic implementation. The aim of the greedy constructive heuristic that we propose here consists of providing starting solutions having already a good quality, and obtained with an high speed too, which can bring to an improvement of the overall performance of the MQTC heuristic deployed afterwards.

We are given as input $n \geq 4$ different objects and the corresponding symmetric distance matrix D containing the $n \times n$ pairwise distances among those objects. The algorithm makes use also of another distance matrix D' among a set N' of $n' \geq 4$ objects, with $n' \leq n$, which will be used iteratively from our optimization routine to reduce the dimensionality n of the original set of objects in N. Initially, matrix D' is set equal to D, i.e. the sets of objects N and N' are equivalent. At this stage, another graph t', which will be used during the algorithm iterations as a support solution, is initialized to null, i.e. $t' \leftarrow \emptyset$. Then the core of our greedy heuristic begins by selecting objects from N' to be included in the support solution t'. At this purpose we greedily select from N' the objects that have the shortest minimum pairwise distance from D'. Say the two objects a and b in N' have this shortest distance, that is $D'_{(a,\ b)} \le D'_{(c,\ d)}, \forall (c,\ d) \in N'$. Note that in case of ties for the object pairs having the shortest distance in D', the routine simply selects an object pair at random within this set. Afterwards, these nodes a and b are connected to the support solution graph t'. The following three cases are possible:

- None of the two objects a and b are already connected in the partial solution t', and therefore they are joined together by means of a terminal node;
- One of the two objects is already included as a leaf node in t', and therefore the other object b is linked to the subgraph in t' containing a by means of a transition node (i.e. the dotted internal node in the figure). Please note that node b requires to be included in the partial solution t' by a link with a new transition node since, being it a leaf, if it would be included by a link with a terminal node instead, we would not be able to add any further nodes afterwards;
- Both objects a and b are already included in the partial solution t' but they belong to two different subgraphs, and therefore these two subgraphs containing respectively the two objects are connected together by means of a cross node.

Afterwards, we apply a routine, referred to as *distance matrix reduction*, to merge the added nodes a and b together to form another object, say x, by reducing in this way the dimension of N' of one unit, i.e. $n' = n' - 1$. The distance matrix D' is recomputed accordingly by removing the distances of the two objects a and b with all the other objects in N', and adding the distances of the new node x with the other objects, which are calculated as the averages distances, respectively of a and b, with the other nodes in N'. That is $D'(x,\ y) = \frac{D'(a,\ y) + D'(b,\ y)}{2}$, for all objects $y \in N'$, $y \ne a, b$. The rationale behind this procedure is to greedily merge together highly connected objects which may bring higher values of the quartet cost function of the subgraphs inferred in the partial solution t'.

This greedy procedure is repeated iteratively until a fully connected unrooted binary tree t' is obtained, i.e. $t' \in \Gamma$, which is equivalent also in getting a reduced distance matrix D' with a size $n' = 4$ (i.e. it would not be possible to reduce further the corresponding set N' since it only contain four objects). Then the support solution t' is assigned to the output full unrooted binary tree t, which is produced as final outcome of the algorithm.

3.2   Reduced Variable Neighbourhood Search

Variable Neighbourhood Search (VNS) is a popular metaheuristic for solving hard combinatorial optimization problems based on dynamically changing neighbourhood structures during the search process [14]. VNS does not follow a trajectory, but it searches for new solutions in increasingly distant neighbourhoods of the current solution, jumping only if a better solution is found. Reduced Variable Neighbourhood Search (RVNS) is a variant of the classic VNS algorithm, that has been shown to be successful for many combinatorial problems where local optima with respect to one or several neighbourhoods are relatively close to each other [14]. RVNS is a typical example of a pure stochastic heuristic, akin to a classic Monte-Carlo method, but more systematic [17]. It is useful especially for very large problem instances for which the inner local search within the classic VNS approach is costly, as in the case with quartet clustering.

The Reduced Variable Neighbourhood Search for the MQTC problem starts by selecting an initial full unrooted binary tree $t \in \Gamma$ with $2n - 2$ nodes, obtained by placing the $n \geq 4$ objects to cluster as leaves, with total cost $C(t)$. In the original RVNS implementation in [6], the initial full unrooted binary tree t was selected at random.

Then, the *shaking phase*, which represents the core idea of RVNS, is applied to t. The shaking phase aims to change the neighbourhood structure, $N_k(\cdot)$, when the algorithm is trapped at a local optimum. The new incumbent solution, say t', is generated at random in order to avoid cycling, which might occur if a deterministic rule is used. The simplest and most common choice for the neighbourhood structure consists of setting neighbourhoods with increasing cardinality: $|N_1(\cdot)| < |N_2(\cdot)| < ... < |N_{k_{max}}(\cdot)|$, where k_{max} represents the maximum size of the shaking phase. Let k be the current size of the shaking phase. The algorithm starts by selecting the first neighbourhood ($k \leftarrow 1$) and, at each iteration, it increases the parameter k if a better solution is not obtained ($k \leftarrow k+1$), until the largest neighbourhood is reached ($k \leftarrow k_{max}$). The process of changing neighbourhoods when no improvement occurs diversifies the search. In particular, the choice of neighbourhoods of increasing cardinality yields a progressive diversification of the search process.

For the MQTC problem, a shaking phase of size k consists of the random selection of another full unrooted binary tree t' within the neighbourhood $N_k(t)$ of the current solution t. To obtain t' from $N_k(t)$, the algorithm performs k consecutive *base moves*, where a base move is a single basic modification that each internal node of t can perform with its neighbouring internal nodes. The possible base moves that can be performed depend on the types of internal node pairs [6]. In the case of:

- two transition nodes: either the attached leaves are exchanged, or they are transformed into one cross node and one terminal node connected to the corresponding leaves;
- one terminal node and one transition node: the leaf of the transition node is exchanged with one of the two leaves of the terminal node;
- one terminal node and one cross node: they are transformed into two transition nodes with the two leaves of the terminal node attached;
- one transition node and one cross node: the transition node is moved in one of the other two branches of the cross node;
- two cross nodes: one branch of one cross node is swapped with a branch of the other cross node.

Note that each base move corresponds just to a limited local modification of the structure of the incumbent solution t, which results in most of the coefficients of the corresponding Complete Pseudo-Adjacency matrix C to remain unchanged. In this was there will be no need to recalculate all the coefficients of C, but only recomputing a small subset of it, speeding up consistently this step.

At the beginning of RVNS, the first neighbourhood ($k \leftarrow 1$) is selected and, at each iteration, the parameter k is increased ($k \leftarrow k+1$) whenever the solution obtained is not an improvement of the current best solution (i.e. $C(t') > C(t)$). When $k > 1$, the first base move is performed to a randomly selected internal node and one of its neighbouring internal nodes with respect to the considered distance of rank one. Then, to perform the successive base move, the algorithm selects one of the two internal nodes considered, and another neighbouring internal node that must be different from the two internal nodes already considered, and so on. The procedure is repeated until k consecutive base are performed.

If an improved binary tree t' is produced by the shaking phase ($C(t') < C(t)$), this becomes the best solution to date ($t \leftarrow t'$) and the algorithm restarts from the first neighbourhood ($k \leftarrow 1$) of t. The process of increasing progressively parameter k whenever no improvements are obtained, occurs until the maximum size of the shaking phase, k_{max}, is reached. When this happens, k is re-initialized to the first neighbourhood ($k \leftarrow 1$). The correct setting of k_{max} is an important user task. For the MQTC problem, a simple reactive schema for the efficient tuning of k_{max} has been implemented [6]. At the starting point, k_{max} is set to a small value ($k_{max} = 2$) and is increased ($k_{max} = k_{max}+1$) every i_{update} iterations between two consecutive improvements. For the value of this parameter we use the setting of [6], where $i_{update} = (1.25 \cdot 10^5)/n^2 + 50$. Throughout the execution of the algorithm, the best solution to date is stored as the binary tree t, which will be produced as output of the algorithm when the user termination condition (e.g. a maximum allowed CPU time) is reached.

4   Computational Results

In order to evaluate the algorithms, we performed experiments to compare them in terms of quality of produced solutions and computational running time. For evaluating solution quality, we used both the cost function $C(\cdot)$, already defined previously, and of another metric, referred to as *normalized tree benefit score*, $S(\cdot) \in [0, 1]$ [2, 6], which is a more intuitive performance measure of the goodness of quartet clustering. Given the set N of $n \geq 4$ documents to cluster, let m be the *best (minimal) cost*, calculated as the sum of the $\binom{n}{4}$ minimum costs of each set of four objects in N, and let M be the *worst (maximal) cost*, calculated as the sum of the $\binom{n}{4}$ maximum costs of each set of four objects in N. The normalized tree benefit score $S(t)$ of a full unrooted binary tree $t \in \Gamma$ is obtained by rescaling and normalizing in $[0, 1]$ the cost function $C(t)$, i.e. $S(t) = \frac{M-C(t)}{M-m} \in [0, 1]$. While a lower cost function $C(t)$ results in a better solution t, conversely a higher normalized tree benefit score means a better clustering quality.

Our experimental algorithms comparison was made upon classic MQTC problem datasets, already used in previous studies in the literature (see e.g. [2,4,6]). They are briefly described in the following, but for more details the reader is referred to [2,4,6].

- Data constructed artificially to have none inconsistency, that is data for which the exact solutions are known in advance and have been built to have normalized tree benefit score equal to one. The construction mechanism is described in detail in [2,6]. These data aim at testing whether the quartet-based tree reconstruction is reliable and accurate on clean consistent data with known solutions. They consist of ten different problem instances ranging from a number of objects $n = 10$ to 100. They are referred to as: *artificial*.
- Example of natural data concerning a study in genomics with DNA sequences of different placental mammalian species. The distance matrices from the genomic data were computed by using an automated software method described by Cilibrasi and Vitányi [2,4], who downloaded the whole mitochondrial genomes of the placental mammalian species from the GenBank Database on the World Wide Web. They consists of three sets of data with $n = 10$, $n = 24$, and $n = 34$, and are referred to as: *nature*.

Table 1 show the results of our experimental comparison of the algorithms on the considered datasets. The heuristics are identified with the following abbreviations: *Greedy*, for the greedy constructive heuristic described in Sect. 3.1; $RVNS_{rand}$, for the original implementation of the Reduced Variable Neighbourhood Search (Sect. 3.2) with initial solution selected at random; $RVNS_{greedy}$, for the new Reduced Variable Neighbourhood Search implementation where the initial solution is selected by using the greedy constructive heuristic, *Greedy*. All the algorithms were implemented in C++ under the Microsoft Visual Studio 2015 framework, and were deployed on an Intel Quad-Core i5 64-bit microprocessor at 2.30 GHz with 16 GB RAM.

As stopping condition for the RVNS-based metaheuristics it was considered a maximum allowed CPU time (*max-CPU-time*). In particular, as also used in [2,6], we set *max-CPU-time* to one hour. Selection of the maximum allowed CPU time as the stopping criterion was made in order to have a direct comparison among the RVNS metaheuristics with respect to the quality of their solutions. Instead, for the *Greedy* algorithm it was not necessary to set any stopping criterion since, being a constructive heuristic, it automatically ends when a feasible solution, i.e. a fully connected unrooted binary tree, is obtained.

Looking at Table 1, the first column shows the number of objects, n, characterizing the different datasets (*artificial*, *nature*, *geographical*) while the remaining columns give the computational results in terms of clustering quality (i.e. cost function values $C(\cdot)$, *cost*, and normalized tree benefit scores $S(\cdot)$, *score*), and computational running time in seconds (*time*) for the different algorithms. The performance of an heuristic can be considered better than another if it obtains a lower cost function value, or more intuitively a larger normalized tree benefit score. In case of ties, an algorithm is consider better than another if it was faster.

Table 1. Computational results of the compared algorithms (*Greedy*, *RVNS_rand*, and *RVNS_greedy*) in terms of cost function values (*cost*, normalized tree benefit scores (*score*, and computational running times in seconds (*time*) for the considered datasets.

size n	Greedy			$RVNS_{rand}$			$RVNS_{greedy}$		
	cost	score	time	cost	score	time	cost	score	time
artificial									
10	210.8000	0.89500	0.004	202.4000	1.00000	0.040	202.4000	1.00000	0.005
20	3301.2500	0.95270	0.003	3231.8500	1.00000	0.421	3231.8500	1.00000	0.006
30	15013.8656	0.92419	0.002	14559.4658	1.00000	0.861	14559.4658	1.00000	0.023
40	45245.7000	0.92400	0.006	43449.7500	1.00000	8.413	43449.7500	1.00000	0.554
50	104709.0800	0.85641	0.008	97207.8400	1.00000	10.606	97207.8400	1.00000	0.055
60	196608.3789	0.89166	0.012	186787.0474	1.00000	38.724	186787.0474	1.00000	0.176
70	360831.5064	0.88295	0.022	338182.8198	1.00000	38.858	338182.8198	1.00000	0.243
80	561067.4125	0.79739	0.028	509526.6875	1.00000	66.880	509526.6875	1.00000	0.193
90	806291.9392	0.89048	0.037	769344.2770	1.00000	101.512	769344.2770	1.00000	0.287
100	1232141.4400	0.92249	0.059	1178538.2000	1.00000	115.013	1178538.2000	1.00000	10.292
nature									
10	349.0720	0.99979	0.002	349.0720	0.99979	0.006	349.0720	0.99979	0.039
24	18649.3360	0.98524	0.002	18637.3390	0.99588	2.083	18637.3390	0.99588	0.232
34	82934.2444	0.98323	0.005	82922.0360	0.98792	10.610	82922.0360	0.98792	0.542
geographical									
13	476.4124	0.74265	0.002	439.7529	0.96843	0.270	439.7529	0.96843	0.016
22	4644.7111	0.82426	0.004	4377.1741	0.93507	3.140	4377.1741	0.93507	0.029
24	6839.0909	0.79911	0.004	6422.0182	0.92459	3.290	6422.0182	0.92459	0.044
25	8876.1814	0.79267	0.004	7827.1264	0.98760	2.840	7827.1264	0.98760	0.046
35	29209.7541	0.81774	0.004	26332.7485	0.98367	10.750	26332.7485	0.98367	0.205
37	28298.8724	0.79559	0.006	26846.2316	0.91973	32.940	26846.2316	0.91973	0.318

From the results showed in the table, we can immediately denote that *Greedy* was much faster of several orders of magnitude than the original RVNS implementation, *RVNS_rand*, although in most of the cases it obtained solutions with worst quality. This is an understandable result since *RVNS_rand* is an explorative metaheuristic that runs for a longer time, *max-CPU-time*, while *Greedy* instead stops immediately when a feasible solution is reached. But when *Greedy* was then embedded inside the RVNS metaheuristic in order to produce initial good-quality solutions in *RVNS_greedy*, a very powerful metaheuristic was obtained. Indeed, as it can be seen in the table, *RVNS_greedy* retained the high-speed feature from *Greedy*, but also the characteristics of good-quality solutions that is proper of the RVNS approach for the given problem. Indeed, looking at the performance of both the RVNS algorithms, the obtained solutions were comparable with respect to clustering quality, but *RVNS_greedy* was much faster. Note that for the *artificial* datasets without inconsistencies, both RVNS implementations were able to reach optimality, *RVNS_greedy* being much faster, while this was not achieved by *Greedy*.

Summarizing, the novel RVNS implementation with the greedy constructive heuristic used for selecting the initial starting solutions resulted to be the best performing method in our computational experiments in terms of both quartet clustering quality and, especially, computational running time.

5   Conclusions

In this paper we proposed some improved heuristics for quartet clustering, a novel hierarchical clustering approach based on the minimum quartet tree cost (MQTC) problem, which is NP-hard and whose goal is to derive an optimal tree from the total number of possible combinations of quartet topologies on some input objects n, where optimality means that the sum of the dissimilarities of the embedded (or consistent) quartet topologies is minimal.

In particular we provided the details of a new basic greedy heuristic that is characterized by a very high speed. Although the performance of this simple method in terms of quartet clustering quality, evaluated by means of a defined cost function and of a normalized tree benefit score, was not as good as that of the best solution method reported in the literature, i.e. a Reduced Variable Neighbourhood Search (RVNS) metaheuristic, this greedy method was used to considerably improve the performance of the RVNS by using it to construct initial good-quality solutions instead of randomly selected solutions.

This produces a very efficient solution approach to the problem, as demonstrated by our experiments on the comparison of the considered algorithms on a set of well-known MQTC datasets and by the reported computational results, which represents an advancement of the state-of-the-art on the solution methods used for quartet clustering.

Acknowledgements. The author Dr. Sergio Consoli wants to dedicate this work with deepest respect to the memory of Professor Kenneth Darby-Dowman, a great scientist, an excellent manager, the best supervisor, a wonderful person, a real friend.

References

1. Berry, V., Jiang, T., Kearney, P., Li, M., Wareham, T.: Quartet cleaning: improved algorithms and simulations. In: Nešetřil, J. (ed.) ESA 1999. LNCS, vol. 1643, pp. 313–324. Springer, Heidelberg (1999). https://doi.org/10.1007/3-540-48481-7_28
2. Cilibrasi, R., Vitányi, P.M.B.: Clustering by compression. IEEE Trans. Inf. Theory **51**(4), 1523–1545 (2005)
3. Cilibrasi, R., Vitányi, P.M.B.: The google similarity distance. IEEE Trans. Knowl. Data Eng. **19**(3), 370–383 (2007)
4. Cilibrasi, R., Vitányi, P.M.B.: A fast quartet tree heuristic for hierarchical clustering. Pattern Recogn. **44**(3), 662–677 (2011)
5. Cilibrasi, R., Vitányi, P.M.B., de Wolf, R.: Algorithmic clustering of music based on string compression. Comput. Music J. **28**(4), 49–67 (2004)
6. Consoli, S., Darby-Dowman, K., Geleijnse, G., Korst, J., Pauws, S.: Heuristic approaches for the quartet method of hierarchical clustering. IEEE Trans. Knowl. Data Eng. **22**(10), 1428–1443 (2010)
7. Consoli, S., Stilianakis, N.I.: A VNS-based quartet algorithm for biomedical literature clustering. Electron. Notes Discrete Math. **47**, 13–20 (2015)
8. Consoli, S., Stilianakis, N.I.: A quartet method based on variable neighborhood search for biomedical literature extraction and clustering. Int. Trans. Oper. Res. **24**(3), 537–558 (2017)

9. Costa, L.R., Aloise, D., Mladenović, N.: Less is more: basic variable neighborhood search heuristic for balanced minimum sum-of-squares clustering. Inf. Sci. **415–416**, 247–253 (2017)
10. Diestel, R.: Graph Theory. Springer, New York (2000)
11. Felsenstein, J.: Evolutionary trees from DNA sequences: a maximum likelihood approach. J. Mol. Evol. **17**(6), 368–376 (1981)
12. Furnas, G.W.: The generation of random, binary unordered trees. J. Classif. **1**(1), 187–233 (1984)
13. Granados, A., Cebrian, M., Camacho, D., Rodriguez, F.B.: Reducing the loss of information through annealing text distortion. IEEE Trans. Knowl. Data Eng. **23**(7), 1090–1102 (2011)
14. Hansen, P., Mladenović, N.: Variable neighbourhood search. Comput. Oper. Res. **24**, 1097–1100 (1997)
15. Jiang, T., Kearney, P., Li, M.: A polynomial time approximation scheme for inferring evolutionary trees from quartet topologies and its application. SIAM J. Comput. **30**(6), 1942–1961 (2000)
16. Li, M., Vitányi, P.M.B.: An Introduction to Kolmogorov Complexity and Its Applications, 2nd edn. Springer, New York (1997)
17. Mladenović, N., Petrović, J., Kovačević-Vujčić, V., Čangalović, M.: Solving spread spectrum radar polyphase code design problem by tabu search and variable neighbourhood search. Eur. J. Oper. Res. **151**(2), 389–399 (2003)
18. Mladenović, N., Todosijević, R., Urošević, D.: Less is more: basic variable neighborhood search for minimum differential dispersion problem. Inf. Sci. **326**, 160–171 (2016)
19. Steel, M.A.: The complexity of reconstructiong trees from qualitative characters and subtrees. J. Classif. **9**, 91–116 (1992)
20. Strimmer, K., von Haeseler, A.: Quartet puzzling: a quartet maximum-likelihood method for reconstructing tree topologies. Mol. Biol. Evol. **13**(7), 964–969 (1996)

On the k-Medoids Model
for Semi-supervised Clustering

Rodrigo Randel[1]($\boxtimes$), Daniel Aloise[1], Nenad Mladenović[2], and Pierre Hansen[3]

[1] École Polytechnique de Montréal, Montréal, Canada
{rodrigo.randel,daniel.aloise}@polymtl.ca
[2] Mathematical Institute, Serbian Academy of Science and Arts, Belgrade, Serbia
nenad@mi.sanu.ac.rs
[3] HEC Montréal, Montréal, Canada
pierre.hansen@hec.ca

Abstract. Clustering is an automated and powerful technique for data analysis. It aims to divide a given set of data points into clusters which are homogeneous and/or well separated. A major challenge with clustering is to define an appropriate clustering criterion that can express a good separation of data into homogeneous groups such that the obtained clustering solution is meaningful and useful to the user. To circumvent this issue, it is suggested that the domain expert could provide background information about the dataset, which can be incorporated by a clustering algorithm in order to improve the solution. Performing clustering under this assumption is known as semi-supervised clustering. This work explores semi-supervised clustering through the k-medoids model. Results obtained by a Variable Neighborhood Search (VNS) heuristic show that the k-medoids model presents classification accuracy compared to the traditional k-means approach. Furthermore, the model demonstrates high flexibility and performance by combining kernel projections with pairwise constraints.

Keywords: k-medoids · Semi-supervised clustering ·
Variable Neighborhood Search

1 Introduction

In unsupervised machine learning, no information is known in advance about the input data. In this learning category, the objective is usually to provide the best description of the input data by looking at the similarities/dissimilarities between its elements. Clustering is one of the main unsupervised machine learning techniques. It addresses the following general problem: given a set of data objects $O = \{o_1, \ldots, o_n\}$, find subsets, namely *clusters*, which are homogeneous and/or well separated [1]. Homogeneity means that objects in the same cluster must be similar and separation means that objects in different clusters must differ one from another. The dissimilarity (or similarity) d_{ij} between a pair of

A. Sifaleras et al. (Eds.): ICVNS 2018, LNCS 11328, pp. 13–27, 2019.
https://doi.org/10.1007/978-3-030-15843-9_2

objects (o_i, o_j) is usually computed as a function of the objects' attributes, such that d values (usually) satisfy: (i) $d_{ij} = d_{ji} \geq 0$, and (ii) $d_{ii} = 0$. Note that dissimilarities do not need to satisfy triangle inequalities, i.e., to be distances.

Despite its concise definition, the clustering problem can have significant variations, depending on the specific model used and the type of data to be clustered. The clustering criterion used plays a crucial role in the clustering obtained. For example, the homogeneity of a particular cluster can be expressed by its *diameter* defined as the maximum dissimilarity between two objects within the same cluster, while the separation of a cluster can be expressed by the *split* or the minimum dissimilarity between an object inside the cluster and another outside.

When considering dissimilarity measures, the definitions above yield two families of clustering criteria: those to be *maximized* for separation and those to be *minimized* for homogeneity. In general, these criteria are expressed in the form of thresholds, min-sum or max-sum for a set of clusters. Thus, for instance, the diameter minimization problem corresponds to minimizing for a set of clusters the maximum diameter found among them, while in the split maximization, one seeks to maximize the minimum split found in the clustering partition. The clustering criterion used is also determinant to the computational complexity of the associated clustering problem. For example, split maximization is polynomially solvable in time $O(n^2)$, while diameter minimization is NP-hard already in the plane for more than two clusters [2].

In order to overcome this difficulty and improve the result of the data clustering, it has been suggested that the domain expert could provide, whenever possible, auxiliary information regarding the data distribution, thus leading to better clustering solutions more in accordance to his knowledge, beliefs, and expectations. The clustering process driven by this side-information is called *Semi-Supervised Clustering* (SSC). SSC has become an essential tool in data mining due to the continuous increase in the volume of generated data [3].

The most common types of side-information are pairwise constraints such as *must-link* and *cannot-link* [4]. A *must-link* constraint between two objects implies that they must be assigned to the same cluster, whereas a *cannot-link* constraint that they must be allocated in different clusters. In this paper, we make an in-depth analysis of the use of the k-medoids model for the SSC problem. We also propose a new Variable Neighborhood Search (VNS) [5] algorithm that uses a location-allocation heuristic and takes into consideration pairwise constraints.

The paper is organized as follows. The next section presents the related works to this research. Section 3 describes the k-medoids model for the SSC problem. Section 4 describes the two-stage local descent algorithm proposed, and in Sect. 5 a VNS algorithm is presented for optimizing the described model. Computational experiments that demonstrate the effectiveness of our methodology in a set of benchmark data sets are reported in Sect. 6. Finally, the conclusions are presented in Sect. 7.

2   Related Works

Algorithms that make use of constraints as *must-link* and *cannot-link* in clustering became widely studied and developed after the COP-Kmeans algorithm of Wagstaff and Cardie's work [6]. The algorithm is based on modifying the unsupervised original k-means algorithm by adding a routine to prevent an object from changing cluster if any of the *must-link* or *cannot-link* constraints are violated.

The model optimized by COP-Kmeans consider that objects $o_i \in O$ correspond to points p_i of a s-dimensional Euclidean space, for $i = 1, \ldots, n$. The objective is to find k clusters such that the sum of squared Euclidean distances from each point to the centroid of the cluster to which it belongs is minimized while respecting a set of pairwise constraints. The set $\mathcal{ML}$ is formed by the pairs of points (p_i, p_j) such that p_i and p_j must be clustered together, whereas the set $\mathcal{CL}$ contains the pair of points (p_i, p_j) such that p_i and p_j must be assigned to different clusters.

The *semi-supervised minimum sum-of-squared clustering* (SSMSSC) model is mathematically expressed by:

$$\min_{x,y} \quad \sum_{i=1}^{n} \sum_{j=1}^{k} x_{ij} \| p_i - y_j \|^2 \tag{1}$$

subject to

$$\sum_{j=1}^{k} x_{ij} = 1, \quad \forall i = 1, \ldots, n \tag{2}$$

$$x_{ij} - x_{wj} = 0, \quad \forall (p_i, p_w) \in \mathcal{ML}, \quad \forall j = 1, \ldots, k \tag{3}$$

$$x_{ij} + x_{wj} \leq 1, \quad \forall (p_i, p_w) \in \mathcal{CL}, \quad \forall j = 1, \ldots, k \tag{4}$$

$$x_{ij} \in \{0, 1\}, \quad \forall i = 1, \ldots, n; \quad \forall j = 1, \ldots, k. \tag{5}$$

The binary decision variables x_{ij} express the assignment of point p_i to the cluster j whose centroid is located at $y_j \in \mathbb{R}^s$. Constraints (2) guarantee that each data point is assigned to exactly one cluster. Constraints (3) refer to the must-link constraints, and constraints (4) to the cannot-link ones.

The simplicity and pioneering of COP-Kmeans have made it a basic algorithm for many later works. Some examples are: semi-supervised clustering using combinatorial Markov random fields [7]; adaptive kernel method [8]; clustering by probabilistic constraints [9]; and density-based clustering [10].

A relevant work involving clustering under pairwise constraints was conducted by Xia [11]. The global optimization method proposed in that work is an adaptation of the Tuy's cutting planes method [12]. The algorithm is proved to obtain optimal solutions in exponential time in the worst case, and hence, it cannot be used for practical purpose for larger data mining tasks. Xia [11] reported a series of experiments where the algorithm is halted before convergence. The obtained clustering results were superior other algorithms based on COP-Kmeans.

Restricting the solution space through the explicit use of pairwise constraints is not the only possible approach for SSC. Many works have been published to propose mechanisms using distance metric learning to explore these side-information. Among them, a well-known algorithm is the *Semi-Supervised-Kernel-kmeans* [13] that enhances the similarity matrix obtained from the application of a kernel function by adding a term that brings closer together *must-link* objects while driving away *cannot-link* objects. The algorithm defines a similarity matrix $\mathbf{S} = \mathcal{K} + W + \sigma I$, where $\mathcal{K}$ is a kernel matrix, W is the matrix responsible to include the pairwise constraints into the distance metric, and σ is the term that multiplies an identity matrix I to ensure that $\mathbf{S}$ is semi-definite positive. The kernel-k-means algorithm [14] is then executed over $\mathbf{S}$ in an unsupervised manner (see [13] for details).

3   Proposed Model

Another classical representative-based clustering model is the k-medoids whose objective is to partition the points into exactly k clusters so that the sum of distances between each point and the central object (i.e., the *medoid*) of their respective cluster is minimized.

The input of the k-medoids model is a distance matrix, D, with each entry d_{ij} providing the dissimilarity between points p_i and p_j. It can be mathematically formulated in its semi-supervised version as:

$$\min \quad \sum_{i=1}^{n}\sum_{j=1}^{n} x_{ij}d_{ij} \tag{6}$$

subject to

$$\sum_{j=1}^{n} x_{ij} = 1, \quad \forall i = 1, ..., n \tag{7}$$

$$x_{ij} - x_{wj} = 0 \quad \forall (p_i, p_w) \in \mathcal{ML}, \quad \forall j = 1, ..., n \tag{8}$$

$$x_{ij} + x_{wj} \leq 1 \quad \forall (p_i, p_w) \in \mathcal{CL}, \quad \forall j = 1, ..., n \tag{9}$$

$$x_{ij} \leq y_j \quad \forall i = 1, ..., n, \forall j = 1, ..., n \tag{10}$$

$$\sum_{j=1}^{n} y_j = k \tag{11}$$

$$x_{ij} \in \{0, 1\} \quad \forall i = 1, ..., n, \forall j = 1, ..., n, \tag{12}$$

$$y_j \in \{0, 1\} \quad \forall j = 1, ..., n, \tag{13}$$

where y_j is equal to 1 if p_j is selected as the medoid of cluster j, and 0 otherwise. Constraints (10) assure that points can only be assigned to selected medoids, and constraint (11) defines that k medoids must be selected. The resulting model

(6)–(13) is named thereafter the **Semi-Supervised K-Medoids Problem** (SSKMP).

The possibility of defining the matrix D allows the objective function of the model to be flexible to use different measures to express the dissimilarities between points and medoids. The k-medoids model can be used to cluster metric data, as well as more generic data with notions of similarity/dissimilarity. For this reason, one of the main features of k-medoids is its vast list of applications [15].

When comparing the k-means model with the k-medoids model, Steinley [16] listed three important advantages in using the later for clustering:

1. Although both models work with a center-based approach, the k-means model defines the central element as the centroid of the cluster, while in the k-medoids this element is taken directly from the data set. This feature allows, for example, to identify which is the most representative element of each cluster.
2. The k-medoids, in its formal definition, usually consider the Euclidean distance to measure the dissimilarity between points and medoids, instead of the quadratic one considered in k-means. As a consequence, the k-medoids is generally more robust to outliers and noise present in the data [17].
3. While k-means only uses quadratic distance and may need to constantly recompute the distances between points and centroids every time centroids are updated, the k-medoids run over any distance matrix, even those for which there exist triangle inequality violations and which are not symmetric.

4   Local Descent Algorithm for SSKMP

Several heuristics methods have already been proposed to solve the original k-medoids problem. A very popular one is the *interchange* heuristic introduced in [18]. This local descent method searches, in each iteration, for the best pair of medoids (one to be inserted in the current solution, and another to be removed) that leads to the best-improving solution if swapped. If such pair exists, the swap is performed, and the procedure is repeated. Otherwise, the algorithm stops and the best solution found during this descent path is returned. An efficient implementation of this procedure, called *fast-interchange*, was proposed by Whitaker [19]. However, this method was not widely used (possibly due to an error in the article) until Hansen and Mladenović [20] corrected it and successfully applied it as a subroutine of a VNS heuristic. After, Resende and Werneck [21] proposed an even more efficient implementation by replacing one of the data structures present in the implementation of Whitaker [19] with two new data structures. Although the implementation suggested in [21] has the same worst-case complexity, $O(n^2)$, it is significantly faster and, to the best of the authors' knowledge, is the best implementation for the heuristic *interchange* already published.

In this paper, the method proposed in [21] is used as local descent procedure for our algorithm in order to refine a given SSKMP solution, but with a slight modification to ensure that pairwise constraints are respected.

4.1   Handling Must-Link Constraints

The following strategy is proposed to respect *must-link* constraints:

If must-link constraints connect a set of points, they can all be merged into a single point, which is enough to represent them all.

This assumption relies on the fact that all these points need to be together in the final partition, and aggregating them is just an efficient shortcut for assigning them to the same cluster repeatedly times.

Figure 1 illustrates this process on a set of *must-link* constraints given by $\mathcal{ML} = \{(p_1, p_2), (p_4, p_6), (p_2, p_6)\}$. It is possible to replace the set $\mathcal{ML}$ by an equivalent set $\mathcal{ML}' = \{(p_1, p_2), (p_1, p_4), (p_1, p_6)\}$, with p_1 as the root point for all other linked points p_2, p_4 and p_6 (Fig. 1b). This aggregation creates a so-called *super-point* and is showed in Fig. 1c where all points involved in that *must-link* constraint are all represented by the super-point p_1. Note that the super-point could have been aggregated over p_2, p_4 or p_6 instead of p_1 without prejudice.

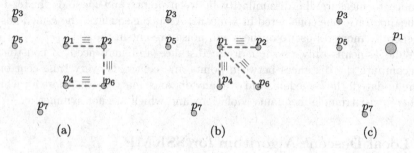

(a)   (b)   (c)

Fig. 1. Illustration of a *super-point* aggregation.

However, since the points involved in *must-link* constraints can be aggregated and viewed as a single point, then it is also necessary to update the dissimilarity d_{ij} of a super-point p_i, and all medoids $j = 1, \ldots, n$, as the sum of dissimilarities of all points that compose it. Let $H(p_i) = \{p_h \in P \mid (p_i, p_h) \in \mathcal{ML}\}$ be the set of points that are part of the super-point p_i. The cost $d_{ij} = d_{ji}$, for each $j = 1, \ldots, n$ is then calculated as the sum of dissimilarities considering all aggregated points, i.e.,

$$d_{ij} = \sum_{h : p_h \in H(p_i)} d_{hj} \quad j = 1, \ldots, n \tag{14}$$

For the example in Fig. 1, d_{13} is updated as: $d_{13} = d_{13} + d_{23} + d_{43} + d_{63}$.

The super-point aggregation is a quick step that can be entirely performed during the preprocessing stage of the algorithm. It also helps to reduce the dimension of the original data set once the points are merged. Consequently, the more *must-link* constraints are provided by the expert, the best is the performance of our algorithmic approach.

4.2 Handling Cannot-Link Constraints

Once all *must-link* constraints are respected, the local descent algorithm only concerns violated *cannot-link* constraints. A solution is said to be infeasible if there exists any pair $(p_i, p_w) \in \mathcal{CL}$ such that p_i and p_w are assigned to the same cluster. In order to avoid that, the algorithm is divided into two stages:

1. **Stage 1.** In this first stage, the *cannot-link* constraints are temporarily neglected, and the local descent algorithm proceeds to improve the current best solution.
2. **Stage 2.** For each new improved solution found in stage 1, there is a chance of this solution be infeasible, so the algorithm invokes a routine able to restore its feasibility (concerning the *cannot-link* constraints).

In summary, the approach of our local descent algorithm is to allow an efficient search to be executed in the direction of the best possible solution (regardless of the *cannot-link* constraints), whereas the solutions obtained during the descent search path are turned into feasible solutions. Thus, the algorithm relies upon the possibility of restoring the feasibility of solutions generated in the first stage of the algorithm. The key point of our strategy is to guide the search exclusively by the gradient of the objective function, disregarding the *cannot-link* constraints.

Let **s** be a solution for the problem with its k selected medoids, i.e., $\mathbf{s} = \{j | y_j = 1\}$. Let us denote $X(i) = \{j | x_{ij} = 1\}$ the cluster of point p_i. We also define the set $E(i) = \{h | (p_i, p_h) \in \mathcal{CL}\}$ as the set of points that cannot be clustered together with p_i, and $B(i) = \{j \in \mathbf{s} | \exists h \in E(i), X(h) = j\}$ as the set of clusters in **s** that are *blocked* to p_i since it contains at least one point from $E(i)$.

The feasibility routine is presented in Algorithm 1. It is called whenever a new infeasible solution **s** is obtained by the algorithm. Let $\phi_1(i) \in \mathbf{s}$ be the closest medoid in **s** from point p_i. Remark that after stage 1, since *cannot-link* constraints are not considered, every point p_i, for $i = 1, \ldots, n$, is assigned to its closest medoid, i.e., $X(i) = \phi_1(i)$.

The restore function works as follows: first, between lines 1–6, a new set R is built to contain all points p_i that are involved in *cannot-link* constraints (i.e., $E(i) \neq \emptyset$). The loop of lines 7–14 proceeds by removing the cannot-link violations. The order in which the points are examined determines the solution obtained or even if the method is able to restore feasibility. Therefore, set R is shuffled at the beginning of that loop at line 8. Then, the algorithm iterates in the loop of lines 9–13 searching for assignments that can make the solution feasible (condition $X(i) \in B(i)$) or that can improve its cost (condition $X(i) \neq \phi_1(i)$). The rationale behind the second condition is that p_i might have been allocated to a farther cluster in a previous iteration of the restoration routine because its closest medoid was not available for assignment due to a cannot-link constraint. Note that Algorithm 1 needs to keep that B updated. This is performed every time after a point p_i is assigned from a medoid p_ℓ to medoid p_j, blocking this medoid for each point in $h \in E(i)$, and maybe removing ℓ from their sets $B(h)$, depending on the presence of any other point in $E(h)$ assigned to medoid p_ℓ.

Algorithm 1. Restore feasibility function

1: $R \leftarrow \emptyset$
2: **for** $i = 1, ..., n$ **do**
3:     **if** $E(i) \neq \emptyset$ **then**
4:         $R \leftarrow R \cup \{i\}$
5:     **end if**
6: **end for**
7: **repeat**
8:     shuffle(R)
9:     **for all** $i \in R$ **do**
10:         **if** $X(i) \in B(i)$ **or** $X(i) \neq \phi_1(i)$ **then**
11:             Assign p_i to the closest medoid $j \in$ **s** such that $j \notin B(i)$
12:         **end if**
13:     **end for**
14: **until** no assignment is made

Algorithm 1 is assured of finishing although a feasible solution is not guaranteed. Indeed, the decision problem of whether a clustering problem is feasible given a set $\mathcal{CL}$ of cannot-link constraints is NP-complete [22]. In that case, the obtained solution is simply discarded.

5   Variable Neighborhood Search for SSKMP

Variable Neighborhood Search (VNS) metaheuristic [23] has been successfully applied to many clustering problems (e.g. [24–27]). The neighborhood structure adopted in our VNS algorithm is based on *swapping* selected medoids of a solution **s** by others non-selected medoids outside **s**. In this sense, v_{max} neighborhoods are defined, where the v-th neighborhood of **s**, $N_v(\mathbf{s})$, contains all solutions obtained after replacing v medoids $j \in$ **s** with others v not-selected medoids $l \notin$ **s**.

The Algorithm 2 presents the complete framework of our VNS algorithm. It starts by preprocessing the *must-link* constraints via the *super-point* concept (lines 1). Following that, the algorithm constructs an initial feasible solution (line 2) obtained in a series of three steps: (i) an initial solution $\mathbf{s}_b$ is built by randomly selecting k initial medoids and assigning each point to its closest medoid; (ii) the restore feasibility function is applied for $\mathbf{s}_b$; (iii) if $\mathbf{s}_b$ is still infeasible, the algorithm proceeds and replace $\mathbf{s}_b$ by the first feasible solution found during the VNS. We assume that the problem is always feasible, i.e., the sets $\mathcal{ML}$ and $\mathcal{CL}$ allows to obtain a feasible solution for the SSC under consideration. The algorithm considers that infeasible solutions have infinity cost.

Next, the algorithm starts the VNS block (loop 3–15) that chooses a random neighbor solution (line 6) and applies the two-stage local descent method described in Sect. 4 to possibly improve it (line 10). However, notice that after a random solution $\mathbf{s}^r$ is chosen in the neighborhood $N_v(\mathbf{s})$ of our VNS algorithm, an allocation step must follow to re-assign the points that were allocated to the replaced medoids (removed medoids of **s**) to their new closest medoid. However, this process does not take into consideration the *cannot-link* constraints, and then, $\mathbf{s}^r$ might be infeasible. To overcome this situation, we also invoke the restore function for $\mathbf{s}^r$ before proceeding to the local search procedure (line 8). If the best feasible solution found in the descent path has a better cost than $\mathbf{s}_b$, then it is stored in $\mathbf{s}_b$ (line 12). The algorithm repeats this process until a defined stopping criterion is met.

Algorithm 2. VNS for SSC k-medoids

1: Apply the *super-point* concept, merging points interconnected by *must-link* constraints into super-points;
2: Find an initial feasible solution $\mathbf{S}_b$;
3: **repeat**
4:     $v \leftarrow 1$;
5:     **repeat**
6:         Choose a random neighbor solution $\mathbf{S}^r \in N_v(\mathbf{s})$;
7:         **if** $\mathbf{S}^r$ is infeasible **then**
8:             Call the restore feasibility function for $\mathbf{S}^r$.
9:         **end if**
10:         Apply the local descent method from $\mathbf{S}^r$, obtaining a local minimum S_f
11:         **if** cost of $\mathbf{S}_b >$ cost of $\mathbf{S}_f$ **then**
12:             $\mathbf{S}_b \leftarrow \mathbf{S}_f$; $v \leftarrow 1$;
13:         **end if**
14:         $v \leftarrow v + 1$;
15:     **until** $v = v_{max}$
16: **until** a stopping criterion is met

---

6   Experiments

This work explores the results from three different perspectives. First, model SSKMP is analyzed concerning its accuracy performance when compared with the traditional SSC model, SSMSSC. Next, the VNS performance is tested using a set of benchmark datasets for SSC problem. Third, the flexibility of SSKMP is explored in combination with distance metric learning.

Computational experiments were performed on an Intel i7-6700 CPU with a 3.4 GHz clock and 16 Gigabytes of RAM. The algorithms were implemented in C++ and compiled by gcc 6.3.

6.1   Model Accuracy

First of all, it is essential to keep in mind that it is impossible to determine whether a model is better than another with respect to all possible data sets (see Kleinberg's impossibility theorem [28]). The SSMSSC and SSKMP are comparable models given that (i) both are representative-based; and (ii) the data sets used in the experiments are considered as points in the Euclidean space. It was decided to compare the models regarding accuracy using the *Adjusted Rand Index* (ARI) [29], which can measure how close the clustering result is to the ground-truth classification obtained in the UCI repository [30].

We first compare the models using the ARI results reported by Xia [11]. As done in her work, we ran the VNS algorithm 100 times and reported the average ARI value. We also defined the stop criterion as the average CPU time used by Xia's algorithm. In all experiments we used the v_{max} parameter equal to 10. Table 1 presents this the of these two models for 12 benchmark data sets. For each of them, column n indicates the number of points and k the number of clusters. In the following, we present results for two configurations of $\mathcal{ML}$ and $\mathcal{CL}$ used in [11]. The first two columns refer to the number of *must-link* and *cannot-link* constraints, and the last two refer to the ARI index values obtained by each model concerning the ground-truth partition.

Table 1. Datasets configurations and ARI results for SSMSSC and SSKMP

Instance	n	k	Configuration 1				Configuration 2			
			$|\mathcal{ML}|$	$|\mathcal{CL}|$	ssmssc	sskmp	$|\mathcal{ML}|$	$|\mathcal{CL}|$	ssmssc	sskmp
Soybean	47	4	4	24	0.55	**0.60**	8	4	0.62	0.62
Protein	116	6	18	12	**0.31**	0.25	26	18	**0.32**	0.25
Iris	150	3	12	12	0.74	**0.75**	16	8	0.75	**0.76**
Wine	178	3	44	26	0.44	**0.45**	72	44	0.45	0.45
Ionosphere	351	2	52	36	0.16	0.16	122	64	0.14	**0.15**
Control	600	6	60	30	**0.54**	0.50	90	60	**0.53**	0.51
Balance	625	3	156	94	**0.32**	0.24	218	126	**0.43**	0.25
Yeast	1484	10	296	178	0.16	0.16	520	296	0.17	0.17
Optical	3823	10	496	306	**0.70**	0.68	689	420	**0.71**	0.69
Statlog	4435	6	444	222	0.53	0.53	666	444	**0.54**	0.53
Page	5473	5	548	274	0.01	**0.03**	1024	820	0.01	**0.03**
Magic	19020	2	1902	952	0.05	**0.18**	2854	1902	0.04	**0.16**

We note from Table 1 that both models present quite similar results and comparable clustering performances. For the 24 tests cases, each model had nine times each the best ARI, and for six data sets, they had the same ARI value. Moreover, even when the ARI indices were not equal, the difference in values was marginal.

6.2   VNS Performance

This section is dedicated to evaluating the VNS performance for optimizing the SSKMP model. To obtain the optimal solution for the tested datasets, we used

Table 2. Performance results for VNS and CPLEX.

Instance	Configuration	f_{opt}	t_{opt}	vns	$\overline{vns}$	$\overline{t_{vns}}$	restore
Soybean	1	1.138047e+02	0.12	0%	0%	0.00	11%
	2	1.156629e+02	0.16	0%	0%	0.00	10%
Protein	1	1.269331e+03	2.46	0%	0%	0.01	10%
	2	1.262633e+03	0.93	0%	0%	0.00	15%
Iris	1	9.835843e+01	8.85	0%	0%	0.01	7%
	2	9.962796e+01	7.94	0%	0%	0.00	6%
Wine	1	1.749303e+04	10.07	0%	0%	0.01	7%
	2	1.907746e+04	17.16	0%	0%	7.08	7%
Ionosphere	1	8.172423e+02	61.44	0%	0%	5.13	9%
	2	8.384550e+02	53.58	0%	0.02%	65.04	9%
Control	1	2.693438e+04	135.20	0%	0%	0.13	5%
	2	2.693937e+04	124.34	0%	0%	0.12	5%
Balance	1	1.466425e+03	881.94	0%	0.002%	110.42	4%
	2	1.471803e+03	816.61	0%	0%	91.51	4%
Yeast	1	2.523202e+02	166622.71	0%	0%	97.78	3%
	2	2.605097e+02	42557.74	0%	0.004%	124.44	3%

the solver CPLEX 12.6. This restricted our sample in this experiment because CPLEX was not able to solve data sets Optical, Statlog, Page and Magic in a reasonable amount of time (less than 50 h).

Table 2 shows the results of our computational experiments. For each configuration, we executed the algorithm 10 times using 300 s as time limit. Columns f_{opt} and t_{opt} provide optimal solution values and the time needed by CPLEX to obtain it, respectively. The column vns reports the gap between the optimal solution and the best solution found by the VNS from the 10 distinct executions. In the sequel, columns $\overline{vns}$ and $\overline{t_{vns}}$ report the average values for the same 10 execution of the algorithm. The column restore presents the average percentage of time required by the restore feasibility function during the execution.

Firstly, we justify the importance of having a heuristic approach to the problem since the time to optimally solve it increases exponentially as the number of points scales (t_{opt}). On the other hand, for all the 16 test cases, the VNS was able to find the optimal solutions using much less time. For the test where CPLEX took the longest time to solve, 46 h for Yeast configuration 1, the VNS only needed, on average, 98 s to obtain a solution with the same cost. Furthermore, only in three scenarios, the VNS was not able to obtain the best solution in all ten executions, but still, the gaps are tiny.

From the results reported in column restore, we verify that the restore feasibility function does not require much computational time (7% on average) for the instances used in [11]. Besides, the amount of time is reduced for larger datasets, which is expected as the local descent procedure starts to demand more computational resources to perform the search.

6.3   Model Flexibility

One of the main advantages of using SSKMP is the ability to work with a general dissimilarity matrix D as input. This feature not only allows the model to work with many different metric systems but also provides great flexibility to define a clustering criterion. For example, it is possible to use the *distance metric learning* technique without a single modification with our algorithm. Take for instance the *Semi-Supervised-Kernel-Kmeans* (SS-Kernel-k-means) algorithm [13], which defines the similarity matrix $\mathbf{S} = \mathcal{K} + W + \sigma I$, aggregating the kernel matrix $\mathcal{K}$ and constraints matrix W (metric learning). Then, we can easily transform **s** into a dissimilarity matrix D (e.g. subtract each entry by the maximum element in **s**) and use it as input for SSKMP. Furthermore, having the distance metric modification in the input does not preclude the use of pairwise constraints in combination, which has already been proven to be a good approach [31].

Consider the synthetic data set **Two Circles** showed in Fig. 2, which presents 200 points in the Euclidean plane, with 100 points in each class. This dataset has an inner circle and a surrounding outer circle. The Fig. 3 presents the ARI results for our proposed VNS and the SS-Kernel-k-means algorithm using the two circles instance. Both algorithms were executed 100 times starting from a random initial solution, and the average ARI was reported. As suggested in [13], we used an exponential kernel ($exp(-\|x - y\|^2/2\sigma)$) for SS-Kernel-$k$-means to separate the two classes in the mapped space linearly. We also included the algorithm vns+ which combines the distance metric learning and solution space restriction due to pairwise constraints into the model optimized by our proposed VNS. The time limit used for both VNS algorithms was the average time needed by SS-Kernel-k-means to finish one execution.

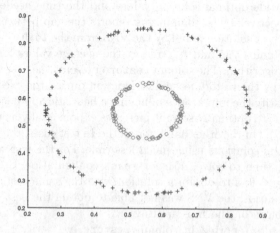

Fig. 2. Two circles synthetic data set.

We note from Fig. 3 that the VNS algorithms based on model SSKMP outperformed the typical kernel approach, both reaching the maximum ARI value.

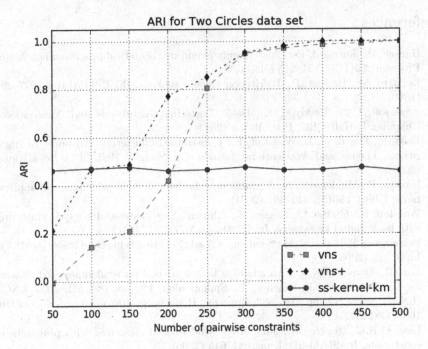

Fig. 3. ARI performance for two circles data set.

We highlight that the VNS algorithm improved its accuracy performance by adding the distance metric learning mechanism, reaching the maximum ARI value with 20% fewer constraints than the VNS algorithm that uses only the pairwise constraints. We also observed that the SS-Kernel-k-means algorithm was not able to improve ARI as the number of pairwise constraints increased. We believe that the kernel-based algorithm is more sensitive to initialization besides not being able to escape from local optima. In contrast, the VNS was proved robust, making powerful use of a priori information.

7 Conclusion

This paper proposed a VNS heuristic for assessing the performance of the k-medoids model for semi-supervised clustering. Experiments showed that the new model had similar classification performance when compared with the traditional k-means model. The VNS algorithm was validated in a series of comparative experiments against CPLEX, presenting solutions very close to the optimal ones (never exceeding 0.02% in average) using much less CPU time. Moreover, the flexibility of the k-medoids model was tested regarding the addition of a dissimilarity matrix generated by a kernel function with distance metric learning. The VNS that combined the kernel trick with the explicit use of pairwise constraints presented the best accuracy performance among the algorithms compared.

References

1. Hansen, P., Jaumard, B.: Cluster analysis and mathematical programming. Math. Program. **79**(1–3), 191–215 (1997)
2. Delattre, M., Hansen, P.: Bicriterion cluster analysis. IEEE TPAMI **4**, 277–291 (1980)
3. Aggarwal, C.C., Reddy, C.K.: Data Clustering: Algorithms and Applications. Chapman & Hall/CRC, Boca Raton (2013)
4. Basu, S., Davidson, I., Wagstaff, K.: Constrained Clustering: Advances in Algorithms, Theory, and Applications, 1st edn. Chapman & Hall/CRC, Boca Raton (2008)
5. Hansen, P., Mladenović, N.: Variable neighborhood search: principles and applications. EJOR **130**(3), 449–467 (2001)
6. Wagstaff, K., Cardie, C., Rogers, S., Schroedl, S.: Constrained k-means clustering with background knowledge. In: ICML, pp. 577–584 (2001)
7. Bekkerman, R., Sahami, M.: Semi-supervised clustering using combinatorial MRFs. In: ICML (2006)
8. Yan, B., Domeniconi, C.: An adaptive kernel method for semi-supervised clustering. In: Fürnkranz, J., Scheffer, T., Spiliopoulou, M. (eds.) ECML 2006. LNCS (LNAI), vol. 4212, pp. 521–532. Springer, Heidelberg (2006). https://doi.org/10.1007/11871842_49
9. Law, M.H.C., Topchy, A., Jain, A.K.: Model-based clustering with probabilistic constraints. In: SIAM-SDM, pp. 641–645 (2005)
10. Ruiz, C., Spiliopoulou, M., Menasalvas, E.: Density-based semi-supervised clustering. Data Min. Knowl. Disc. **21**(3), 345–370 (2010)
11. Xia, Y.: A global optimization method for semi-supervised clustering. Data Min. Knowl. Disc. **18**(2), 214–256 (2009)
12. Tuy, H.: Concave programming under linear constraints. Soviet Math. **5**, 1437–1440 (1964)
13. Kulis, B., Basu, S., Dhillon, I., Mooney, R.: Semi-supervised graph clustering: a kernel approach. Mach. Learn. **74**(1), 1–22 (2009)
14. Schölkopf, B., Smola, A., Müller, K.R.: Nonlinear component analysis as a kernel eigenvalue problem. Neural Comp. **10**(5), 1299–1319 (1998)
15. Christofides, N.: Graph Theory: An Algorithmic Approach (Computer Science and Applied Mathematics). Academic Press Inc., Orlando (1975)
16. Steinley, D.: K-medoids and other criteria for crisp clustering. In: Handbook of Cluster Analysis. Chapman and Hall/CRC Handbooks of Modern Statistical Methods. CRC Press (2015)
17. Kaufman, L., Rousseeuw, P.J.: Partitioning around medoids (program PAM). In: Finding Groups in Data: An Introduction to Cluster Analysis, pp. 68–125 (1990)
18. Teitz, M.B., Bart, P.: Heuristic methods for estimating the generalized vertex median of aweighted graph. Oper. Res. **16**(5), 955–961 (1968)
19. Whitaker, R.: A fast algorithm for the greedy interchange for large-scale clustering and median location problems. INFOR **21**(2), 95–108 (1983)
20. Hansen, P., Mladenović, N.: Variable neighborhood search for the P-median. Locat. Sci. **5**(4), 207–226 (1997)
21. Resende, M.G.C., Werneck, R.F.: A fast swap-based local search procedure for location problems. ANOR **150**(1), 205–230 (2007)
22. Davidson, I., Ravi, S.: Clustering with constraints: feasibility issues and the k-means algorithm. In: SIAM-SDM, pp. 138–149 (2005)

23. Mladenović, N., Hansen, P.: Variable neighborhood search. Comput. Oper. Res. **24**(11), 1097–1100 (1997)
24. Costa, L.R., Aloise, D., Mladenović, N.: Less is more: basic variable neighborhood search heuristic for balanced minimum sum-of-squares clustering. Inf. Sci. **415**, 247–253 (2017)
25. Hansen, P., Ruiz, M., Aloise, D.: A VNS heuristic for escaping local extrema entrapment in normalized cut clustering. Pattern Recog. **45**(12), 4337–4345 (2012)
26. Hansen, P., Mladenović, N.: J-means: a new local search heuristic for minimum sum of squares clustering. Pattern Recog. **34**(2), 405–413 (2001)
27. Santi, É., Aloise, D., Blanchard, S.J.: A model for clustering data from heterogeneous dissimilarities. EJOR **253**(3), 659–672 (2016)
28. Kleinberg, J.: An impossibility theorem for clustering. In: Advances in Neural Information Processing Systems, pp. 463–470 (2003)
29. Hubert, L., Arabie, P.: Comparing partitions. J. Classif. **2**(1), 193–218 (1985)
30. Lichman, M.: UCI machine learning repository (2013)
31. Bilenko, M., Basu, S., Mooney, R.J.: Integrating constraints and metric learning in semi-supervised clustering. In: Proceedings of the Twenty-First International Conference on Machine Learning. ICML 2004, pp. 81–88. ACM, New York (2004)

Complexity and Heuristics for the Max Cut-Clique Problem

Mathias Bourel, Eduardo Canale, Franco Robledo, Pablo Romero, and Luis Stábile[✉]

Instituto de Matemática y Estadística, IMERL Facultad de Ingeniería, Universidad de la República, Montevideo, Uruguay
{mbourel,canale,frobledo,promero,lstabile}@fing.edu.uy

Abstract. In this paper we address a metaheuristic for an combinatorial optimization problem. For any given graph $\mathcal{G} = (V, E)$ (where the nodes represent items and edges correlations), we want to find the clique $\mathcal{C} \subseteq V$ such that the number of links shared between $\mathcal{C}$ and $V - \mathcal{C}$ is maximized. This problem is known in the literature as the Max Cut-Clique (MCC).

The contributions of this paper are three-fold. First, the complexity of the MCC is established, and we offer bounds for the MCC using elementary graph theory. Second, an exact Integer Linear Programming (ILP) formulation for the MCC is offered. Third, a full GRASP/VND methodology enriched with a Tabu Search is here developed, where the main ingredients are novel local searches and a Restricted Candidate List that trades greediness for randomization in a multi-start fashion. A dynamic Tabu list considers a bounding technique based on the previous analysis.

Finally, a fair comparison between our hybrid algorithm and the globally optimum solution using the ILP formulation confirms that the globally optimum solution is found by our heuristic for graphs with hundreds of nodes, but more efficiently in terms of time and memory requirements.

Keywords: Combinatorial optimization problem · Max Cut-Clique · ILP · GRASP · VND · Tabu Search

1 Motivation

The MCC has an evident application to product-placement in Market Basket Analysis (MBA), sometimes known as *affinity analysis* [1]. For instance, the manager of a supermarket must decide how to locate the different items in the different compartments. In a first stage, it is essential to determine the correlation between the different pairs of items, for psychological/attractive reasons. Then, the priceless/basic products (bread, rice, milk and others) could be hidden on the back, in order to give the opportunity for other products in a large corridor (and candies should be at hand by kids as well). Observe that the MCC appears in

© Springer Nature Switzerland AG 2019
A. Sifaleras et al. (Eds.): ICVNS 2018, LNCS 11328, pp. 28–40, 2019.
https://doi.org/10.1007/978-3-030-15843-9_3

the first stage, while marketing/psychological aspects play a key role in a second stage for product-placement in a supermarket.

This work is focused on a specific combinatorial optimization methodology to assist product placement; however, related applications could be found. The problem under study is called Max Cut-Clique (MCC), and it was introduced by Martins [5]. For any given graph $\mathcal{G} = (V, E)$ (where the nodes are items and links represent correlation), we want to find the clique $\mathcal{C} \subseteq V$ such that the number of links shared between $\mathcal{C}$ and $V - \mathcal{C}$ is maximized.

In [5], the author states that the MCC is presumably hard, since related problems such as $MAX\text{-}CUT$ and $MAX\text{-}CLIQUE$ are both $\mathcal{NP}$-Complete. To the best of our knowledge, there is no formal proof available for the hardness of the MCC in the published scientific literature. Nevertheless, the MCC is systematically addressed by the scientific community with metaheuristics and exact solvers that run in exponential time.

A recent work in the field develops an Iterated Local Search for the MCC [6]. As far as we know, this work belongs to the state-of-the-art techniques for the MCC. The authors find optimal solutions for most instances under study, and suggest a rich number of applications.

The contributions of this paper can be summarized in the following items:

1. The $\mathcal{NP}$-Completeness of MCC is established (Subsect. 2.1).
2. Bounds for both the globally optimum solution and the clique size are produced (Subsect. 2.2).
3. A hybrid GRASP/VND heuristic enriched with Tabu Search is developed to address the MCC (Sect. 3).
4. An exact Integer Linear Programming (ILP) formulation for the MCC is proposed (Sect. 4).
5. The performance of our approach is studied (Sect. 5).
6. A discussion of applications for product-placement is included (Sect. 6).

2   Analysis and Complexity

In this section, the computational complexity for the MCC is established. We formally prove that the corresponding decision version for the MCC belongs to the class of $\mathcal{NP}$-Complete decision problems (Subsect. 2.1). Then, we find bounds for the MCC using elementary graph theory (Subsect. 2.2).

It is worth to remark that the hardness promotes the development of heuristics, and these bounds will enrich our GRASP/VND heuristic with a dynamic Tabu List.

2.1   Complexity

We formally prove that the MCC is at least as hard as $MAX\text{-}CLIQUE$. First, we describe both decision problems and the decision versions for the MCC:

Definition 1 (*MAX-CLIQUE*).

> *GIVEN: a graph $G = (V, E)$ and a real number K.*
> *QUESTION: is there a clique $C \subseteq V$ such that $|C| \geq K$?*

For convenience, we describe the MCC as a decision problem. Let us denote $\delta(C)$ to the cut produced by a node-set C, or the objective value for the MCC whenever C is a clique.

Definition 2 (*MCC*).

> *GIVEN: a graph $G = (V, E)$ and a real number K.*
> *QUESTION: is there a clique $C \subseteq G$ such that $|\delta(C)| \geq K$?*

Theorem 1. *The MCC belongs to the class of $\mathcal{NP}$-Complete problems.*

Proof. We prove that the MCC is at least as hard as MAX-$CLIQUE$. Consider a simple graph $G = (V, E)$ with order $n = |V|$ and size $m = |E|$. Let us connect m leaf-nodes hanging to every single node $v \in V$ (observe that there are $m \times n$ such nodes). The resulting graph is called H. If we find a polynomial-time algorithm for MCC, then we can produce the max cut-clique in H. But observe that the max cut-clique C in H must belong to G. If C has cardinality c, then the cut-clique has precisely $c \times m$ hanging nodes. By construction, the cut-clique must maximize the number of hanging nodes, since the whole size $|E| = m$ is added to the cut by a single addition of a node in the clique. As a consequence, c must be the MAX-$CLIQUE$. We proved that the MCC is at least as hard as MAX-$CLIQUE$, as desired. Since MCC belongs to the set of $\mathcal{NP}$ decision problems, it belongs to the $\mathcal{NP}$-Complete class. ∎

2.2 Bounds for *MCC*

Observe that the globally optimum for the MCC could be attained by more than one clique. Let us denote by C_{min} the minimum cardinality clique such that $|\delta(C_{min})| = OPT$, the optimal value for the MCC, and $c_{min} = |C_{min}|$.

Definition 3. *A finite sequence $\{a_i\}_{i=0}^{n}$ is strictly unimodal if there exists some index $k_0 \in \{0, \ldots, n\}$ such that $a_0 < a_1 < \cdots < a_{k_0}$ and $a_{k_0} \geq a_{k_0+1} > a_{k_0+2} > \cdots > a_n$.*

Lemma 1. *Consider a connected graph G with degree-sequence $(\delta_1, \ldots, \delta_n)$, where for convenience we consider $\delta_1 \leq \delta_2 \leq \cdots \leq \delta_n$. Then, the following finite sequence $\{f(k)\}_{k=0}^{n}$ is strictly unimodal, where*

$$f(k) = -k(k-1) + \sum_{i=1}^{k} \delta_{n-i+1}, \forall k \in \{0, 1, \ldots, n\}. \tag{1}$$

Proof. The difference between consecutive terms, $\Delta_k = f(k) - f(k-1)$, is:

$$\Delta_k = -k(k-1) + \sum_{i=1}^{k} \delta_{n-i+1} + (k-1)(k-2) - \sum_{i=1}^{k-1} \delta_{n-i+1} = -2(k-1) + \delta_{n-k+1}.$$

Since $-2(k-1)$ and δ_{n-k+1} are monotonically decreasing sequences, being the former strictly decreasing, $\{\Delta_k\}_{k \geq 0}$ must be strictly decreasing as well. Furthermore, since $\Delta_1 = f(1) - f(0) = \delta_n > 0$ and $\Delta_n = -2(n-1) + \delta_1 < 0$, there exists some index k_0 such that: $f(0) < f(1) < \cdots < f(k_0)$ and $f(k_0) \geq f(k_0 + 1) > f(k_0 + 2) > \cdots > f(n)$, as desired. ∎

Lemma 2. *The following inequalities hold for any clique $\mathcal{C}$:*

$$- |\mathcal{C}|(|\mathcal{C}| - 1) + \sum_{i=1}^{|\mathcal{C}|} \delta_i \leq |\delta(\mathcal{C})| \leq f(|\mathcal{C}|) \tag{2}$$

Proof. The sum $|\delta(\mathcal{C})| + |\mathcal{C}|(|\mathcal{C}| - 1) = \sum_{v_i \in \mathcal{C}} \delta_i$ is greater (smaller) than the sum of the $|\mathcal{C}|$ smallest (greatest) degrees. ∎

In the following, we will provide an upper-bound for OPT, the globally optimum value for the MCC, and bounds for the size c_{min} of the minimum cardinality clique $\mathcal{C}_{min}$, in terms of the auxiliary sequence $\{f(k)\}_{k=0}^{n}$.

Theorem 2 (*Upper-Bound for MCC*). *If OPT denotes the optimal value for the MCC and f is maximized at k_0, then $OPT \leq f(k_0)$.*

Proof. If we are given an arbitrary clique $\mathcal{C}$, the incident edges to some $v \in \mathcal{C}$ either belong to the clique or to the cut. Then:

$$|\delta(\mathcal{C})| = \sum_{v_i \in \mathcal{C}} \delta_i - |\mathcal{C}|(|\mathcal{C}| - 1) \leq f(|\mathcal{C}|) \leq f(k_0), \tag{3}$$

where Lemmas 2 and 1 were considered in the last two inequalities. Since the inequalities hold for every clique, in particular we get that $OPT \leq f(k_0)$. ∎

Theorem 3 (*Bounds for c_{min}*). *If $\{k_0, k_1\} = \mathrm{argmax} f(k)$ with $k_0 \leq k_1$, then the following inequalities hold for any clique $\mathcal{C}$:*

$$\max\{k \leq k_0 : f(k) \leq |\delta(\mathcal{C})|\} \leq c_{min} \leq \min\{k \geq k_1 : f(k) \leq |\delta(\mathcal{C})|\}. \tag{4}$$

Proof. Let $\mathcal{C}'$ be a clique such that $f(|\mathcal{C}'|) \leq |\delta(\mathcal{C})|$ and $|\mathcal{C}'| \leq k_0$. Since $|\delta(\mathcal{C})| \leq OPT \leq f(c_{min})$ and f is strictly increasing in $[1, k_0]$, then $|\mathcal{C}'| \leq c_{min}$. Taking maximum on $|\mathcal{C}'| \leq k_0$, we obtain the first inequality.

The reasoning for the second inequality is analogous. ∎

Corollary 1. *If k_0 and k_1 are as in previous Theorem, then the the following inequalities hold:*

$$\max\{k \leq k_0 : f(k) \leq \delta_n\} \leq c_{min} \leq \min\{k \geq k_1 : f(k) \leq \delta_n\} \tag{5}$$

Proof. Apply Theorem 3 with the clique $\mathcal{C} = \{v_n\}$. ∎

3   Methodology

GRASP, VND and Tabu Search are well known metaheuristics that have been successfully used to solve many hard combinatorial optimization problems. GRASP is an iterative multi-start process which operates in two phases [7]. In the Construction Phase a feasible solution is built whose neighborhood is then explored in the Local Search Phase [7]. The second phase is usually enriched by means of different variable neighborhood structures. For instance, VND (Variable Neighborhood Descent) explores several neighborhood structures in a deterministic order. Its success is based on the simple fact that different neighborhood structures do not usually have the same local minimum. Thus local optima can be escaped by applying some deterministic rule for altering the neighborhoods [3]. Tabu Search is a strategy to prevent local search algorithms getting trapped in previously visited solutions. It accepts non-improving moves and uses a penalization mechanism called Tabu List [2,4]. The reader is invited to consult the comprehensive Handbook of Metaheuristic for further information [8].

Here, we develop a GRASP/VND methodology enriched with Tabu Search in order to avoid getting trapped in previous visited solutions. In the following, the Pseudo-code of our Hybrid Metaheuristic (HM) for the max cut-clique is presented (see Algorithm 1). It follows the traditional two-phase GRASP template enriched with a VND (Lines 4–5).

A Tabu Search strategy is included in order to enhance feasible solutions. The tabu list $\mathcal{T}$ stores tabu nodes (Line 6), discarding previous solutions. Essentially, the most frequent nodes involved in all solutions after the second phase (VND) are not considered for further solutions during θ iterations, whenever we reach θ^{max} consecutive iterations without improvement. Most frequent nodes are selected if they appear more than ϕ times since the last tabu list refresh. The real numbers ϕ and θ are uniformly chosen at random in the interval $[1, \theta^{max}]$, being θ^{max} a parameter of the algorithm. The specific GRASP phases for the MCC are described in detail in the following subsections.

Algorithm 1. HM PSEUDO-CODE

    Input: α, θ^{max}, maxIter, $\mathcal{G}$
    Output: $\mathcal{C}^*$
1: $\mathcal{C}^* \leftarrow \emptyset$
2: $\mathcal{T} \leftarrow \emptyset$
3: **for** iter = 1 to maxIter **do**
4:     $\mathcal{C} \leftarrow$ CLIQUE$(\alpha, \mathcal{T}, \mathcal{G})$
5:     $\mathcal{C} \leftarrow$ VND$(\mathcal{C}, \mathcal{T}, \mathcal{G})$
6:     $\mathcal{T} \leftarrow$ UPDATE$(\mathcal{T}, \theta^{max}, \mathcal{C})$ ▷ Tabu List
7:     **if** $|E'(\mathcal{C})| > |E'(\mathcal{C}^*)|$ **then**
8:         $\mathcal{C}^* \leftarrow \mathcal{C}$
9: **return** $\mathcal{C}^*$

3.1 Construction Phase - *Clique*

The construction phase of the proposed algorithm is depicted in Algorithm 2. Let us denote by $\mathcal{C}$ the clique under construction, $\delta(U)$ and $\Delta(U)$ the minimum and maximum degree of the node-set U. The clique $\mathcal{C}$ is initially empty (Line 1), and a multi-start process is considered (Line 2). A Restricted Candidate List (*RCL*) is defined in Line 3. Observe that the RCL includes nodes with the highest degree, and α trades greediness for randomization. During the *While* loop of Lines 4–11, a singleton $\{i\}$ is uniformly picked from the RCL (Line 5), and the maximum clique $\mathcal{C}'$ is built using the nodes from the set $\mathcal{C} \cup \{i\}$, specifically, $[\mathcal{C} \cap N(i)] \cup \{i\}$, being $N(i)$ the neighbor-set of node i (see Line 6). The best solution is updated if necessary (Lines 7–8). Observe that the process is finished only if we meet $MAX\_ATTEMPTS$ without improvement (Lines 9–11). The reader can appreciate that the output $\mathcal{C}$ is the best feasible clique during the whole process (Line 12).

Algorithm 2. CLIQUE

    Input: $\alpha, \mathcal{T}, \mathcal{G}$
    Output: $\mathcal{C}$
1: $\mathcal{C} \leftarrow \emptyset$
2: $improving = MAX\_ATTEMPTS$
3: $RCL \leftarrow \{v \in V - \mathcal{C} : |E'(v)| \geq \Delta(V - \mathcal{C}) - \alpha(\Delta(V - \mathcal{C}) - \delta(V - \mathcal{C}))\}$
4: **while** $improving > 0$ **do**
5:    $i \leftarrow selectRandom(RCL)$
6:    $\mathcal{C}' \leftarrow [\mathcal{C} \cap N(i)] \cup \{i\}$
7:    **if** $|E'(\mathcal{C}')| > |E'(\mathcal{C})|$ **then**
8:       $\mathcal{C} \leftarrow \mathcal{C}'$
9:       $improving \leftarrow MAX\_ATTEMPTS$
10:   **else**
11:      $improving \leftarrow improving - 1$
12: **return** $\mathcal{C}$

3.2 Local Search Phase - *VND*

The goal is to combine a rich diversity of neighborhoods in order to obtain an output that is locally optimum solution for every feasible neighborhood. Five neighborhood structures are considered to build a VND [3]. **Add**, **Swap**, and **Aspiration** are taken from a previous ILS [6]. However, our VND is enriched with 2 additional neighborhood structures, named **Remove** and **Cone**. The following neighborhood take effect whenever the resulting cut-clique is increased:

- **Remove**: a singleton $\{i\}$ is removed from a clique $\mathcal{C}$.
- **Add**: a singleton $\{i\}$ is added from a clique $\mathcal{C}$.
- **Swap**: if we find $j \notin \mathcal{C}$ such that $\mathcal{C} - \{i\} \subseteq N(j)$, we can include j in the clique and delete i (swap i and j).

- **Cone**: generalization of Swap for multiple nodes. The clique $\mathcal{C}$ is replaced by $\mathcal{C} \cup \{i\} - \mathcal{A}$, being $\mathcal{A}$ the nodes from $\mathcal{C}$ that are non-adjacent to i.
- **Aspiration**: this movement offers the opportunity of nodes belonging to the Tabu List to be added.

Observe that the dynamic Tabu list works during the potential additions during **Add**, **Swap** and **Cone**. On the other hand, **Aspiration** provides diversification with an *opportunistic unchoking* process: it picks nodes from the Tabu List instead. For the remaining four local searches, there is an efficient way to determine whether there is an improvement with respect to some neighbor-set. Specifically, the Test Lemmas 3 to 6 are useful to determine the improvements for **Remove**, **Add**, **Swap** and **Cone** movements, respectively. We call Aspiration Test to Lemma 4 but applied in a different domain (specifically, the candidate nodes must belong to the Tabu List).

Lemma 3 (Remove). $|\delta(\mathcal{C} - \{i\})| > |\delta(\mathcal{C})|$ *iff* $|\delta(i)| < 2(|\mathcal{C}| - 1)$.

Proof.

$$
\begin{aligned}
|\delta(\mathcal{C} - \{i\})| &= |\delta(\mathcal{C})| + |\mathcal{C}| - 1 - (|\delta(i)| - (|\mathcal{C}| - 1)) \\
&= |\delta(\mathcal{C})| + |\mathcal{C}| - 1 - |\delta(i)| + |\mathcal{C}| - 1 \\
&= |\delta(\mathcal{C})| + 2(|\mathcal{C}| - 1) - |\delta(i)| \\
&> |\delta(\mathcal{C})|,
\end{aligned}
$$

where the last inequality holds iff $2(|\mathcal{C}| - 1) - |\delta(i)| > 0$.   ∎

Lemma 4 (Add). $|\delta(\mathcal{C} \cup \{i\})| > |\delta(\mathcal{C})|$ *iff* $|\delta(i)| > 2|\mathcal{C}|$.

Proof.

$$
\begin{aligned}
|\delta(\mathcal{C} \cup \{i\})| &= |\delta(\mathcal{C})| - |\mathcal{C}| + |\delta(i)| - |\mathcal{C}| \\
&= |\delta(\mathcal{C})| + |\delta(i)| - 2|\mathcal{C}| \\
&> |\delta(\mathcal{C})|,
\end{aligned}
$$

where the last inequality holds iff $|\delta(i)| > 2|\mathcal{C}|$.   ∎

Lemma 5 (Swap). $|\delta(\mathcal{C} - \{j\} \cup \{i\})| > |\delta(\mathcal{C})|$ *iff* $|\delta(i)| > |\delta(j)|$.

Proof.

$$
\begin{aligned}
|\delta(\mathcal{C} - \{j\} \cup \{i\})| &= |\delta(\mathcal{C})| - |\delta(j)| + 2(|\mathcal{C}| - 1) + |\delta(i)| - 2(|\mathcal{C}| - 1) \\
&= |\delta(\mathcal{C})| - |\delta(j)| + |\delta(i)| \\
&> |\delta(\mathcal{C})|,
\end{aligned}
$$

where the last inequality holds iff $|\delta(i)| > |\delta(j)|$.   ∎

Lemma 6 (Cone). $|\delta(\mathcal{C} - \mathcal{A} \cup \{i\})| > |\delta(\mathcal{C})|$ *iff* $|\delta(i)| > |\delta(\mathcal{A})| - 2|\mathcal{C} - \mathcal{A}|(|\mathcal{A}| - 1)$.

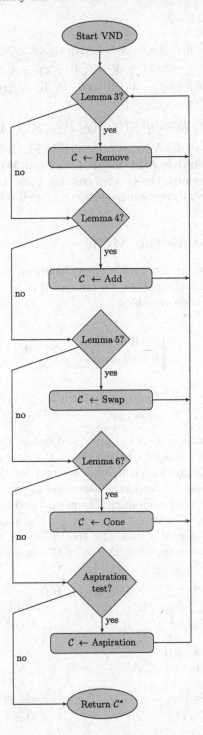

Fig. 1. Flow diagram for the local search phase - VND.

Proof.

$$|\delta(\mathcal{C} - \mathcal{A} \cup \{i\})| = |\delta(\mathcal{C})| + |\mathcal{A}||\mathcal{C} - \mathcal{A}| - (|\delta(\mathcal{A})| - |\mathcal{A}||\mathcal{C} - \mathcal{A}|) - 2|\mathcal{C} - \mathcal{A}| + |\delta(i)|$$
$$= |\delta(\mathcal{C})| + 2|\mathcal{A}||\mathcal{C} - \mathcal{A}| - |\delta(\mathcal{A})| - 2|\mathcal{C} - \mathcal{A}| + |\delta(i)|$$
$$= |\delta(\mathcal{C})| + 2|\mathcal{C} - \mathcal{A}|(|\mathcal{A}| - 1) - |\delta(\mathcal{A})| + |\delta(i)||\delta(\mathcal{C} - \mathcal{A} \cup \{i\})|$$
$$> |\delta(\mathcal{C})|$$

where the last inequality holds iff $|\delta(i)| > |\delta(\mathcal{A})| - 2|\mathcal{C} - \mathcal{A}|(|\mathcal{A}| - 1)$. ∎

The Flow Diagram of our VND is presented in Fig. 1. The ordered sequence of local searches are **Remove, Add, Swap, Cone** and **Aspiration** moves. Once an improvement is obtained, the process restarts from the beginning. Observe that, in the output, a locally optimum solution under all neighborhood structures is met.

4   Exact Method for the MCC

In this section, we present an exact method based on a mathematical formulation. Due to combinatorial nature, we addressed it by integer programming, using the following decision variables:

$$w_i = \begin{cases} 1 & \text{if node } i \in \mathcal{C} \\ 0 & \text{otherwise} \end{cases}, \forall i \in V$$

$$w_{(i,j)} = \begin{cases} 1 & \text{if edge } (i,j) \in E(\mathcal{C}) \\ 0 & \text{otherwise} \end{cases}, \forall (i,j) \in E$$

An integer programming model is presented below. Constraint (1) and (2) state that both nodes i, j belong to the clique $\mathcal{C}$ if and only if $(i,j) \in E(\mathcal{C})$. Recall that Theorem 3 provides a feasible interval for the size of the clique, c_{min}. Constraints (3) and (4) determine lower and upper bounds Lb and Ub for the size of the clique, found combining Theorem 3 and the best output of our GRASP/VND heuristic. Constraints (5) and (6) just state that w_i and $w_{(i,j)}$ are binary variables. The goal is maximize the cut-clique, which is precisely the difference between the sum-degree minus twice the number of internal links.

$$\max \sum_{i \in V} d_i \times w_i - 2 \times \sum_{(i,j) \in E} w_{(i,j)}$$

$$\text{s.a. } 2w_{(i,j)} \leq w_i + w_j \qquad\qquad \forall (i,j) \in E \ (1)$$
$$w_i + w_j - 1 \leq w_{(i,j)} \qquad\qquad \forall i, j \in V \quad (2)$$

$$\sum_{i \in V} w_i \geq Lb \qquad\qquad\qquad\qquad\qquad (3)$$
$$\sum_{i \in V} w_i \leq Ub \qquad\qquad\qquad\qquad\qquad (4)$$

$$w_{(i,j)} \in \{0,1\} \qquad\qquad\qquad\quad \forall (i,j) \in E \ (5)$$
$$w_i \in \{0,1\} \qquad\qquad\qquad\qquad \forall i \in V \quad (6)$$

5 Computational Results

In order to test the performance of the algorithm we carried out a fair comparison with respect to the ILP model implemented using IBM CPLEX 12.8. Both algorithms are executed on a Home-PC (Intel Core i7, 2.4 GHz, 8 GB RAM). The graphs under study were obtained from the SteinLib[1] and DIMACS.

Table 1 reports the performance of both algorithms for each instance. All HM algorithm instances were tested using a single run with one-hundred iterations

Table 1. HM versus ILP for the MCC.

Instances			GRASP/VND			ILP				
Name	n	Density	$|E'(\mathcal{C})|$	$|\mathcal{C}|$	Time (s)	Lb	Ub	$|E'(\mathcal{C})|$	$|\mathcal{C}|$	Time (s)
i080-001	80	0.039	**13**	2	0.7120	2	8	**13**	2	0.66
i080-002	80	0.039	**13**	2	1.4779	2	9	**13**	2	0.62
i080-011	80	0.11	**38**	4	3.0396	3	16	**38**	4	0.94
i080-044	80	0.2	**80**	5	1.1495	4	26	**80**	5	0.71
i080-045	80	0.2	**74**	4	0.5748	4	25	**74**	4	0.90
i080-111	80	0.11	**35**	4	0.4141	3	15	**35**	4	0.88
i080-112	80	0.11	**39**	3	0.3191	3	19	**39**	3	0.64
i080-131	80	0.05	**16**	2	1.1908	2	10	**16**	2	0.73
i080-132	80	0.05	**15**	3	0.5146	2	9	**15**	3	0.69
i080-142	80	0.2	**74**	4	1.0306	4	25	**74**	4	0.70
i080-143	80	0.2	**80**	4	0.8546	4	26	**80**	4	0.85
i160-001	160	0.019	**15**	2	0.6351	2	10	**15**	2	2.84
i160-002	160	0.019	**14**	2	2.8054	2	9	**14**	2	2.44
i160-011	160	0.064	**44**	3	4.6177	3	20	**44**	3	24.53
i160-044	160	0.2	**180**	5	5.7298	5	47	**180**	5	11.23
i160-045	160	0.2	**173**	5	3.8451	5	42	**173**	5	14.92
i160-111	160	0.064	**50**	4	5.9956	4	18	**50**	4	5.59
i160-112	160	0.064	**46**	4	0.47	3	18	**46**	4	5.71
i160-131	160	0.025	**19**	3	2.5908	3	10	**19**	3	2.80
i160-132	160	0.025	**22**	3	2.3052	3	12	**22**	3	2.95
i160-142	160	0.2	**183**	5	3.7192	5	45	**183**	5	10.63
i160-143	160	0.2	**170**	5	05.097	5	44	**170**	5	10.57
mc11	400	0.0095	**6**	2	0.7292	2	2	**6**	2	6.43
c-fat200-1	200	0.077	**81**	9	0.1860	9	17	**81**	9	29.57
c-fat200-2	200	0.163	**306**	17	0.8388	17	34	**306**	17	88.79
c-fat200-5	200	0.426	**1892**	43	13.0593	43	86	**1892**	43	1717.39
c-fat500-1	500	0.036	**110**	10	4.77459	10	20	**110**	10	1198.31
c-fat500-2	500	0.073	**380**	19	14.1875	19	38	**380**	19	1822.08
c-fat500-5	500	0.186	2304	48	121.32	48	95	2304	48	10800
c-fat500-10	500	0.374	8930	94	33.298	94	188	8930	94	10800

---

[1] The dataset can be found in the URL http://steinlib.zib.de/steinlib.php.

and $\alpha = \frac{1}{2}$, $MAX\_ATTEMPTS = \lfloor \frac{|V|}{10} \rfloor$, $\theta^{max} = 4$. Lower and upper bounds Lb and Ub were obtained for each topology under study using Corollary 1.

The values remarked using bold letters from column $|\delta(\mathcal{C})|$ indicate that the best solution was reached according to the output from the ILP solver.

Following the terminology, $|\delta(\mathcal{C})|$, $|\mathcal{C}|$ and $Time$ represent maximum cut-clique size found, best solution, and the CPU time for the best solution found. Lb, Ub columns are reported for the ILP solver which represents the lower an upper bound for the ILP model. Under ILP, $Time$ give the time to reach the optimum value or the best lower bound to the optimum when the optimum is not attained within the given time limit (10800 s).

Table 2. Performance of the local search phase.

Instances			GRASP/VND						
Name	n	Density	Remove (%)	Add (%)	Swap (%)	Cone (%)	Aspiration (%)	#moves	mp (%)
i080-001	80	0.039	0	38	57	5	0	154	30.949
i080-002	80	0.039	3	32	59	7	0	147	27.571
i080-011	80	0.11	1	48	48	3	0	158	18.647
i080-044	80	0.2	0	54	46	0	0	239	17.390
i080-045	80	0.2	0	56	44	0	0	217	20.585
i080-111	80	0.11	0	59	39	3	0	176	22.572
i080-112	80	0.11	0	48	37	5	0	244	16.261
i080-131	80	0.05	1	30	60	9	0	151	25.983
i080-132	80	0.05	0	32	68	0	0	136	23.402
i080-142	80	0.2	0	50	50	0	0	202	19.762
i080-143	80	0.2	0	52	48	0	0	253	18.467
i160-001	160	0.019	0	22	73	5	0	143	29.452
i160-002	160	0.019	0	20	80	0	0	135	26.427
i160-011	160	0.064	0	57	40	3	0	186	21.803
i160-044	160	0.2	0	60	40	0	0	251	18.079
i160-045	160	0.2	0	53	47	0	0	211	16.358
i160-111	160	0.064	0	61	38	1	0	181	23.816
i160-112	160	0.064	0	55	45	0	0	154	23.940
i160-131	160	0.025	0	39	58	2	0	168	23.710
i160-132	160	0.025	0	42	54	3	0	179	26.588
i160-142	160	0.2	0	57	42	0	0	250	18.146
i160-143	160	0.2	0	59	41	0	0	196	17.664
mc11	400	0.0095	0	100	0	0	0	46	50.000
c-fat200-1	200	0.077	0	100	0	0	0	376	13.159
c-fat200-2	200	0.163	0	100	0	0	0	475	6.338
c-fat200-5	200	0.426	0	99	1	0	0	138	0.505
c-fat500-1	500	0.036	0	100	0	0	0	391	12.965
c-fat500-2	500	0.073	0	100	0	0	0	278	3.736
c-fat500-5	500	0.186	2	98	0	0	0	132	3.252
c-fat500-10	500	0.374	0	100	0	0	0	11	0.132

The reader can appreciate from Table 1 that our GRASP/VND algorithm meets the best solution in all cases. The globally optimum for all the instances under study is formally proved using the ILP formulation. Furthermore, our GRASP/VND approach presents consistently smaller CPU times for graphs with large size.

Table 2 shows the performance of the VND algorithm. The activity of every single local search is studied. Swap and Add movements show to be more effective, while Remove and Cone take effect few times. Aspiration has no effect, but it works for dense graphs.

In order to understand the global effectiveness of our VND scheme, a *mid-point test* is performed. The columns Remove, Add, Swap, Cone and Aspiration show the percentage of each kind of movement applied over one-hundred executions of the VND local search phase. The column #moves states the amount of movement applied during these iterations. The column entitled *mp* displays the average gap in percentage between the best solution found in each local search phase with respect to the feasible solution obtained from the construction phase over one-hundred iterations. The reader can appreciate that the VND effect is notorious, since the cut-clique is roughly half the optimum in most cases using only the Construction Phase.

It is worth to remark that we further studied the performance of our GRASP/VND methodology versus a state-of-the-art ILS heuristic for the MCC, detailed in [6]. We could find optimality in all the reported instances which achieved optimality, and we found the best feasible solutions so far in the remaining cases, with identical results offered in [6].

6   Conclusions and Trends for Future Work

Several business models can be represented by Market Basket Analysis (MBA). A relevant marketing approach is to find a subset of items that are strongly correlated with the others. This intuition is formalized by means of a combinatorial optimization problem, called Max Cut-Clique (MCC).

In this paper, the $\mathcal{NP}$-Completeness of MCC is established. This fact promotes the development of heuristics and bounds. As a consequence, we offered bounds for both the globally optimum solution and the size of the minimum cardinality clique with maximum cut. Then, a GRASP/VND methodology enriched with Tabu Search is developed to address the MCC. A fair comparison with an exact ILP formulation confirms the optimality of our approach for hundreds of nodes. Furthermore, the computational effort is reduced for the heuristic under large-sized graphs. The movements Swap and Add have the largest activity for the instances under study. The experiments shows that our GRASP/VND heuristic is competitive with state-of-the-art solutions for the MCC. Further analysis should be done to determine the best order for the VND in terms of computational efficiency.

As future work we would like to implement our solution into a real-life product-placement scenario. In a first stage, we need historical information to determine the links between pairs of items. The physical location of the items must be determined using a complementary geometrical problem with constraints. The solution could consider multi-constrained clustering in order to include categories for the items, or other Machine Learning techniques to determine profiles for the customers, according to the product under study. After the real implementation, the feedback of sales in a period is a valuable metric of success.

Acknowledgements. This work is partially supported by Project 395 CSIC I+D *Sistemas Binarios Estocásticos Dinámicos*.

References

1. Aguinis, H., Forcum, L.E., Joo, H.: Using market basket analysis in management research. J. Manag. **39**(7), 1799–1824 (2013)
2. Amuthan, A., Thilak, K.D.: Survey on tabu search meta-heuristic optimization. In: 2016 International Conference on Signal Processing, Communication, Power and Embedded System (SCOPES), pp. 1539–1543, October 2016
3. Duarte, A., Mladenović, N., Sánchez-Oro, J., Todosijević, R.: Variable neighborhood descent. In: Martí, R., Panos, P., Resende, M. (eds.) Handbook of Heuristics, pp. 1–27. Springer, Cham (2016). https://doi.org/10.1007/978-3-319-07153-4_9-1
4. Glover, F., Laguna, M.: Tabu Search. Kluwer Academic Publishers, Norwell (1997)
5. Martins, P.: Cliques with maximum/minimum edge neighborhood and neighborhood density. Comput. Oper. Res. **39**(3), 594–608 (2012)
6. Martins, P., Ladrón, A., Ramalhinho, H.: Maximum cut-clique problem: ILS heuristics and a data analysis application. Int. Trans. Oper. Res. **22**(5), 775–809 (2015)
7. Resende, M.G.C., Ribeiro, C.C.: Optimization by GRASP - Greedy Randomized Adaptive Search Procedures. Computational Science and Engineering. Springer, New York (2016). https://doi.org/10.1007/978-1-4939-6530-4
8. Salhi, S.: Handbook of metaheuristics. J. Oper. Res. Soci. **65**(2), 320 (2014). (2nd edition)

A VNS Approach to Solve Multi-level Capacitated Lotsizing Problem with Backlogging

Jerzy Duda(✉) and Adam Stawowy

Department of Applied Computer Science, Faculty of Management,
AGH University of Science and Technology,
Gramatyka St. 10, 30-067 Krakow, Poland
{jduda,astawowy}@zarz.agh.edu.pl

Abstract. In this paper a multi-level capacitated lotsizing problem with machine-capacity-constraint and backlogging is studied. The main objective is to minimize the total cost which includes the inventory and delaying costs of produced items. Since the problem under study is NP-hard, a variable neighborhood search (VNS) combined with CPLEX solver is proposed as a solution approach. Neighborhood is changed according to VNS scheme employing four different functions and is locally optimized for a set of partial MIP problems that can be easily solved.

Finally, extensive computational tests demonstrate that the proposed search algorithm can find good quality solutions for all examined problems. The objective values obtained by the proposed algorithm are comparable to the results of state-of-the art, much more complicated algorithms.

Keywords: Multi-level capacitated lotsizing ·
Variable neighborhood search · Hybrid approach

1 Introduction

The multi-level capacitated lotsizing problem (MLCLSP) is a generalization of the classical capacitated lotsizing problem (CLSP). CLSP is a large-bucket model which determines the lotsizes of produced items but not the sequence of the lots. The setup times may be needed to produce some lots and all machines have limited capacities. MLCLSP extends CLSP by assuming that the manufacturing process is performed at multi stages machines, according to the BOM list of MRP system. The goal is to determine production plan for final products and their components to prevent delays in the delivery to the customers and to minimize the inventory costs. The multi-level capacitated lotsizing problem was originally introduced by Billington et al. [2]. In such problem the planning horizon is finite and it is divided into T discrete time periods (e.g. day or weeks). The plan has

This work is supported by AGH UST statutory research no. 11/11.200.327.

A. Sifaleras et al. (Eds.): ICVNS 2018, LNCS 11328, pp. 41–51, 2019.
https://doi.org/10.1007/978-3-030-15843-9_4

to be prepared for I items (finished products) with external demand d known for each period or internal demand resulting from the preceding production (for sub-products). Original formulation of MLCLSP required demand to be met without any delay, however, later an extended formulation allowed for backlogging of the finished products, which greatly complicates solution of the problem.

In this paper we study MLCLSP with backlogging and setup times for product families. The aim of the paper is to develop the model and appropriate algorithm of its solution able to achieve optimized production plans for multi-level system within few minutes, which is enough for the planners in industrial practice (plans are prepared usually at the beginning of each working week but sometimes need to be rearranged, if some unexpected event occurs like order cancellation or machine breakdown). We propose a hybrid approach combing variable neighborhood search with CPLEX solver. The paper has the following structure. Section 2 presents the literature review. Definition of the problem and notation are described in Sect. 3. Section 4 gives the details on proposed heuristic approach. The computational experiments are summarized in Sect. 5, and the conclusions are drawn in Sect. 6.

2   State of the Art

There are many studies reported in the literature for capacitated lotsizing problems. A comprehensive review on CLSP and MLCLSP formulations as well as solution approaches are presented in [3] and [7]. For practical reasons, this section gives a limited review of the methods used to solve considered problem: we focus only on approaches that are close related to our work.

Wu et al. [14] developed the LugNP (Lower and upper bound guided Nested Partitions) framework using two new MIP reformulations for capacitated multi-level lotsizing problems with backlogging. LugNP effectively combines lower and upper bounding techniques with NP method, incorporating an efficient partitioning and sampling strategy, guiding the search to the most promising region of the solution space. Extensive computational tests demonstrated the quality of proposed framework.

Zhao et al. [15] used a variable neighborhood decomposition search (VNDS) integrated with a general-purpose CPLEX MIP solver - a method suggested by Lazic et al. [8]. The VNDS algorithm consequently fixes values for hard variables on the basis of solution achieved with MIP relaxation and then VNS algorithm is used. The heuristic is able to solve problem with 40 items, 16 periods, and 6 resources.

Seeanner et al. [9] proposed the similar approach which combines the principles of variable neighborhood decomposition search and fix-and-optimize (FO) heuristic. They provided results for real world problems with hundreds of items and 12 time periods. However, the computation time was an one hour for a single instance of the problem, and the gap between the obtained results and reference values was also significant.

Toledo et al. [13] presented a hybrid multi-population genetic algorithm (HMPGA) which combines a multi-population genetic algorithm using FO heuristic and MIP submodels solved by a CPLEX solver. HMPGA evolves three populations of individuals hierarchically structured in a tree. Each individual is represented as a 0–1 matrix $F*T$, where F is the number of families and T is the number of periods (1 - if a setup of family f in period t happens, 0 - otherwise). The fix-and-optimize heuristic tries to improve the current best individual, sequentially fixing and optimizing binary variables, using period and family rolling horizon windows. A total of four test sets from the MULTILSB (Multi-Item Lot-Sizing with Backlogging - the same we use in this research) library were solved and the results have shown that HMPGA had a better performance for most of the test sets than two competitive approaches (Akartunali and Miller's Heuristic [1] and Lower and upper bound guided Nested Partitions heuristic [14]).

More recently, Chen [4] developed a VNS and FO approach that iteratively solves a series of sub-problems of the model until no better solution can be found. Each sub-problem re-optimizes a determined subset of binary decision variables based on the interrelatedness of binary variables in the constraints, while the values of the other binary variables are fixed. Then CPLEX branch and bound algorithm is used to solve the relaxed problem. The largest instances considered by the author had a size of 100 items and 16 periods, and it took more than 10 min to solve them.

Similar approach for multi-item capacitated lotsizing with time windows and setup times was applied by Erromdhani et al. [6]. However, due to the problem nature, in the FO stage they optimized matrix of product setups, not the family setups as it was in [4]. The maximum number of items in computational experiments was 30.

Other types of lot-sizing problems were studied by Sifaleras and Konstantaras. In [10] the authors used Variable Neighborhood Descent (VND) algorithm with eight different neighborhood structure types to solve multi-item dynamic lotsizing problem in a closed-loop supply chain. In [11] they used General VNS with four different neighborhoods to solve reverse logistics multi-item dynamic lotsizing problem. In both cases they considered instances up to 300 items and 52 periods.

3   Problem Formulation

Many formulations of MLCLSP with backlogging have been presented; a basic formulation of the problem can be expressed as follows:

Indices

$i = 1, \dots, n, \dots, I$ produced items, first n items are the finished products
$t = 1, \dots, T$ production periods
$m = 1, \dots, M$ machines
$f = 1, \dots, F$ families of products

Data

d_{it}	demand for item i in period t
a_{mi}	capacity used to produce item i on machine m
C_{mt}	total capacity of machine m in period t
h_i	holding cost of item i
bc_i	backlog cost of item i, $i = 1, \ldots, n$
c_{if}	1, if product i belongs to family f
r_{ij}	quantity of product i required to produce product j
st_{mf}	setup time for family f on machine m

Variables

x_{it}	production of item i in period t
y_{it}	1, if there is setup resulting from production of item i in period t
w_{ft}	1, if there is setup resulting from production of family f in period t
s_{it}	stock holding of item i in period t
b_{it}	backlog of item i in period t

Minimize

$$\sum_{i=1}^{n}\sum_{t=1}^{T} bc_i \cdot b_{it} + \sum_{i=1}^{I}\sum_{t=1}^{T} h_i \cdot s_{it} \tag{1}$$

Subject to:

$$x_{it} + b_{it} - b_{it-1} + s_{it-1} - s_{it} = d_{it}, \quad i = 1, \ldots, n, \ t = 1, \ldots, T \tag{2}$$

$$x_{it} + s_{it-1} - s_{it} = \sum_{j=1}^{i-1} r_{ij} \cdot x_{jt}, \quad i = n+1, \ldots, I, \ t = 1, \ldots, T \tag{3}$$

$$x_{it} \le y_{it} \cdot d_{tT}^{i}, \quad i = 1, \ldots, I, \ t = 1, \ldots, T \tag{4}$$

$$y_{it} \cdot c_{if} \le w_{ft}, \quad i = 1, \ldots, I, \ t = 1, \ldots, T \tag{5}$$

$$\sum_{i=1}^{I} a_{mi} \cdot x_{it} + \sum_{f=1}^{F} st_{mf} \cdot w_{ft} \le C_{mt}, \quad m = 1, \ldots, M, \ t = 1, \ldots, T \tag{6}$$

$$x_{it}, s_{it}, b_{it} \ge 0, \quad y_{it}, w_{ft} \in \{0, 1\} \tag{7}$$

The objective function (1) minimizes the sum of backlogging and holding costs of all items (backlog is valid only for the finished products). Equation (2) balances stock levels, backlogs and current production with the demand for the finished products, while Eq. (3) balances stock and current production with the demand resulting from production of preceding items. Constraint (4) ensures that there is a setup ($y_{it} = 1$), if item i is produced in period t and simultaneously limits the number of produced items to the cumulated demand in interval $[k, t]$ $d^i_{kt} = \sum_{l=k}^{t} d_{il}$. Lot size upper bound for the product i in the period t can be further limited as shown in [1]. Constraint (5) ensures that there is a setup ($w_{ft} = 1$), if family f is produced in period t. Constraint (6) limits the production of items and setup of the machines to the capacity of machine m available in period t. Decision variables x, s, b are continuous nonnegative numbers, while setup decision variables are binary (7).

4 VNS Algorithm

4.1 Idea of the Algorithm

At first we planned to extend our VNS algorithm for CLSP [5] so that it would take into account many levels of products. The matrix $\mathbf{x}$ representing the lotsizes for the products in every period will be disturbed in accordance with the VNS rules and locally optimized by linear programming model simplifying original model of the problem. The problem however turned out to have so much different characteristics from the previously analyzed variant of CLSP that the obtained results were highly unsatisfying.

We had to focus not on the $\mathbf{x}$ matrix, but on either the matrix of product setups $\mathbf{y}$, as it was in [6] or the matrix of family setups $\mathbf{w}$, as it was proposed in [13]. After series of experiments we decided to employ the latter matrix as the representation of the solution.

An exemplary representation used later for computational experiments is presented in Fig. 1. Columns contain 17 families, rows represent 16 subsequent periods of time.

	f_1	f_2	f_3	f_4	f_5	f_6	f_7	f_8	f_9	f_{10}	f_{11}	f_{12}	f_{13}	f_{14}	f_{15}	f_{16}	f_{17}
t_1	0	0	1	1	1	1	1	0	0	0	1	1	1	1	1	1	0
t_2	0	0	1	1	1	1	1	0	0	0	1	1	1	1	1	1	0
t_3	0	0	1	1	1	1	0	0	0	0	1	1	0	1	1	0	0
t_4	0	0	1	1	1	1	1	0	0	0	1	1	1	1	1	1	0
t_5	0	0	1	1	1	1	0	0	0	0	1	1	0	1	1	0	0
t_6	0	0	1	1	1	1	1	0	0	0	1	1	1	1	1	1	0
t_7	0	0	1	1	1	1	0	0	0	0	1	1	0	1	1	0	0
t_8	0	0	1	1	1	1	1	0	0	0	1	1	1	1	1	1	0
t_9	0	0	1	1	1	1	0	0	0	0	1	1	0	1	1	0	0
t_{10}	0	0	1	1	1	1	1	0	0	0	1	1	1	1	1	1	0
t_{11}	0	0	1	1	1	1	0	0	0	0	1	1	0	1	1	0	0
t_{12}	0	0	1	1	1	1	1	0	0	0	1	1	1	1	1	1	0
t_{13}	0	0	1	1	1	1	0	0	0	0	1	1	0	1	1	0	0
t_{14}	0	0	1	1	1	1	1	0	0	0	1	1	1	1	1	1	0
t_{15}	0	0	1	1	1	1	0	0	0	0	1	1	0	1	1	0	0
t_{16}	0	0	1	1	1	1	1	0	0	0	1	1	1	1	1	1	0

Fig. 1. Representation of the solution

The main procedure of the VNS algorithm is shown in Fig. 2. At the beginning each cell in the solution matrix **w** is initialized with the value 1 with the probability of 90%. Next, for a given number of iterations (max_{iter}) current neighborhood is changed using only two functions: $AddW()$ and $SubW()$. After that the local search procedure based on MIP CPLEX Solver is used to find the best solution in the neighborhood. This neighborhood is next slightly changed with the functions $FlipW()$ and $UnifyW()$ to further exploit it unless no improvement is achieved in 10 consequent attempts.

```
Initialize(w)
w_best:=w
for iter:=1 to max_iter
   w:=w_best
   if rnd() < 0.5 then      // Shaking stage
      w := AddW(w)
   else
      w := SubW(w)
   do
      k:=0
      w* := LocalMIPSearch(w)    //Descent local search
      if f(w*) < f(w_best) then
         w_best := w*
      else
         k := k+1
         w := FlipW(w)     // Additional neighborhood perturbation
      if rnd() < 0.3 then
         w := UnifyW(w)
   while k < 10
next iter
```

Fig. 2. Main VNS procedure

4.2 Neighborhood Perturbations

We use four functions to alter the **w** matrix. Function $AddW()$ changes the value from 0 to 1 for a randomly chosen family and a randomly chosen period. If setups for the selected family has been already planned for all periods, another family is drawn. In other words it forces one family to have one more setup planned. Similarly, $SubW()$ removes one setup from a randomly selected family. If there are no setups planned for the selected family, family number is redrawn, so that one family has one less planned setup.

Function $FlipW()$ simply changes value 0 in a randomly chosen cell to value 1 or vice versa, depending on its original content. Such a function was used as a mutation operator by Toledo et al. [13] in their genetic algorithm.

The most complex function is $UnifyW()$, which purpose is to even the number of setups in the families containing one or two setups fewer than the randomly chosen family. The application of this function results from the observation that in many cases the setup structure have to be propelled for the families of products and their sub-products.

4.3   Local Search Algorithm

The local search algorithm use a CPLEX solver to solve a greatly simplified version of the original problem. All families - except one - in setup matrix **w** are fixed and such problem is solved within maximum 3 s. (usually immediately). The procedure is repeated for all the families in the random order or (with probability of 10%) in the reverse order of the families. The local search algorithm is shown in Fig. 3. If the solution found by solving MIP for one of the families is better than the current best, the best solution is updated.

```
function LocalMIPSearch(w)
    fam := 1...F
    if rnd() < 0.9 then
        fam := MixOrder(fam)
    else
        fam := ReverseOrder(fam)
    foreach f in fam
        w = solveMIP(f)
        if f(w) < f(w_best) then
            w_best := w
    return w
end function
```

Fig. 3. Local MIP search procedure

5   Computational Experiment and Results Analysis

In order to evaluate our VNS approach we used two out of four tests originally proposed by Simpson and Erengue [12] for MLCLSP without backlogging and next extended by Akartunali and Miller that introduced penalty for backlogging of finished products [1]. Currently it is the only one benchmark available but at the same time containing the most difficult cases to solve. We choose SET01 containing instances for which upper bounds are well defined and confirmed by the best algorithms currently available (i.e. the easiest to solve) and SET03 containing the instances for which those algorithms provide significantly different results (i.e. the hardest to solve). Each set contains 30 instances with 78 items, 6 machines and 17 families.

Table 1. Comparison of the results for SET01 instances.

SET01	LugNP AMH HMPGA		VNS			Deviation			
	Best LAH	Worst LAH	Best VNS	Avg VNS	Sd VNS	bL-bV	bL-aV	wL-bV	wL-aV
1	22382.45	22460.73	22382.45	22822.30	291.31	0.000	0.019	−0.003	0.016
2	27584.79	27584.79	27926.91	28200.56	139.81	0.012	0.022	0.012	0.022
3	25187.25	25239.43	25300.22	25627.15	203.49	0.004	0.017	0.002	0.015
4	26334.72	26436.92	26789.12	26974.66	129.71	0.017	0.024	0.013	0.020
5	25145.49	25254.62	25319.36	25882.55	283.31	0.007	0.028	0.003	0.024
6	26667.42	26770.84	27040.81	27248.87	124.99	0.014	0.021	0.010	0.018
7	24123.78	24218.39	24313.06	24723.61	337.70	0.008	0.024	0.004	0.020
8	29640.42	29645.94	29969.76	30262.30	253.09	0.011	0.021	0.011	0.020
9	20971.19	21362.68	21345.80	21515.59	125.88	0.018	0.025	−0.001	0.007
10	22580.00	22647.53	22563.12	23178.29	406.28	−0.001	0.026	−0.004	0.023
11	12955.57	12955.58	12955.57	13124.47	176.65	0.000	0.013	0.000	0.013
12	26831.25	26831.26	26831.25	26971.12	172.32	0.000	0.005	0.000	0.005
13	23127.84	23127.84	23127.84	23170.99	19.83	0.000	0.002	0.000	0.002
14	25035.84	25035.84	25161.34	25380.37	180.16	0.005	0.014	0.005	0.014
15	14118.11	14118.11	14118.11	14719.36	441.12	0.000	0.041	0.000	0.041
16	17400.12	17540.20	17515.66	17981.46	294.89	0.007	0.032	−0.001	0.025
17	22996.13	23007.51	22996.13	23307.41	206.87	0.000	0.013	0.000	0.013
18	12973.77	12973.77	12973.77	12999.90	43.58	0.000	0.002	0.000	0.002
19	16349.58	16502.94	16562.39	16864.13	157.11	0.013	0.031	0.004	0.021
20	17158.59	17158.59	17158.59	17482.52	236.19	0.000	0.019	0.000	0.019
21	12421.19	12421.19	12421.19	12687.50	265.69	0.000	0.021	0.000	0.021
22	40158.34	40188.74	40216.82	40392.27	165.98	0.001	0.006	0.001	0.005
23	30605.70	30605.70	30605.70	31076.74	289.95	0.000	0.015	0.000	0.015
24	32035.02	32190.36	32024.81	32305.71	111.63	−0.000	0.008	−0.005	0.004
25	52959.94	52989.21	52959.94	53361.99	280.73	0.000	0.008	−0.001	0.007
26	41221.51	41221.51	41553.05	41977.92	268.81	0.008	0.018	0.008	0.018
27	43289.36	43319.73	43289.36	43563.94	268.32	0.000	0.006	−0.001	0.006
28	40993.46	41019.84	40993.46	41084.61	71.05	0.000	0.002	−0.001	0.002
29	25322.35	25492.58	25322.35	25432.04	101.91	0.000	0.004	−0.007	−0.002
30	70863.66	70863.66	70984.24	71150.79	136.03	0.002	0.004	0.002	0.004
Average deviation						0.004	0.016	0.002	0.014

We then compared the results of VNS approach with the best known algorithms proposed in the literature to solve MLCLSP with backlogging. In the order of their publication time they are: Akartunali and Miller Heuristic (AMH) proposed in [1], Lower and upper bound guided Nested Partitions (LugNP) proposed by Wu et al. [14] and a hybrid multi-population genetic algorithm (HMPGA) proposed by Toledo et al. [13]. All algorithms have been run for 100 sec. for the instances in SET01 and for 300 s. for those in set SET03.

Tables 1 and 2 show the comparison for SET01 and SET03, respectively. Best LAH column contains the best result achieved by the best algorithm out of LugNp, AHM and HMPGA, while Worst LAH column contains the result

Table 2. Comparison of the results for SET03 instances.

SET01	LugNP AMH HMPGA		VNS			Deviation			
	Best LAH	Worst LAH	Best VNS	Avg VNS	Sd VNS	bL-bV	bL-aV	wL-bV	wL-aV
1	186680.49	195149.43	191387.30	209939.89	8808.20	0.025	0.111	−0.020	0.070
2	212852.65	236041.47	225310.26	229930.70	3254.25	0.055	0.074	−0.048	−0.027
3	199569.69	235541.55	216103.06	222468.79	4031.02	0.077	0.103	−0.090	−0.059
4	205775.33	225455.50	226820.97	231067.34	3130.64	0.093	0.109	0.006	0.024
5	205079.78	215570.37	214297.28	225388.88	6249.56	0.043	0.090	−0.006	0.044
6	205737.56	221315.06	219692.19	228520.77	5077.96	0.064	0.100	−0.007	0.032
7	196613.20	205742.46	205016.64	215799.17	6519.64	0.041	0.089	−0.004	0.047
8	221449.75	246816.32	240602.09	244125.31	2603.04	0.080	0.093	−0.026	−0.011
9	182662.20	189237.06	195412.15	203968.33	3983.39	0.065	0.104	0.032	0.072
10	188378.86	206095.40	203747.33	212488.50	6202.33	0.075	0.113	−0.012	0.030
11	128756.14	135132.01	143547.75	150481.94	3545.43	0.103	0.144	0.059	0.102
12	200361.47	217015.46	215685.05	222474.07	4388.37	0.071	0.099	−0.006	0.025
13	197661.06	217793.79	212410.95	219793.41	4327.45	0.069	0.101	−0.025	0.009
14	198324.37	206490.19	215134.21	218734.72	2158.66	0.078	0.093	0.040	0.056
15	128153.34	150579.48	147818.73	154606.02	5410.67	0.133	0.171	−0.019	0.026
16	140947.21	146981.65	148427.99	154414.70	4605.40	0.050	0.087	0.010	0.048
17	186345.86	208868.28	204722.77	215460.26	5622.27	0.090	0.135	−0.020	0.031
18	98976.15	114150.49	116622.27	125073.74	3583.24	0.151	0.209	0.021	0.087
19	143961.87	161507.94	156964.17	164340.70	4011.59	0.083	0.124	−0.029	0.017
20	162960.82	164285.30	184593.85	187562.63	2374.04	0.117	0.131	0.110	0.124
21	121932.44	154392.30	138171.62	147151.40	6163.14	0.118	0.171	−0.117	−0.049
22	244366.01	274202.11	274066.46	278636.34	3541.05	0.108	0.123	0.000	0.016
23	211899.91	229468.79	228175.50	243597.15	7501.89	0.071	0.130	−0.006	0.058
24	245491.89	253504.00	262608.85	273941.80	6099.97	0.065	0.104	0.035	0.075
25	326629.29	331890.64	339733.28	346333.16	4715.18	0.039	0.057	0.023	0.042
26	278748.97	290192.54	290044.75	295980.47	3690.88	0.039	0.058	−0.001	0.020
27	291300.89	306675.79	311446.25	319883.48	5029.53	0.065	0.089	0.015	0.041
28	224659.18	225729.92	240805.18	250221.02	5440.39	0.067	0.102	0.063	0.098
29	188074.48	197801.38	202261.93	208023.84	3672.16	0.070	0.096	0.022	0.049
30	394691.55	415301.14	405878.57	415275.59	6397.24	0.028	0.050	−0.023	0.000
Average deviation						0.074	0.109	−0.001	0.037

achieved by the algorithm that occurred to be the worst for the particular instance. Columns Best VNS and Avg VNS present the best and the average result (out of 10 runs) for the VNS approach and a standard deviation for those 10 runs. Last four columns compare the results achieved by the best and the worst out of LugNp, AHM and HMPGA algorithms (denoted as bL and wL, respectively) to the best and average results obtained by the VNS approach (denoted as bV and aV, respectively).

We can observe that for more than a half of SET01 instances the best results achieved by the VNS approach do not differ from the ones achieved by the best algorithms known in the literature (the average deviation is 0.4% when compared to the best algorithm and 0.2% when compared to the worst LAH algorithm). When the average VNS results are compared, the deviation raises to 1.6% and 1.4% respectively, which can be acceptable, taking into account the simplicity of the proposed approach.

Moreover, for two instances (10 and 24) the VNS approach was able to find better results than provided by LugNp, AHM and HMPGA algorithms. The results achieved by the VNS approach for the instances in SET03 are not so outstanding, if we compare them to the best results provided by the three competitive algorithms (in this case the best algorithm was always HMPGA). The best results achieved by VNS approach are worse on average by 7.4%, while the average VNS results are 10.9% behind HPMGA. However, if we compare the VNS results with LugNp and AHM algorithms, the best of VNS results is on average slightly better (by 0.1%) and the average VNS result is 3.7% worse, however, still being better in 4 cases.

The observations presented above show the potential of the proposed VNS approach and, at the same time, the need for further progress, especially for difficult to solve instances (e.g. instances 11 and 20 in SET03). However, this will most likely complicate the algorithm and then it will limit its applicability to other types of MLCLSP.

6   Conclusions and the Future Work

Multilevel capacitated lot sizing problem is one of the hardest and most practical problems that is faced in a wide variety of production systems. The paper proposes an efficient VNS approach integrated with fix and optimize heuristic for MLCLSP to minimize the delaying and inventory costs.

The computational experiments presented in the paper prove that the proposed VNS based approach can be well applied to the multi-level capacitated lotsizing problem and gives only slightly less satisfactory results than much more complicated algorithms. However, for some particularly difficult cases (like two instances in the SET03 described in the computational experiments) further improvement of the VNS algorithm is desirable.

It is worth to notice that the proposed approach due to its simplicity can be easily applied for solving any lotsizing and scheduling optimization problem with setups dependent on the families of products. In the future research, we plan to extend our approach first of all to include a smarter neighborhood generator. It might be also interesting to check whether the approach proposed by Erromdhani et al. [6] for the multi-item lotsizing problem with product-dependent setups would be beneficial also for the MLCLSP with family-dependent setups. Eventually, we plan to study a multilevel capacitated lotsizing problem in which lot sizes for products are expressed as integer values what has practical applications in many industrial cases and is much harder to solve than the problem with continuous values.

References

1. Akartunali, K., Miller, A.J.: A heuristic approach for big bucket multi-level production planning problems. Eur. J. Oper. Res. **193**(2), 396–411 (2009)
2. Billington, P., McClain, J., Thomas, L.: Mathematical programming approaches to capacity-constrained MRP systems: review, formulation and problem reduction. Manag. Sci. **29**(10), 1126–1141 (1983)
3. Buschkuehl, L., Sahling, F., Helber, S., Tempelmeier, H.: Dynamic capacitated lot-sizing problems: a classification and review of solution approaches. OR Spectr. **32**(2), 231–261 (2010)
4. Chen, H.: Fix-and-optimize and variable neighborhood search approaches for multi-level capacitated lot sizing problems. Omega **56**, 25–36 (2015)
5. Duda, J., Stawowy, A.: A variable neighborhood search for multi-family capacitated lot-sizing problem. Electron. Notes Discrete Math. **66**, 119–126 (2018)
6. Erromdhani, R., Jarboui, B., Eddaly, M., Rebai, A., Mladenovic, N.: Variable neighborhood formulation search approach for the multi-item capacitated lot-sizing problem with time windows and setup times. Yugoslav J. Oper. Res. **27**(3), 301–322 (2017)
7. Jans, R., Degraeve, Z.: Meta-heuristics for dynamic lot sizing: a review and comparison of solution approaches. Eur. J. Oper. Res. **177**(3), 1855–1875 (2007)
8. Lazic, J., Hanafi, S., Mladenovic, N., Urosevic, D.: Variable neighbourhood decomposition search for 0–1 mixed integer programs. Comput. Oper. Res. **37**(6), 1055–1067 (2010)
9. Seeanner, F., Almada-Lobo, B., Meyr, H.: Combining the principles of variable neighborhood decomposition search and the fix and optimize heuristic to solve multi-level lot-sizing and scheduling problems. Comput. Oper. Res. **40**(1), 303–317 (2013)
10. Sifaleras, A., Konstantaras, I.: General variable neighborhood search for the multiproduct dynamic lot sizing problem in closed-loop supply chain. Electron. Notes Discrete Math. **47**, 69–76 (2015)
11. Sifaleras, A., Konstantaras, I.: Variable neighborhood descent heuristic for solving reverse logistics multi-item dynamic lot-sizing problems. Comput. Oper. Res. **78**, 385–392 (2017)
12. Simpson, N., Erengue, S.: Modeling multiple stage manufacturing systems with generalized costs and capacity issues. Nav. Res. Logist. **52**(6), 560–570 (2005)
13. Toledo, C., de Oliveira, R., Franca, P.: A hybrid multi-population genetic algorithm applied to solve the multi-level capacitated lot sizing problem with backlogging. Comput. Oper. Res. **40**(4), 910–919 (2013)
14. Wu, T., Shi, L., Geunes, J., Akartunali, K.: An optimization framework for solving capacitated multi-level lot-sizing problems with backlogging. Eur. J. Oper. Res. **214**(2), 428–441 (2011)
15. Zhao, Q., Xie, C., Xiao, Y.: A variable neighborhood decomposition search algorithm for multilevel capacitated lot-sizing problem. Electron. Notes Discrete Math. **39**, 129–135 (2012)

How to Locate Disperse Obnoxious Facility Centers?

Jesús Sánchez-Oro[1] , J. Manuel Colmenar[1]([⊠]) , Enrique García-Galán[1] ,
and Ana D. López-Sánchez[2]

[1] Rey Juan Carlos University, C/ Tulipán s/n, 28933 Móstoles, Madrid, Spain
{jesus.sanchezoro,josemanuel.colmenar,enrique.garciag}@urjc.es
[2] Pablo de Olavide University, Ctra. Utrera Km 1, 41013 Sevilla, Spain
adlopsan@upo.es

Abstract. The bi-objective obnoxious p-median problem has not been extensively studied in the literature yet, even having an enormous real interest. The problem seeks to locate p facilities but maximizing two different objectives that are usually in conflict: the sum of the minimum distance between each customer and their nearest facility center, and the dispersion among facilities, i.e., the sum of the minimum distance from each facility to the rest of the selected facilities. This problem arises when the interest is focused on locating obnoxious facilities such as waste or hazardous material, nuclear power or chemical plants, noisy or polluting services like airports. To address the bi-objective obnoxious p-median problem we propose a variable neighborhood search approach. Computational experiments show promising results. Specifically, the proposed algorithm obtains high-quality efficient solutions compared to the state-of-art efficient solutions.

Keywords: Location problem · Obnoxious p-median problem ·
Multi-objective optimization · Variable neighborhood search

1 Introduction

The importance of locating centers no matter the nature of them is crucial to manage any company either private or public. In our field, a *location problem* can be defined as an optimization problem that seeks to place one or more centers or facilities having into account a given set of customers or demand points [9].

According to [5], location problems can be classified into four categories regarding the objective function criteria: *facility location problems*, which seek to find a place to locate a facility in order to minimize the total cost between demand points and facilities; *p-median problems*, which determine the locations of p facilities in order to minimize the total cost between demand points and facilities; *p-center problems*, which minimize the maximum distance between each demand point and its assigned facility; and *covering problems* whose objective is to find the minimum number of facilities to cover all the demand points

© Springer Nature Switzerland AG 2019
A. Sifaleras et al. (Eds.): ICVNS 2018, LNCS 11328, pp. 52–63, 2019.
https://doi.org/10.1007/978-3-030-15843-9_5

or to maximize the number of demand points covered by a given number of facilities. All those problems can be considered with or without a demand value in facilities and/or in demand points. In those cases they are known as *capacitated* or *uncapacitated* problems, respectively [4, 15]. Furthermore, location problems can be considered on the *discrete* space, when facilities can be only placed at specific locations [12], or *continuous* space, in which facilities can be placed at any location of a given region [1]. This work deals with an uncapacitated discrete facility location problem.

The problem that is considered in this paper is known as the bi-objective obnoxious p-median problem, *Bi-OpM*, firstly introduced in [3]. It mainly consists on locating a set of obnoxious facilities on a landscape shared with customers (also known as demand points). The term obnoxious referring to a facility is used when it is desired to locate it as far as possible from the demand points. This situation appears when the interest is to locate facilities such as waste or hazardous material, nuclear power or chemical plants, noisy or polluting services like airports. Besides, the facilities should be properly distributed to avoid the situation where several obnoxious facilities are close to each other.

The *Bi-OpM* can be formally stated as follows. Let I be a set of customers, and J a set of candidate facility centers, where $|I| = n$ and $|J| = m$, and let d store the distances among all the considered nodes. The aim of the *Bi-OpM* is to locate a set P candidate facilities, having $|P| = p$ and $p < m$, while maximizing two objective functions: (f_1), the distance from each demand point to the facilities, computed as the sum of the minimum distances between each demand point and the nearest facility; and (f_2), the dispersion among the facilities, computed as the sum of the minimum distances from each facility to the rest of the selected facilities. More precisely, these objective functions can be described in the following way:

$$\max \; f_1 = \sum_{i \in I} \min d_{ij} : j \in P$$

$$\max \; f_2 = \sum_{j \in P} \min d_{jk} : k \in P, j \neq k$$

$$\text{s.t.} \;\; P \subseteq J$$

$$|P| = p$$

Some authors name facilities in P as *open* facilities and facilities in $J \backslash P$ as *closed* or *unopened* facilities.

On the other hand, it is important to emphasize that we are dealing with a multi-objective optimization problem. Hence, the definition of an efficient solution is the one for which no single-objective function value can be improved without deteriorating another objective function value. It is said that a solution P^* dominates another solution P if P^* is not worse than P in all the objectives, and P^* is better than P in at least one objective. Similarly, we say that P^* weakly dominates P if P^* is not worse than P in all the objectives [2]. Formally, as we are maximizing the objectives, a solution P^* dominates another solution P, if $f_i(P^*) \geq f_i(P)$ for all $i = 1, 2$ and $f_i(P^*) > f_i(P)$ for at least one $i = 1, 2$. According to this, we will say that a solution is *efficient* if there is no other solution that dominates it. The Pareto front, also known as the efficient frontier, is the set of efficient solutions. Our purpose then is to find a good approximation to the Pareto front, denoted as PF from now on.

As stated before, the *Bi-OpM* was first introduced in [3]. The authors proposed a Multi-Objective Memetic Algorithm (MOMA) defining two new variants of the crossover and mutation operators and studying three local search strategies applied in the MOMA. Furthermore, they performed a comparison using two multi-objective state-of-the-art methods, specifically the Non-dominated Sorting Genetic Algorithm II, (NSGA-II, [6]), and the Strength-Pareto Evolutionary Algorithm 2 (SPEA2, [16]), and also adding single-objective Genetic Algorithm (GA) which combines the objectives under study through a weighted sum of their values.

In this paper, a Variable Neighborhood Search algorithm (VNS) is adapted to solve the considered multi-objective optimization problem. Another contribution of this paper is to compare our algorithm against the best algorithm proposed in the literature so far, [3].

The rest of the paper is organized as follows. Section 2 describes our VNS proposal and details how the algorithm has being adapted and implemented to solve this bi-objective optimization problem. Section 3 presents the computational results where a experimentation and analysis of the results is shown. Finally, Sect. 4 summarizes the paper and discusses future work.

2   VNS Algorithm

To solve the *Bi-OpM* problem we propose a VNS approach that considers all the features of this bi-objective optimization problem. VNS is a metaheuristic framework originally introduced by [13] that relies on the idea of systematic changes in the neighborhood structures. The adaptability of the methodology has resulted in several variants in recent years (see [11] for a recent survey on the methodology), which has led to several successful applications for a variety of difficult optimization problems, such as those in [7] and [14]. In this work, VNS is adapted to solve a bi-objective optimization problem.

We propose a Basic VNS (BVNS) algorithm which combines deterministic and stochastic changes of neighborhood in order to obtain high quality solutions. Multi-objective VNS was originally proposed recently, see [8]. However, we follow a different approach in this paper, which is briefly described in Algorithm 1.

BVNS requires from two input parameters: a set of non-dominated solutions PF and the largest neighborhood to be explored, k_{max}. Starting from the first neighborhood (step 1), the method iterates until reaching the maximum predefined neighborhood $k_{\max}$ (steps 2–16). At each iteration two different phases are applied to every solution from the incumbent set of non-dominated solutions. In particular, the solution is randomly perturbed in the current neighborhood k using the *Shake* procedure (step 5). The proposed *Shake* algorithm consists in

Algorithm 1. BVNS(PF, k_{max})

```
1:  k ← 1
2:  while k ≠ k_max do
3:      PF' ← PF
4:      for all P ∈ PF' do
5:          P' ← Shake(P, k)
6:          PF' ← Insert& Update(P')
7:          P'' ← LocalSearch(P')
8:          PF' ← Insert& Update(P'')
9:          if PF' ≠ PF then                    ▷ Improve in the Pareto front
10:             k ← 1
11:             PF ← PF'
12:         else
13:             k ← k + 1
14:         end if
15:     end for
16: end while
17: return PF
```
---

randomly interchanging k assigned facilities with k candidate locations that do not belong to the current solution yet, generating solution P', which is added to the updated Pareto front, PF' (step 6). It is worth mentioning that the shake method performs random movements (whose sizes depend on the current neighborhood) which are not considered in the local search (i.e., interchange $k \leq p$ facilities simultaneously, while the local search only interchanges a single facility). Furthermore, the shake method accepts solutions of lower quality that will eventually let us explore further regions of the search space, while the local search only considers improved solutions.

A local search method is responsible of locally improving the perturbed solution P', obtaining solution P'' (step 7). Notice that every feasible solution generated during the search is a candidate solution for entering in the set of non-dominated solutions. The method *Insert & Update* (steps 6 and 8) performs this verification, inserting the solution if it is non-dominated by others already in the set, removing those solutions dominated by the new one. Regarding this behavior, any modification in the Pareto front is considered as an improvement since a new non-dominated solution has been included in it. Therefore, if the Pareto front has been modified, the search starts again from the first neighborhood (step 9), updating the incumbent Pareto front. Otherwise, the method explores the next neighborhood (step 13) until reaching the largest considered neighborhood. BVNS ends returning the set of non-dominated solutions generated in the search.

2.1   Constructive Method

The initial solution for the VNS algorithm is the set of non-dominated solutions PF conformed with the solutions generated by a constructive procedure inspired by the Greedy Randomized Adaptive Search Procedure (GRASP) methodology [10]. For this work, we have decided to use a semi-greedy procedure that combines greediness (intensification) and randomness (diversification) by means of a parameter α. The procedure generates a predefined number of initial solutions for both objective functions f_1 and f_2 that are evaluated for entering in the set of non-dominated solutions. Therefore, the output of the constructive phase is a set of non-dominated solutions PF, which acts as the input Pareto front for the VNS algorithm.

Algorithm 2 details the constructive method proposed for the Bi-OpM, which is generalized for any objective function, f_i. The input for the method is comprised of the set of candidate locations to host a facility J, the parameter α which controls the greediness/randomness of the method, and the objective function under consideration f_i. The method starts by randomly selecting a candidate location from the available ones, including it in the solution under construction (steps 1–2). Then, a Candidate List (CL) is created with the remaining candidate locations (step 3). The method iteratively selects new candidates until p locations has been selected (step 4). In each iteration, the minimum and maximum values of the objective function among all the candidates are evaluated (steps 5–6). After that, the Restricted Candidate List (RCL) is created (step 8) with the most promising candidates, i.e., those whose objective function value is larger or equal than a threshold th (step 7). On the one hand, if $\alpha = 0$, the construction is totally greedy (it only considers the facilities that produce the greatest increase in the objective function value). On the other hand, if $\alpha = 1$, then all facilities are included in the RCL so the construction is totally random. The next vertex to be added to the solution is selected at random from the RCL (step 9), updating the solution under construction and the CL (steps 10–11). The method ends when the solution has exactly p locations selected.

Algorithm 2. Construct(J, α, f_i)

1: $v \leftarrow Random(J)$
2: $P \leftarrow \{v\}$
3: $CL \leftarrow J \setminus \{v\}$
4: **while** $|P| < p$ **do**
5:     $g_{\min} \leftarrow \min_{v \in CL} f_i(P \cup \{v\})$
6:     $g_{\max} \leftarrow \max_{v \in CL} f_i(P \cup \{v\})$
7:     $th \leftarrow g_{\max} - \alpha \cdot (g_{\max} - g_{\min})$
8:     $RCL \leftarrow \{v \in CL \ : \ f_i(P \cup \{v\}) \geq th\}$
9:     $v' \leftarrow Random(RCL)$
10:     $P \leftarrow P \cup \{v'\}$
11:     $CL \leftarrow CL \setminus \{v'\}$
12: **end while**
13: **return** P

2.2 Local Search

The problem under consideration tries to optimize two different objective function, so a traditional approach would propose a different local search method for each objective function. Instead, we propose a single local search method that aggregates both objective functions in a unique search. Parameter β controls the influence of each objective function in the aggregated function f_a. More formally,

$$f_a \leftarrow \beta \cdot f_1 + (1 - \beta) \cdot f_2$$

Varying the value of β parameter will result in exploring different regions of the search space, potentially increasing the number of solutions included in the set of non-dominated solutions.

The local search method traverses all the selected facilities and tries to exchange it with every candidate location. The search follows a first improvement approach in order to reduce the computational effort of the method. In particular, every time an improvement move is found, it is performed and the search starts again.

The value of β must vary in order to obtain a more dense set of non-dominated solutions. In the context of the BVNS algorithm, we consider a random value of β in the range 0–1 in each local search phase, in order to explore a wider portion of the search space.

3 Computational Results

This section presents and discusses the results of the experimental experience conducted in this paper. In order to perform a fair comparison against the most competitive algorithm proposed in the literature, we have solved the same set of instances considered in [3]. Specifically, the previous paper presents eight instances where the number of nodes (indicated as $|V| = |I \cup J|$) ranges from 400 to 900, the number of demand points and facilities ($|I|$ and $|J|$) varies from 200 to 450 and the number of open facilities (p) is between 25 and 225. We show in Table 1 these features for all the instances, which are available at http://www.optsicom.es/biopm/. Our experiments were run on a computer provided with an Intel i7 2600 processor running at 3.4 GHz, 4 GB RAM and Ubuntu 16.04. The algorithms were implemented using Java 8.

After some preliminary computational experiments where several values of α and k_{max} were tested, we decided to use a random value for the α parameter and $k_{max} = 0.3 \cdot p$, since they obtained the best performance in our tests. Besides, a number of 100 iterations of the constructive method generated the initial Pareto front that our BVNS requires as input parameter.

Firstly, we show the results of our VNS proposal in terms of the hypervolume [17], which is the metric that was presented in [3]. In the previous paper, a total of seven different algorithms where compared. For the sake of the space, we will compare the results of our VNS proposal against three of the seven approaches from the state of the art. These algorithms are the NSGA-II [6],

Table 1. Description of the instances.

Instance	$|V|$	$|I|$	$|J|$	p
pmed17-p25	400	200	200	25
pmed20-p50	400	200	200	50
pmed22-p62	500	250	250	62
pmed28-p75	600	300	300	75
pmed33-p87	700	350	350	87
pmed36-p100	800	400	400	100
pmed39-p112	900	450	450	112
pmed40-p225	900	450	450	225

which is a classical multi-objective evolutionary algorithm, and the two variants of the memetic algorithm that obtained the best results in [3]: dominance-based local search, DBLS, and alternate objective local search, AOLS. Table 2 compares the performance of our VNS proposal showing the hypervolume values for all these algorithms and instances, the average hypervolume value normalized to the best hypervolume obtained for each instance (Avg. Norm. Hyp.), and the average normalized deviation between the hypervolume and the best hypervolume value (Avg. Dev.).

As it can be seen in the table, DBLS and AOLS obtain the best hypervolume values for two of the instances respectively, which are the four smaller ones. However, VNS reaches the best hypervolume in the four larges instances, obtaining also the best average normalized hypervolume and the best average deviation.

Notice that the results for NSGA-II, DBLS and AOLS, extracted from [3], correspond to the set of non-dominated solutions obtained after 30 runs of each algorithm. In the case of VNS, the results correspond to one single run. Therefore, it is clear that the efficiency of our VNS proposal is higher in relation to the other algorithms.

We have accounted for the number of efficient points obtained by each algorithm, which are shown in Table 3. In this case, our VNS proposal obtains the higher number of efficient points in all but the smallest instance. Therefore, VNS shows again a more efficient behavior in the optimization process, considering again that the results of VNS come from one single run.

In addition to the hypervolume and the number of efficient points, we depict in Fig. 1 the efficient points obtained in the experiments described before. From Fig. 1(a) to (h), the instances are sorted by size, from the smallest to the largest. As shown in the picture, the four fronts in the smallest instance, `pmed17.p25`, are almost completely overlapped. However, as the size of the instance grows, the front obtained with the NSGA-II algorithm is shifted downwards, while the other three algorithms maintain a good performance with similar shapes. This trend is more clear in the case of the efficient points of the largest instances,

Table 2. Hypervolume of final non-dominated fronts. Best values are depicted in bold font.

Instance	NSGA-II	DBLS	AOLS	VNS
pmed17.p25	8692436	8706765	**8710887**	8597575
pmed20.p50	9455493	**10042017**	10020130	9930426
pmed22.p62	10782181	**12565434**	12503057	12482833
pmed28.p75	8552938	10360761	**10383320**	10330591
pmed33.p87	8246628	10464880	10496109	**11011544**
pmed36.p100	9050694	11962494	11925733	**12413823**
pmed39.p112	7925756	11301612	11275309	**11707564**
pmed40.p225	7779073	10830924	10750521	**11978974**
Avg. Norm. Hyp.	0.8032	0.9726	0.9709	**0.9955**
Avg. Dev.	0.1968	0.0274	0.0291	**0.0045**

Table 3. Number of efficient points of final non-dominated fronts. Best values are depicted in bold font.

Instance	NSGA-II	DBLS	AOLS	VNS
pmed17.p25	84	85	**88**	68
pmed20.p50	41	136	121	**146**
pmed22.p62	17	140	127	**193**
pmed28.p75	15	115	102	**218**
pmed33.p87	24	103	108	**284**
pmed36.p100	12	89	82	**278**
pmed39.p112	8	104	76	**294**
pmed40.p225	22	133	123	**313**
Avg.	27.88	113.13	103.38	**224.25**

displayed in Fig. 1(e) to (h). In those instances, the front generated by our VNS proposal outperforms both the memetic and the NSGA-II approaches.

We have also obtained, for each instance, the complete set of non-dominated solutions after executing all the algorithms, that is, the final PF for each instance. This way, we have also measured the contribution of each algorithm to the final PF. Table 4 shows the ratio for each algorithm. We can see that VNS is the best contributor to the final PF but in the two smallest instances. However, it contributes in more that a 94% in the four largest instances, reaching the 100% in pmed36.p100. Again, it is worth mentioning that NSGA-II, DBLS and AOLS were executed 30 times, whereas VNS was run just once, and reached an average contribution of 71.27%.

Finally, we compare the execution time of the analyzed algorithms. Table 5 presents the time spent by VNS on the single run that produced the results

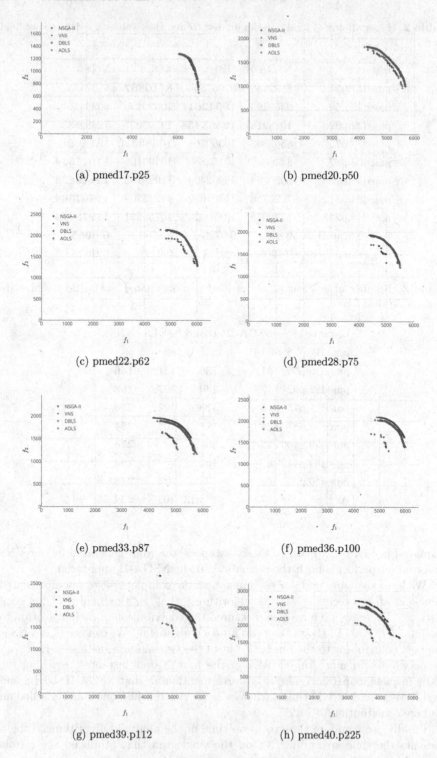

(a) pmed17.p25

(b) pmed20.p50

(c) pmed22.p62

(d) pmed28.p75

(e) pmed33.p87

(f) pmed36.p100

(g) pmed39.p112

(h) pmed40.p225

Fig. 1. Trade-off between f_1 and f_2.

Table 4. Contribution of each algorithm to the final PF. Best values are depicted in bold font.

Instance	NSGA-II	DBLS	AOLS	VNS
pmed17.p25	0.9545	0.9659	**0.9773**	0.0114
pmed20.p50	0	**0.3963**	0.25	0.3537
pmed22.p62	0	0.3108	0.0676	**0.6532**
pmed28.p75	0	0.1458	0.0833	**0.7708**
pmed33.p87	0	0.0036	0.0143	**0.9821**
pmed36.p100	0	0	0	**1**
pmed39.p112	0	0.0169	0	**0.9831**
pmed40.p225	0	0.0494	0.0031	**0.9475**
Avg.	0.1193	0.2361	0.1744	**0.7127**

previously analyzed, and the average execution time of NSGA-II, DBLS and AOLS. This average time was obtained after the 30 runs that produced the results shown before. Hence, despite that VNS is slower in one case (pmed28-p75), it is important to report that it obtains better results in one single run than the other algorithms after 30 runs.

Table 5. CPU time (secs). One single run of VNS versus average execution time of NSGA-II, DBLS and AOLS. Best values are depicted in bold font.

Instance	NSGA-II	DBLS	AOLS	VNS
pmed17-p25	1759.2	540.0	658.5	**21.8**
pmed20-p50	4995.3	933.9	1214.7	**410.1**
pmed22-p62	7990.8	1414.2	1768.2	**970.1**
pmed28-p75	13491.6	**1839.9**	2365.8	1856.1
pmed33-p87	19639.2	2706.9	3361.5	**1825.9**
pmed36-p100	27532.5	4367.7	3691.5	**1809.8**
pmed39-p112	29739.9	6099.6	3855.0	**2162.0**
pmed40-p225	114285.9	11440.8	8475.9	**3608.7**

4   Conclusions and Future Research

This paper generalizes the Variable Neighborhood Search algorithm (VNS) to solve a bi-objective optimization problem known as the bi-objective obnoxious p-median problem, *Bi-OpM*. To that end, the VNS approach is designed to take into account two conflicting objectives: to maximize the sum of the distances to the nearest demand point to each obnoxious facility and, to maximize the dispersion of obnoxious facilities. The interest of this problem appears because the *Bi-OpM* fits in many realistic situations where it is desired to locate facilities as far as possible from the demand points and among them.

Computational results show the superiority of the proposed algorithm over the state-of-art algorithms to solve the *Bi-OpM* so far on the same set of instances. Results obtained by the VNS algorithm outperform, in most of the instances, the three considered algorithms: NSGA-II, DBLS, and AOLS, spending less computational time.

As future work, it would be interesting to solve an extension of the *Bi-OpM* which will include an additional objective function. The new multi-objective obnoxious facility location problem that we will address, seeks to maximize the sum of the minimum distances between each demand point and its nearest facility and maximize the sum of the minimum distances between two facilities but also to minimize the number of demand points affected (or covered) by the facilities.

References

1. Carlsson, J.G., Jia, F.: Continuous facility location with backbone network costs. Transp. Sci. **49**(3), 433–451 (2014)
2. Coello, C.A.C., Lamont, G.B., Veldhuizen, D.A.V.: Evolutionary Algorithms for Solving Multi-Objective Problems. Genetic and Evolutionary Computation. Springer, New York (2006)
3. Colmenar, J., Martí, R., Duarte, A.: Multi-objective memetic optimization for the bi-objective obnoxious p-median problem. Knowl.-Based Syst. **144**, 88–101 (2018)
4. Cornuéjols, G., Nemhauser, G., Wolsey, L.: The uncapacitated facility location problem. In: Mirchandani, P.B., Francis, R.L. (eds.) Discrete Location Theory, pp. 119–171. Wiley-Interscience, New York (1990)
5. Dantrakul, S., Likasiri, C., Pongvuthithum, R.: Applied p-median and p-center algorithms for facility location problems. Expert Syst. Appl. **41**(8), 3596–3604 (2014)
6. Deb, K., Pratap, A., Agarwal, S., Meyarivan, T.: A fast and elitist multiobjective genetic algorithm: NSGA-II. IEEE Trans. Evol. Comput. **6**(2), 182–197 (2002)
7. Duarte, A., Pantrigo, J., Pardo, E., Sánchez-Oro, J.: Parallel variable neighbourhood search strategies for the cutwidth minimization problem. IMA J. Manag. Math. **27**(1), 55 (2016)
8. Duarte, A., Pantrigo, J.J., Pardo, E.G., Mladenovic, N.: Multi-objective variable neighborhood search: an application to combinatorial optimization problems. J. Glob. Optim. **63**(3), 515–536 (2015)
9. Farahani, R.Z., Hekmatfar, M.: Facility Location: Concepts, Models, Algorithms and Case Studies. Springer, Heidelberg (2009)
10. Feo, T.A., Resende, M.G.C.: Greedy randomized adaptive search procedures. J. Glob. Optim. **6**(2), 109–133 (1995)
11. Hansen, P., Mladenović, N., Todosijević, R., Hanafi, S.: Variable neighborhood search: basics and variants. EURO J. Comput. Optim. **5**(3), 423–454 (2017)
12. Marín, A.: The discrete facility location problem with balanced allocation of customers. Eur. J. Oper. Res. **210**(1), 27–38 (2011)
13. Mladenović, N., Hansen, P.: Variable neighborhood search. Comput. Oper. Res. **24**(11), 1097–1100 (1997)

14. Sánchez-Oro, J., Sevaux, M., Rossi, A., Martí, R., Duarte, A.: Solving dynamic memory allocation problems in embedded systems with parallel variable neighborhood search strategies. Electron. Notes Discret Math. **47**, 85–92 (2015)
15. Wu, L.Y., Zhang, X.S., Zhang, J.L.: Capacitated facility location problem with general setup cost. Comput. Oper. Res. **33**(5), 1226–1241 (2006)
16. Zitzler, E., Laumanns, M., Thiele, L.: SPEA2: improving the strength pareto evolutionary algorithm for multiobjective optimization. In: Giannakoglou, K., et al. (eds.) Evolutionary Methods for Design, Optimisation and Control with Application to Industrial Problems (EUROGEN 2001), pp. 95–100. International Center for Numerical Methods in Engineering (CIMNE) (2002)
17. Zitzler, E., Thiele, L.: Multiobjective evolutionary algorithms: a comparative case study and the strength pareto approach. IEEE Trans. Evol. Comput. **3**(4), 257–271 (1999)

Basic VNS Algorithms for Solving the Pollution Location Inventory Routing Problem

Panagiotis Karakostas[1] , Angelo Sifaleras[2] ,
and Michael C. Georgiadis[1](✉)

[1] Department of Chemical Engineering, Aristotle University of Thessaloniki,
University Campus, 54124 Thessaloniki, Greece
pkarakost@cheng.auth.gr, mgeorg@auth.gr
[2] Department of Applied Informatics, School of Information Sciences,
University of Macedonia, 156 Egnatia Street, 54636 Thessaloniki, Greece
sifalera@uom.gr

Abstract. This work presents a new variant of the Location Inventory Routing Problem (LIRP), called Pollution LIRP (PLIRP). The PLIRP considers both economic and environmental impacts. A Mixed Integer Programming (MIP) formulation is employed and experimental results on ten randomly generated small-sized instances using CPLEX are reported. Furthermore, it is shown that, CPLEX could not compute any feasible solution on another set of ten randomly generated medium-sized instances, with a time limit of five hours. Therefore, for solving more computationally challenging instances, two Basic Variable Neighborhood Search (BVNS) metaheuristic approaches are proposed. A comparative analysis between CPLEX and BVNS on these 20 problem instances is reported.

Keywords: Variable Neighborhood Search ·
Location Inventory Routing Problem · Green logistics

1 Introduction

In recent years, the efforts to manage the environmental impacts of the logistic activities have been increased. One of the major environment challenges is the global warming. The carbon dioxide (CO_2) emissions are highlighted as its main cause [2,5]. Transportation has been mentioned as the logistic activity with the highest contribution to (CO_2) emissions [2,11]. Also, the combined environmental impact of location-routing activities [10] and inventory-routing activities [2] has already been studied.

More specifically, the amount of the emitted (CO_2) gasses is proportionate to the amount of the consumed fuel [6]. Based on that fact, companies can either

© Springer Nature Switzerland AG 2019
A. Sifaleras et al. (Eds.): ICVNS 2018, LNCS 11328, pp. 64–76, 2019.
https://doi.org/10.1007/978-3-030-15843-9_6

adopt energy efficient vehicles or re-optimize their logistic decisions by taking into account factors affecting the fuel consumption [2], or even adopt a hybrid strategy.

From an economic perspective, the simultaneous tackling of strategic, tactical and operational decisions ensured the efficient performance of the supply chain. The Location Inventory Routing Problem (LIRP) integrates these three decisions [8,13]. However, there is a lack of research about the environmental-related variants of the LIRP. A sustainable closed-loop LIRP proposed by Zhalechian et al. [12] where economic, environmental and social impacts were considered. They formulated a multi-objective stochastic programming model for describing the problem.

In this work, a new green variant of the LIRP, the Pollution Location Inventory Routing Problem (PLIRP) is proposed. A Mixed Integer Programming (MIP) model is presented. In order to solve medium- and large-scaled instances, Basic Variable Neighborhood Search metaheuristic algorithms are developed. The remainder of this work is organized as follows. Section 2 describes the problem and provides its mathematical formulation. In Sect. 3, the proposed solution approach is presented, followed by experimental results in Sect. 4. Finally, Sect. 5 concludes this work and outlines future directions.

2 Problem Statement

This work extends the LIRP presented in [13] by considering fuel consumption and (CO_2) emissions costs, that are influenced by distance, load, speed and vehicles characteristics. The mathematical formulation of the PLIRP integrates the MIP models presented in [13] and [2]. The notations of the proposed model are given in Tables 1 and 2.

Table 1. Sets of the mathematical model

Indices	Explanation
V	Set of nodes
J	Set of candidate depots
I	Set of customers
K	Set of vehicles
H	Set of discrete and finite planning horizon

Table 2. PLIRP model variables and parameters.

Notation	Explanation
f_j	Fixed opening cost of depot j
y_j	1 if j is opened; 0 otherwise
C_j	Storage capacity of depot j
z_{ij}	1 if customer i is assigned to depot j; 0 otherwise
h_i	Unit inventory holding cost of customer i
Q_k	Loading capacity of vehicle k
d_{it}	Period variable demand of customer i
x_{ijkt}	1 if node j is visited after i in period t by vehicle k
q_{ikt}	Product quantity delivered to customer i in period t by vehicle k
w_{itp}	Quantity delivered to customer i in period p to satisfy its demand in period t
c_{ij}	Travelling cost of locations pair (i, j)
a_{vikt}	Load weight by travelling from node v to the customer i with vehicle k in period t
$zz_{v_1 v_2 ktr}$	1 if vehicle k travels from node $v1$ to $v2$ in period t with speed level r
s_r	The value of the speed level r

Table 3 describes the vehicles' parameters and gives their fixed values.

Table 3. Vehicles' parameters.

Parameter	Explanation	Value
ϵ	Fuel-to-air mass ratio	1
g	Gravitational constant (m/s^2)	9.81
ρ	Air density (kg/m^3)	1.2041
CR	Coefficient of rolling resistance	0.01
η	Efficiency parameter for diesel engines	0.45
f_c	Unit fuel cost (e/L)	0.7382
f_e	Unit CO_2 emission cost (e/kg)	0.2793
σ	CO_2 emitted by unit fuel consumption (kg/L)	2.669
$HVDF$	Heating value of a typical diesel fuel (kj/g)	44
ψ	Conversion factor $(g/s$ to $L/s)$	737
θ	Road angle	0
τ	Acceleration (m/s^2)	0
CW_k	Curb weight (kg)	3500
EFF_k	Engine friction factor $(kj/rev/L)$	0.25
ES_k	Engine speed (rev/s)	39
ED_k	Engine displacement (L)	2.77
CAD_k	Coefficient of aerodynamics drag	0.6
FSA_k	Frontal surface area (m^2)	9
$VDTE_k$	Vehicle drive train efficiency	0.4

It should be highlighted that, the values of parameters f_c and f_e are the average price of the petrol prices in 40 European countries, taken from the site www.globalpetrolprices.com in 26th of February in 2018. The value of parameter CW_k can refer to [9]. The rest of the parameters' values are taken by [2].

In order to simplify some parts of the objective function, due to the fuel consumption, the following formulas are utilized.

$$- \lambda = \frac{HVDF}{\psi}$$
$$- \gamma_k = \frac{1}{1000VDTE\eta}$$
$$- \alpha = \tau + gCR\sin\theta + gCR\cos\theta$$
$$- \beta_k = 0.5CAD\rho FSA_k$$

Thus, the mathematical model of the PLIRP is as follows:

$$\min \sum_{j \in J} f_j y_j + \sum_{i \in I} h_i \sum_{t \in H} \left(\tfrac{1}{2}d_{it} + \sum_{p \in H, p<t} w_{itp}(t-p) + \sum_{p \in H, p>t} w_{itp}(t-p+|H|) \right)$$

$$+ \sum_{i \in V} \sum_{j \in V} \sum_{t \in H} \sum_{k \in K} c_{ij} x_{ijkt} + \sum_{i \in V} \sum_{j \in V} \sum_{k \in K} \sum_{t \in H} \left\{ \lambda \left(f_c + (f_e\sigma) \right) \right.$$

$$\left(\sum_{r \in R} \frac{(zz_{ijktr} EFF_k ES_k ED_k c_{ij})}{s_r} + \left(\alpha\gamma_k \left(CW_k x_{ijkt} + a_{ijkt} \right) c_{ij} \right) \right.$$

$$\left. \left. + \left(\beta_k \, \gamma_k \sum_{r \in R} \left(s_r \, zz_{ijktr} \right)^2 \right) \right) \right\}$$

$$\tag{1}$$

Subject to

$$\sum_{r \in R} zz_{ijktr} = 0 \quad \forall i,j \in V, \forall k \in K, \forall t \in H \tag{2}$$

$$\sum_{i \in V} a_{ijkt} - \sum_{i \in V} a_{jikt} = q_{jkt}PW \quad \forall j \in I, \forall k \in K, \forall t \in H \tag{3}$$

$$\sum_{j \in V} x_{ijkt} - \sum_{j \in V} x_{jikt} = 0 \qquad \forall i \in V, \forall k \in K, \forall t \in H \tag{4}$$

$$\sum_{j \in V} \sum_{k \in K} x_{ijkt} \leq 1 \quad \forall t \in H, \forall i \in I \tag{5}$$

$$\sum_{j \in V} \sum_{k \in K} x_{jikt} \leq 1 \quad \forall t \in H, \forall i \in I \tag{6}$$

$$\sum_{i \in I} \sum_{j \in J} x_{ijkt} \leq 1 \quad \forall k \in K, \forall t \in H \tag{7}$$

$$x_{ijkt} = 0 \quad \forall i,j \in J, \forall k \in K, \forall t \in H, i \neq j \tag{8}$$

$$\sum_{i \in I} q_{ikt} \leq Q_k \quad \forall k \in K, \forall t \in H \tag{9}$$

$$\sum_{j \in J} z_{ij} = 1 \quad \forall i \in I \tag{10}$$

$$z_{ij} \leq y_j \quad \forall i \in I, \, \forall j \in J \tag{11}$$

$$\sum_{i \in I} \left(z_{ij} \sum_{t \in H} d_{it} \right) \leq C_j \quad \forall j \in J \tag{12}$$

$$\sum_{u \in I} x_{ujkt} + \sum_{u \in V \setminus \{i\}} x_{iukt} \leq 1 + z_{ij} \quad \forall i \in I, \, \forall j \in J, \, \forall k \in K, \forall t \in H \tag{13}$$

$$\sum_{i \in I} \sum_{k \in K} \sum_{t \in H} x_{jikt} \geq y_j \quad \forall j \in J \tag{14}$$

$$\sum_{i \in I} x_{jikt} \leq y_j \quad \forall j \in J, \, \forall k \in K, \, \forall t \in H \tag{15}$$

$$\sum_{p \in H} w_{itp} = d_{it} \quad \forall i \in I, \, \forall t \in H \tag{16}$$

$$\sum_{t \in H} w_{itp} = \sum_{k \in K} q_{ikp} \quad \forall i \in I, \, \forall p \in H \tag{17}$$

$$q_{ikt} \leq M \sum_{j \in V} x_{ijkt} \quad \forall i \in I, \, \forall t \in H, \, \forall k \in K \tag{18}$$

$$\sum_{j \in V} x_{ijkt} \leq M q_{ikt} \quad \forall i \in I, \, \forall t \in H, \, \forall k \in K \tag{19}$$

$$x_{ijkt} \in \{0,1\} \quad \forall i \in I, \, \forall j \in J, \, \forall t \in H, \, \forall k \in K \tag{20}$$

$$y_j \in \{0,1\} \quad \forall j \in J \tag{21}$$

$$z_{ij} \in \{0,1\} \quad \forall i \in I, \, \forall j \in J \tag{22}$$

$$q_{ikt} \leq \min \left\{ Q_k, \sum_{p \in H} d_{ip} \right\} \quad \forall i \in I, \, \forall j \in J, \forall k \in K \tag{23}$$

$$w_{itp} \leq d_{ip} \quad \forall i \in I, \, \forall t, p \in H \tag{24}$$

The objective function minimizes the sum of facilities opening costs, inventory holding costs, general routing costs and fuel consumption and CO_2 emissions costs. Constraints 2 impose that, only one speed level will be assigned to a vehicle traveling between two nodes in the selected time period. Constraints 3 declare that, the total weight of the incoming flow of product to a selected customer minus the total weight of the outcoming product flow of that customer equals the product weight delivered to that customer in the selected time period with the selected vehicle. Also, they operate as subtour elimination constraints. Constraints 2–9 are related to the routing decisions. As an example Constraints 8 ensure that, a selected vehicle in a selected time period will not travel between two depots. Constraints 10–15 guarantee the feasibility of the location decisions.

For example, Constraints 10 and 11 force a customer to be allocated to a depot, only if that depot is marked as opened. Finally, Constraints 11–19 force the feasibility of the inventory decisions. For instance, Constraints 16 guarantee the satisfaction of a selected customer's demand over the time horizon.

3   Solution Method

Due to the high computational complexity of the PLIRP, two Basic Variable Neighborhood Search metaheuristic algorithms are proposed for solving medium and large scale problem instances.

3.1   Construction Heuristic

Initially, a feasible solution is built by applying a three phase constructive heuristic. In the first phase, a minimum cost criterion procedure is applied for selecting the depots to be opened. Then, the allocation of customers is sequentially scheduled. More specifically, if the total demand of a selected customer does not violate the remaining capacity of the selected opened depot, the customer is allocated to that depot. Otherwise, the customer is allocated to the next opened and capable to service him depot. Finally, the routes are built by applying a random insertion method. It should be clarified that, in this initial solution the scheduled product quantity to be delivered to each customer at each period satisfy its demand for the considered period.

3.2   Basic VNS

The Variable Neighborhood Search is a trajectory-based metaheuristic framework which interchanges two main phases [7]. The first one is the intensification phase, where a local optimum solution is obtained and the second one is the diversification phase. In the last phase the current solution is perturbed for escaping local optimum points. VNS has gained popularity in recent years due to its simplicity and performance [3]. In this work two local search operators are used both in improvement and shaking phase. These neighborhood structures are the following.

- **Inter-route Exchange**. Solutions in this neighborhood are obtained by exchanging two customers located in different routes. These routes could be allocated either on the same or different depots. In the second case, inventory replenishment rescheduling may need to be applied. Figures 1 and 2, illustrate the inter-route exchange move in routes allocated to the same depot. In the first time period an exchange between the third and fourth customers is applied, while in the second and third periods the exchanged customers are the pairs $(1, 4)$ and $(3, 5)$, respectively.

Period_1   Period_2   Period_3

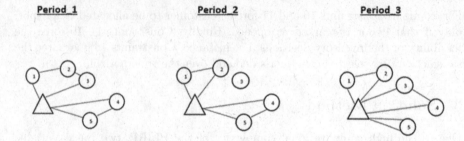

Fig. 1. Routes from the same depot for each time period before the application of the inter-route exchange move.

Period_1   Period_2   Period_3

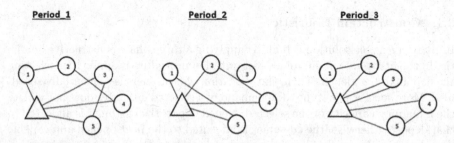

Fig. 2. Routes from the same depot for each time period after the application of the inter-route exchange move.

When the routes are allocated to different depots, the exchange move is applicable only if the two customers are serviced in the same time periods. Figures 3, 4 and 5 illustrate the inter-route exchange between the customers two and three. It is also assumed that, the product quantity delivered to customer three in the third time period exceeds the capacity of the vehicle currently servicing the customer two. Consequently, a replenishment rescheduling is applied as depicted in Fig. 5.

Period_1   Period_2   Period_3

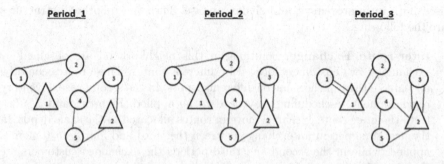

Fig. 3. Routes from different depots for each time period before the application of the inter-route exchange move.

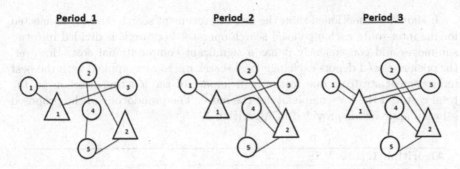

Fig. 4. Routes from different depots for each time period after the application of the inter-route exchange move.

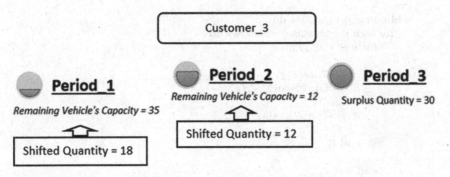

Fig. 5. Shifting the surplus quantity of product of the customer 3 from the third period to other(s).

– **Opened-Closed Depots Exchange.** This neighborhood exchanges a closed depot with an opened one. In each route, allocated to the currently opened depot, a routing reordering procedure is applied.

Figure 6 illustrates an instance of the opened-closed depots' exchange move. More specifically, the first (opened) depot swaps with the second (closed) depot and reordering occurred in the two routes.

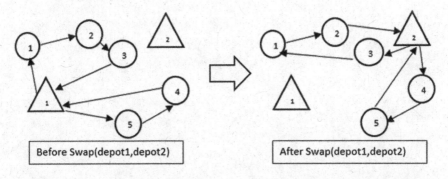

Fig. 6. An illustrated example of the opened-closed depots' exchange move.

It should be mentioned that, the first improvement search strategy is selected for the inter-route exchange local search operator because it is divided into two sub-moves and consequently it has a significant computational cost. However, the opened-closed depots exchange local search operator is applied with the best improvement strategy, due to the high impact of the location decision on the total cost and the low complexity of the move. The pseudocode of the proposed solution approach is provided in Algorithm 1.

Algorithm 1. Basic VNS

---

   procedure BVNS(k_{max}, max_time)

     $S \leftarrow$ Construction_Heuristic

     while $time \leq max\_time$ **do**
        for each neighborhood structure l **do**
           for $k \leftarrow 1, k_{max}$ **do**

               $S' \leftarrow$ Shake(S, k)
               $S'' \leftarrow$ Local_Search(S', l)

               if $f(S'') < f(S)$ **then**
                  $S \leftarrow S''$
               end if

           end for
        end for
     end while

     Return S

Speed Selection Procedure (SSP) examines which speed level has the highest fuel cost decrease for each depot-customer and customer-customer pair in the current solution. It can be used within BVNS and employed after the execution of each local search operator. This version of BVNS is called BVNS_SSP and its pseudocode is summarized at Algorithm 2.

Algorithm 2. Basic VNS with SPP

---

procedure BVNS_SSP(k_{max}, max_time)

    $S \leftarrow$ Construction_Heuristic

    while $time \leq max\_time$ **do**
        for each neighborhood structure l **do**
            for $k \leftarrow 1, k_{max}$ **do**

                $S' \leftarrow$ Shake(S, k)
                $S'' \leftarrow$ Local_Search(S', l)
                $S' \leftarrow$ Speed_Selection(S'')

                if $f(S') < f(S)$ **then**
                    $S \leftarrow S'$
                end if

            end for
        end for
    end while

    return S

In the *Shake* procedure a randomly selected neighborhood structure is applied k times in a current solution, while the *Local_Search* procedure applies the operator specified by l in an incumbent solution both in BVNS and BVNS_SSP.

4   Numerical Results

4.1   Computing Environment and Parameter Settings

The proposed algorithms were implemented in Fortran. The computational experiments ran on a desktop PC running Windows 7 Professional 64-bit with an Intel Core i7-4771 CPU at 3.5 GHz and 16 GB RAM, using Intel Fortran compiler 18.0 with optimization option /O3. The time limit of 60 s was set as the maximum execution time and experimentally k_{max} is set to 18. The mathematical formulation was modeled in GAMS (GAMS 24.9.1) and the problem instances were solved with CPLEX 12.7.1.0 solver with specified time limits (2 h for the small-sized instances and 5 h for the medium-sized instances). CPLEX ran in the same computing environment with Intel Fortran compiler.

4.2   Computational Results

This subsection summarizes the results of the computational tests performed on 20 randomly generated instances, in order to examine the performance of the proposed algorithms. The problem instances are divided into two classes, small-sized instances (up to 20 customers) and medium-sized instances (customers between 20 and 60). Their format follows the format of the instances presented in [13].

Table 4 summarizes the results obtained by the CPLEX solver, BVNS and BVNS with SSP procedure. More specifically, the first column provides the

names of the instances. In the second column the results obtained by CPLEX
are given, while columns three and five show the average results achieved by
BVNS and BVNS_SSP respectively. Columns four and six provide the solution
gap between BVNS and CPLEX results and between BVNS and BVNS_SSP
results respectively. Finally, the solution quality gap between the two proposed
methods are given in column seven.

Table 4. Average computational results on 10 small-sized PLIRP instances

Instance	CPLEX (a)	BVNS (b)	Gap (a-b) %	BVNS_SSP (c)	Gap (a-c) %	Gap (b-c)
4-8-3	22647.63	26895.24	−18.76	26892.47	−18.74	0.01
4-8-5	18282.71	19531.79	−6.83	19617.87	−7.30	−0.44
4-10-3	16929.96	17851.3	−5.44	17887.96	−5.66	−0.21
4-10-5	-	23902.89	-	23895.99	-	0.08
4-15-5	22013.99	23158.8	−5.2	23169.92	−5.25	−0.05
5-9-3	16700.09	17594.66	−5.36	17603.51	−5.41	−0.05
5-12-3	24152.36	30193.34	−25.01	30179.07	−24.95	0.05
5-15-3	16939.71	17719.81	−4.61	15842.7	6.48	10.59
5-18-5	-	19902.78	-	19891.27	-	0.06
5-20-3	24605.637	25132.29	−2.14	25123.32	−2.1	0.04
		Average	−9.17	Average	−7.87	0.093

As it can be seen in Table 4, the CPLEX solver (GAMS) provides 9.17%
better solutions than BVNS and 7.87% better solutions than BVNS_SSP. However, the time limit for the CPLEX was set at two hours, while both BVNS and
BVNS_SSP execute for 60 s. CPLEX is not able to provide any feasible solution
for the medium-sized instances, even with a time limit of five hours. Consequently, Table 5 reports the results achieved by BVNS and BVNS_SSP on the
set of the ten medium-sized instances.

Table 5. Average computational results on ten medium-sized PLIRP instances

Instance	BVNS (a)	BVNS_SSP (b)	Gap (a-b) %
6-22-7	28090.1	28074.69	0.06
6-25-5	22794.17	22747.42	0.21
7-25-5	39927.7	39914.72	0.03
7-25-7	23736.44	23675.7	0.26
8-25-5	26777.71	26773.1	0.02
8-30-7	36648.27	36582.34	0.18
8-50-5	33564.62	33536.73	0.08
8-65-7	27986.69	27988.59	−0.01
9-40-7	23176.14	23190.46	−0.06
9-55-5	23688.55	23697.58	−0.04
		Average	0.07

As it is shown in Table 5, the solutions obtained by BVNS_SSP are 0.07% better than those achieved by BVNS. Table 6 reports the best values achieved by both BVNS and BVNS_SSP in all 20 randomly generated PLIRP instances.

Table 6. Best values on 20 PLIRP instances achieved by BVNS and BVNS_SSP

Instance	BVNS	BVNS_SSP
4-8-3	26894.85	26892.3
4-8-5	19470.34	19617.77
4-10-3	17817.74	17812.94
4-10-5	23901.97	23895.99
4-15-5	23113.12	23134.67
5-9-3	17594.12	17598.73
5-12-3	30193.23	30179.4
5-15-3	17683.65	15816.7
5-18-5	19900.91	19885.77
5-20-3	25102.89	25031.15
6-22-7	28088.87	28074.47
6-25-5	22530.3	22536.06
7-25-5	39926.08	39914.5
7-25-7	23519.59	23214.87
8-25-5	26774.73	26769.87
8-30-7	36584.71	36546.54
8-50-5	33562.31	33491.02
8-65-7	27876.23	27883.81
9-40-7	23146.37	23165.18
9-55-5	23532.72	23596.95

5 Conclusions

This work introduces a new NP-hard combinatorial optimization problem, known as the Pollution Location Inventory Routing Problem which integrates economic and environmental decisions. An MIP formulation of the PLIRP is presented and the optimization solver CPLEX was used for solving small-sized instances. Because of the high complexity of the PLIRP, CPLEX cannot find any feasible solution for medium-sized instances even with a time limit of 5h. For solving more challenging instances, two Basic VNS heuristic algorithms were developed. A future research direction can explore parallel computing techniques [1] for speeding up the solution process. This way, more real-world extensions of this type of problems (e.g., including the utilization of remanufacturing options [4]) can be efficiently addressed.

References

1. Antoniadis, N., Sifaleras, A.: A hybrid CPU-GPU parallelization scheme of variable neighborhood search for inventory optimization problems. Electron. Notes Discret Math. **58**, 47–54 (2017)
2. Cheng, C., Yang, P., Qi, M., Rousseau, L.M.: Modeling a green inventory routing problem with a heterogeneous fleet. Transp. Res. Part E **97**, 97–112 (2017)
3. Coelho, V.N., Santos, H.G., Coelho, I.M., Penna, P.H.V., Oliveira, T.A., Souza, M.J.F., Sifaleras, A.: 5th International Conference on Variable Neighborhood Search (ICVNS 2017). Electron. Notes Discrete Math. **66**, 1–5 (2018)
4. Cunha, J.O., Konstantaras, I., Melo, R.A., Sifaleras, A.: On multi-item economic lot-sizing with remanufacturing and uncapacitated production. Appl. Math. Model. **43**, 678–686 (2017)
5. Demir, E., Bektas, T., Laporte, G.: A review of recent research on green road freight transportation. Eur. J. Oper. Res. **237**, 775–793 (2014)
6. Ehmke, J., Cambell, A., Thomas, B.: Data-driven approaches for emissions-minimized paths in urban areas. Comput. Oper. Res. **67**, 34–47 (2016)
7. Hansen, P., Mladenovic, N., Todosijevic, R., Hanafi, S.: Variable neighborhood search: basics and variants. EURO J. Comput. Optim. **5**, 423–454 (2017)
8. Javid, A., Azad, N.: Incorporating location, routing and inventory decisions in supply chain network design. Transp. Res. Part E **46**, 582–597 (2010)
9. Koç, Ç., Bektaş, T., Jabali, O., Laporte, G.: The fleet size and mix pollution-routing problem. Transp. Res. Part B **70**, 239–254 (2014)
10. Koç, Ç., Bektaş, T., Jabali, O., Laporte, G.: The impact of depot location, fleet composition and routing on emissions in city logistics. Transp. Res. Part B **84**, 81–102 (2016)
11. Leenders, B., Velazquez-Martinez, J., Fransoo, J.: Emissions allocation in transportation routes. Transp. Res. Part D **57**, 39–51 (2017)
12. Zhalechian, M., Tavakkoli-Moghaddam, R., Zahiri, B., Mohammadi, M.: Sustainable design of a closed-loop location-routing-inventory supply chain network under mixed uncertainty. Transp. Res. Part E **89**, 182–214 (2016)
13. Zhang, Y., Qi, M., Miao, L., Liu, E.: Hybrid metaheuristics solutions to inventory location routing problem. Transp. Res. Part E **70**, 305–323 (2014)

Less Is More: The Neighborhood Guided Evolution Strategies Convergence on Some Classic Neighborhood Operators

Vitor Nazário Coelho[1]([⊠]), Igor Machado Coelho[2], Nenad Mladenović[3],
Helena Ramalhinho[4], Luiz Satoru Ochi[1], Frederico G. Guimarães[5],
and Marcone J. F. Souza[6]

[1] Institute of Computation, Universidade Federal Fluminense, Niterói, RJ, Brazil
vncoelho@gmail.com
[2] Department of Computer Science, Universidade do Estado do Rio de Janeiro,
Rio de Janeiro, Brazil
[3] Mathematical Institute, SANU, Belgrade, Serbia
[4] Department of Economics and Business, Universitat Pompeu Fabra,
Barcelona, Spain
[5] Department of Electrical Engineering, Universidade Federal de Minas Gerais,
Belo Horizonte, Brazil
[6] Department of Computer Science, Universidade Federal de Ouro Preto,
Ouro Preto, MG, Brazil

Abstract. This paper extends some explanations about the convergence
of a type of Evolution Strategies guided by Neighborhood Structures, the
Neighborhood Guided Evolution Strategies. Different well-known Neighborhood Structures commonly applied to Vehicle Routing Problems are
used to highlight the evolution of the move operators during the evolutionary process of a self-adaptive Reduced Variable Neighborhood Search
procedure. Since the proposal uses only few components for its search,
we believe it can be seen inside the scope of the recently proposed "Less
Is More Approach".

Keywords: Metaheuristics · Neighborhood structure ·
Reduced VNS · Evolution Strategies ·
Less is more and NP-Hard problems

1 Introduction

This paper extends the explanations regarding the Evolution Strategies applied
for Combinatorial Optimization Problems, recently introduced by Coelho
et al. [2]. When introduced, the method was suggested as a framework that

---

Vitor N. Coelho would like to thank the support given by FAPERJ (grant
E-26/202.868/2016). Marcone J. F. Souza thanks the support given by FAPEMIG
and CNPq (grants CEX-PPM-00676/17 and 307915/2016-6, respectively).

A. Sifaleras et al. (Eds.): ICVNS 2018, LNCS 11328, pp. 77–88, 2019.
https://doi.org/10.1007/978-3-030-15843-9_7

combines a search of Evolution Strategies [1] in a kind of Reduced Variable Neighborhood Search (RVNS) [6] procedure. In this current version, we highlight that it is a simple metaheuristic closely related to the pioneer metaheuristic Simulated Annealing [7]. The similarity borders the use of random moves, however, guided by an evolutionary process instead of a cooling schedule. The use of this basic strategies makes it to be a kind of "Less Is More Approach" [4], a simple metaheuristic framework with few parameters and mechanisms. Simple, but, however, with potential of tackling several complex and large-scale combinatorial problems.

This novel study provides better explanations, acronyms, figures and more details about the proposed metaheuristic. A classic NP-Hard problem, a real-case large-scale Vehicle Routing Problem (VRP), is used as a didactic example. The focus given here is to emphasize the use of important and well known neighborhood structures: swaps, shift, 2-opt and exchange moves. These structures are often used in the resolution of the travelling salesman and vehicle routing problems.

The evolution of the evolutionary operators is discussed with details, highlighting the potential of the proposal in optimizing and adjusting its searching mechanism according to the difficulty in improving the best known solution. Given its main search strategy, based on Neighborhood Structures, we define the procedure as the Neighborhood Guided Evolution Strategies (NGES).

The remainder of this paper is organized as follows: Sect. 2 summarizes the main features used by the method. Section 3 describes the case of study and, in Sect. 4, the results for that specific VRP problem. Finally, Sect. 5 draws some final considerations about the possible insights presented in this study.

2   The Neighborhood Guided Evolution Strategies

The diagram presented in Fig. 1 exemplifies the evolutionary process of the NGES. In summary, a fixed population of size μ guides the evolutionary process, in which, random individuals from it are selected in each generation. Each individual of the population *ind* is composed of two additional mutation vectors, defined as P and A, defined at Sect. 2.2, associated with a solution representation s.

The mutation phase only modifies the operators P_i and A_i of an individual i. That phase does not modify the previously known solution s_i. After perturbing these values, i.e. self-adapting them, a systematic operation of random moves, following the rules and limits expressed in these frequently modified operators, happens. offspring, namely λ, usually, with a population of higher size than parents, are generated and a selection process happens in order to define the next population with size μ.

In order to exemplify and guarantee replicability of the code, Sect. 2.1 exemplifies some C++ code example that could be found at the github repository https://github.com/optframe/optframe/blob/master/ OptFrame/Heuristics/EvolutionaryAlgorithms/NGES.hpp.

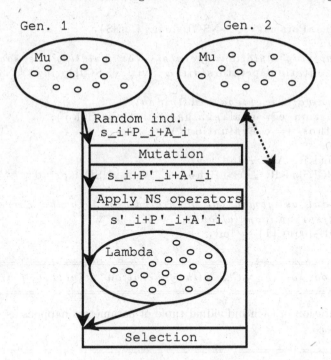

Fig. 1. NGES basic diagram

2.1 Tips for C++ Implementation of the NGES

Generic templates of the pseducodes in C++ are describe below. The first code block explains how the offspring population is obtained from the current population, basically, random individuals are copied and mutated.

```
// ═══════════════════════════════════════════
// Main code of the NGES with offspring generation
// ═══════════════════════════════════════════
for (int l = 0; l < ngesParams.lambda; l++)
{
    // Select a random index from the current population
    int x = rg.rand(ngesParams.mi);
    // Create an offspring
    Solution<R, ADS> filho = pop[x]->sInd;
    // Create the mutation vector
    vector<NGESIndStructure<R, ADS> > vt =
    pop[x]->vEsStructureInd;
    // Additional vector that controls the best order
    // in which neighborhoods are applied.
    // This vector is self adjusted during the search.
    vector<int> vNSOffspring = pop[x]->vNSInd;

    // call method for mutating parameter
```

```
    mutateESParams(vt , vNSOffspring , nNS);

    // applying mutation operators for mutating the son
    applyMutationOperators(filho , vt , vNSOffspring , nNS);

    // Evaluation of each individual
    Evaluation e = eval.evaluateSolution(filho);
    fo_filhos += e.evaluation();

    NGESInd<R, ADS>* ind =
    new NGESInd<R, ADS>(filho , e, vt , vNSOffspring);

    // final assignment of generated son into the current
    // offspring population
    PopOffspring[1] = ind;
}
// ═══════════════════════════════════════════════
// Main code of the NGES with generation selection finishes
// ═══════════════════════════════════════════════
```

The mutation of each individual tuple of parameters happens as specific in the next block.

```
// ═══════════════════════════════════════════════
// Automatic Mutation of NGES parameters
// ═══════════════════════════════════════════════
/* Parameters:
1 -  vector<NGESIndStructure<R, ADS> >& p
2 -  vector<int>& vNSInd
3 -  const int nNS
*/
void mutateESParams(1 , 2 , 3)
{
        double z = rg.rand01();
        if (z <= ngesParams.mutationRate)
        {
                int posX = rg.rand(nNS);
                int posY = rg.rand(nNS);
                if (nNS > 1)
                {
                        while (posY == posX)
                        {
                                posY = rg.rand(nNS);
                        }
                }

                // Swaps only NS order , since preserved for
                // the function applyMutationOperators
                iter_swap(vNSInd.begin() +
                posX, vNSInd.begin() + posY);
        }
```

```
        for (int p = 0; p < nNS; p++)
        {
// sigmaN and sigmaB are special parameters that self-adapt
// the normal and binomial distribution themselves
            p[p].sigmaN += rg.randG(0, 0.1) / 100.0;
            if (p[p].sigmaN < 0)
                p[p].sigmaN = 0;
            if (p[p].sigmaN > 3)
                p[p].sigmaN = 3;
            p[p].pr += rg.randG(0, p[p].sigmaN);

            if (p[p].pr < 0)
                p[p].pr = 0;
            if (p[p].pr > 1)
                p[p].pr = 1;
            p[p].sigmaB += rg.randG(0, 0.1) / 100.0;

            if (p[p].sigmaB < 0)
                p[p].sigmaB = 0;
            if (p[p].sigmaB > 1)
                p[p].sigmaB = 1;
// Negative Binomial is a standard Binomial distribution,
// but with an additional feature
// of generating negative values
            p[p].nap +=
                rg.randNegativeBinomial(p[p].sigmaB, 10);

            if (p[p].nap < 1)
                p[p].nap = 1;
//As described, each move has an upper limit of moves
            if (p[p].nap > ngesParams.maxNS[p])
                p[p].nap = ngesParams.maxNS[p];
        }

}
// ===========================================================
// Automatic Mutation of NGES parameters Ends
// ===========================================================
```

Finally, an adaptive Random Variable Neighborhood Descent happens, using as direction the vectors that were modified in the last step.

```
// ===========================================================
// Apply moves - ARVND
// ===========================================================

/* Parameters:
1 - Solution<R, ADS>& s
2 - const vector<NGESIndStructure<R, ADS> >& p
3 - const vector<int> vNSInd
```

```
4 - const int nNS
*/
void applyMutationOperators (1,2,3,4)
{
        for (int i = 0; i < nNS; i++)
        {
                int param = vNSInd[i]; //Extract index
                double rx = rg.rand01();
                if (rx < p[param].pr)
                for (int a = 0; a < p[param].nap; a++)
                {
                        Move<R, ADS>* mov_tmp =
                        vNS[param]->randomMoveSolution(s);

                        if (mov_tmp->canBeAppliedToSolution(s))
                        {
                                Move<R, ADS>* mov_rev =
                                mov_tmp->applySolution(s);
                                delete mov_rev;
                        }

                        delete mov_tmp;
                }
        }

}
//
// Apply moves - ARVND Ends
//
```

2.2 Basic Principles

Each individual of the population is defined as Eq. 1.

$$ind = \{s, P, A\} \tag{1}$$

Parameters P and A are defined at Eqs. 2 and 3, respectively, and s is the representation of a solution for the problem.

$$P = [p_1, p_2, ..., p_i, ..., p_{neigh}] \tag{2}$$

$$A = [a_1, a_2, ..., a_j, ..., a_{neigh}] \tag{3}$$

The first mutation vector P_i, for an individual i, defines the probability of using moves from a given NS. In this sense, it is the likelihood associated with the application of each NS. A given value $p_i^{neigh} \ \forall \ neigh \in |NS|$ is the current probability of applying a_i^{neigh} moves $m \in NS_{neigh}$, in which $neigh$ is the number of available NS, and $p_i^{neigh} \in [0, 1], p_i^{neigh} \in \mathbb{R}$.

As described above, the number of moves that will be applied, if a random number fits the limits p_i, is the corresponding value a_i found inside the operator A. This operator stores integer values that control the intensity of the step that will be done, once the NS is selected to be applied for modifying the solution s_i. Each position $a_i^{neigh} \in [0, nap^{neigh}], a^{neigh} \in \mathbb{N}$ of this vector limits the number of applications of a given move, with nap^{neigh} representing the maximum number of applications for a given move of type $neigh \in |NS|$.

During the mutation phase, in which each position of the vectors P and A are modified, classical probability distribution functions are used, such as Normal and Binomial distributions. In addition, as described and can be seen in the pseudo-codes presented at Sect. 2.1, two self-adjusted parameters automatically optimize both distributions.

3 Heterogeneous Fleet Vehicle Routing Problem with Multiple Trips

Distribution planning is crucial for most companies that deliver goods. Solving a VRP enable managers to find good routes to deliver their products to a set of dispersed customers. A classical routing problem was first proposed by [5].

Here, VRP with a heterogeneous fleet of vehicles, inspired on a real case of a large distribution company, introduced by [3], is considered. In addition, the problem considers docking constraints, in which some vehicles are unable to serve some particular customers. Objective functions are based on real values provided by a distribution company, which delivers its products to 382 customers and has 169 vehicles of 8 different types.

3.1 Representation and Evaluation of a Solution

A feasible solution s to the HFVRPMT is represented by a set of vectors of routes, respecting each vehicle capacity and attending all costumers that need goods. This solution is evaluated by the sum of the total fixed cost of the vehicles used plus the total cost of the customers visited and the total cost of the distances travelled by the trucks.

3.2 Neighborhood Structures

Six different neighborhood structures are applied to explore the solution space of the VRP dealt in this case study. The first three are intra-route movements while the last two cover inter-route. It is important to note that movements that lead to infeasible solutions are not allowed. The NS are extracted from [3,10], denoted by: NS^{2-opt}, $NS^{Or-opt1}$, $NS^{Or-opt2}$, $NS^{Exchange}$, $NS^{Shift(1,0)}$ and $NS^{Swap(1,1)}$, briefly described bellow:

2-opt Move. A 2-opt move is an intra-route movement that consists in removing two non-adjacent arcs and inserting two new arcs, so that a new route is formed.

Or-optk **Move.** An *Or-optk* move is an intra-route movement that consists in removing k consecutive customers from a given route and reinserting them into another position of the same route. This move is a generalization of the *Or-opt* proposed by [9], in which the removal involves up to three consecutive customers only.

Exchange **Move.** An *Exchange* move is an intra-route movement that consists in exchanging two customers in the same route.

Shift(1, 0) **Move.** A *Shift*(1, 0) move is an inter-route movement that relocates a customer from one route to another. Figure 2 illustrates a *Shift*(1, 0) of customer 6 originally in Route 2 to be the first one in Route 3.

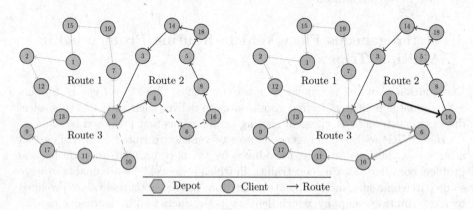

Fig. 2. Example of *Shift*(1, 0) move

Swap(1, 1) **Move.** A *Swap*(1, 1) move is an inter-route movement that exchanges two customers from different routes.

4   Computational Experiments and Analysis

Computational experiments were carried on a Intel Core i7-3537U CPU, 2.00 GHz, with 4 GB of RAM, operating system Ubuntu 14.04.

4.1   Basic Calibration of Population Size

A set of five different instances was used to verify the size of the population that the NGES could perform better. For this purpose, a batch of 30 executions was done and an ANOVA test was conducted in order to analyze the different impact of the population size (μ and $\lambda = 6\mu$). In this ANOVA analyses, all executions were considered as blocking factors of the model, and the different objective function values between the instances were normalized and also considered in the analyses. The null hypotheses was rejected (with $\alpha = 0.05$ and $\beta = 0.8$)

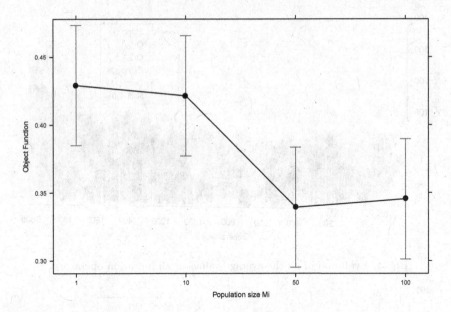

Fig. 3. 95% of confidence effects plot for the NGES population size on five large-scale instances of the HFVRPMT

and a significant difference between the population size was detected. Figure 3 shows an effect plot with 95% of confidence level. No significant difference was detected between the population size 50 and 100 parents. In this sense, the same configuration used for the Open-Pit-Mining Operational Planning Problem (OPMOP), in [2], was kept: $\mu = 100$ and $\lambda = 600$.

4.2 Logic for Setting the Upper Limits for Each Neighborhood Structure

The nap_k limits here were relaxed and left with larger limits compared to the OPMOP. As can be noticed, the classical NS used here can be applied up to 500 times in each application phase. These are, basically, the only limits to be set in this metaheuristic. However, these limits do not strictly represent a critical parameter, as can verify in the evolution of the parameters detailed in the next section.

4.3 NGES Self-adaptive Mechanisms - P and A

The behavior of the mutation operators for two different large scale instances is depicted in Figs. 4, 5 and 6. It is interesting to check the ability of the NGES in increasing the number of applications of the intra-moves of neighborhoods $NS^{Shift(1,0)}$ and $NS^{Swap(1,1)}$ in order to search for new solutions, clearly seen after the first 100 generations. In order to highlight this aspect, Fig. 5 filters these neighborhood for a more clear graph.

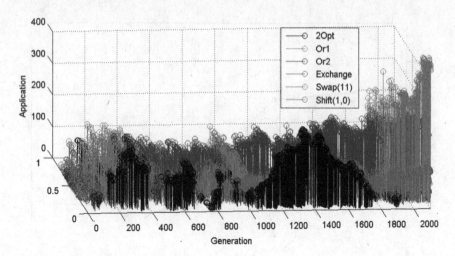

Fig. 4. Evolution of all the average values of all mutation operators

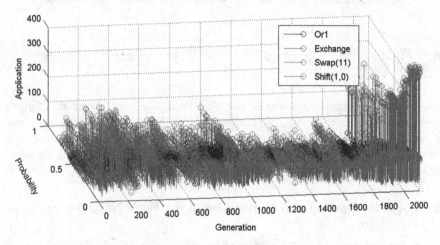

Fig. 5. Evolution of operator highlighting a specific set of NS

Another points that can be seen at Fig. 6 are the high peaks on the probability parameter P for the neighborhood structure $NS^{Or-opt2}$. As should be noticed, these peaks where further investigated and we noticed that they also coincide with the finding of new better solutions, improving the current best solution of the ongoing optimization execution.

5   Final Considerations and Extensions

Analyzing the evolution of the range of probabilities for applying the different NS can help the comprehension of the landscape of an optimization problem. Furthermore, the convergence showed in this study reinforces that the NGES algo-

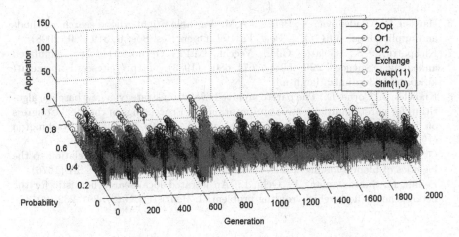

Fig. 6. Self-adaptive operators evolution – Instance HFMVRPMT II

rithm can self-regulate moves application. As could be noticed, when stuck, the operators evolve towards larger probabilities in order to perform larger changes in the solution representation. This represents an effort for escaping from local optima.

In particular, the proposed non sophisticated strategy, basically relying on mutation operations, is an easy to implement metaheuristic but posses sufficient tools for producing successful results in real-world applications.

The method will be extended for handling Multi-objective Optimization Problems. In this sense, it could be an extension of the classic Pareto Archived Evolution Strategy [8]. Basically, the only thing that should be modified is the acceptance criteria which would filter a set of non-dominated solutions. In this sense, the simplicity of the proposal makes its attractive for future improvements and test bed for future adjustments.

References

1. Beyer, H.G., Schwefel, H.P.: Evolution strategies - a comprehensive introduction. Nat. Comput. **1**, 3–52 (2002)
2. Coelho, V.N., et al.: Hybrid self-adaptive evolution strategies guided by neighborhood structures for combinatorial optimization problems. Evol. Comput. **24**(4), 637–666 (2016). https://doi.org/10.1162/EVCO_a_00187
3. Coelho, V.N., Grasas, A., Ramalhinho, H., Coelho, I.M., Souza, M.J.F., Cruz, R.C.: An ILS-based algorithm to solve a large-scale real heterogeneous fleet VRP with multi-trips and docking constraints. Eur. J. Oper. Res. **250**(2), 367–376 (2016). https://doi.org/10.1016/j.ejor.2015.09.047
4. Costa, L.R., Aloise, D., Mladenović, N.: Less is more: basic variable neighborhood search heuristic for balanced minimum sum-of-squares clustering. Inf. Sci. **415–416**, 247–253 (2017). https://doi.org/10.1016/j.ins.2017.06.019
5. Dantzig, G.B., Ramser, J.H.: The truck dispatching problem. Manag. Sci. **6**(1), 80–91 (1959)

6. Hansen, P., Mladenović, N., Pérez, J.A.M.: Variable neighborhood search: methods and applications. 4OR: Q. J. Belg. Fr. Ital. Oper. Res. Soc. **6**, 319–360 (2008)
7. Kirkpatrick, S., Gelatt, C.D., Vecchi, M.P.: Optimization by simulated annealing. Science **220**(4598), 671–680 (1983). http://citeseer.ist.psu.edu/kirkpatrick83optimization.html
8. Knowles, J., Corne, D.: The pareto archived evolution strategy: a new baseline algorithm for pareto multiobjective optimisation. In: Proceedings of the 1999 Congress on Evolutionary Computation, CEC 99, vol. 1, pp. 98–105. IEEE, Washington (1999)
9. Or, I.: Traveling salesman-type combinational problems and their relation to the logistics of blood banking. Ph.D. thesis, Northwestern University, USA (1976)
10. Penna, P., Subramanian, A., Ochi, L.: An iterated local search heuristic for the heterogeneous fleet vehicle routing problem. J. Heuristics **19**(2), 201–232 (2013)

New VNS Variants for the Online Order Batching Problem

Sergio Gil-Borrás[1], Eduardo G. Pardo[1(⊠)], Antonio Alonso-Ayuso[2], and Abraham Duarte[2]

[1] Dept. Sistemas Informáticos, Universidad Politécnica de Madrid, Madrid, Spain
sergio.gil.borras@alumnos.upm.es, eduardo.pardo@upm.es
[2] Department of Computer Science, Universidad Rey Juan Carlos, Móstoles, Spain
{antonio.alonso,abraham.duarte}@urjc.es

Abstract. The Order Batching Problem (OBP) can be considered a family of optimization problems related to the retrieval of goods in a warehouse. The original and most extended version of the problem consists in minimizing the total time needed to collect a group of orders. However, this version has been evolved with many other variants, where the restrictions and/or the objective function might change. In this paper, we deal with the Online Order Batching Problem (OOBP) version, which introduces the novelty to the OBP of considering orders that have arrived to the warehouse once the retrieval of previous orders has started. This family of problems has been deeply studied by the heuristic community in the past. Notice, that solving any variant of the OBP include two important activities: grouping the orders into batches (batching) and determining the route to follow by a picker to retrieve the items within the same batch (routing). We review the most outstanding proposals in the literature for the OOBP variant and we propose a new version of a competitive Variable Neighborhood Search (VNS) algorithm to tackle the problem.

Keywords: Online Order Batching Problem · Batching ·
Variable Neighborhood Search

1 Introduction

The storage of goods in warehouses has associated many tasks such as receiving the goods, storing them or retrieving the products from the shelves of the warehouse, when a new order arrives. Many of those tasks can be enunciated as optimization problems with the aim of saving time, space or work load, among others. The Order Batching Problem (OBP) can be considered a family of optimization problems, more than a single problem, related to the operation of retrieval of goods in a warehouse, when the policy of retrieval is based on order batching.

This work has been partially founded by Ministerio de Economía y Competitividad with grant refs. TIN2015-65460-C2-2-P., MTM2015-63710-P.

The order batching, then, consist in grouping a set of orders together (conforming a batch) and assigning the batch to a person (the picker) who retrieves all the orders within the same batch on a single tour through the warehouse. This policy has been proved to be very effective in contrast with the traditional strict-order picking policy, where each order that arrives to the warehouse is assigned to a picker, who collects exclusively the items from that order on each tour. Some authors point out that it is possible to reduce the travel time up to 35% if the routes followed by the pickers are designed adequately [4]. Additionally, if the batching and routing are considered simultaneously, the save of time can be even larger.

The original and most extended version of the problem, usually known as Order Batching Problem (OBP) consists in minimizing the total time needed to collect a group of orders. However, this version has been evolved with many other variants, where the restrictions and/or the objective function might change. Notice, that solving any variant of the OBP might include two important activities: grouping the orders into batches (batching), and finding the route to follow by the picker to collect the items within the same batch (routing). Additionally, some variants of the OBP also consider a third activity: determining the next batch to be processed (sequencing) once the batches have already been conformed.

In this paper, we deal with the Online Order Batching Problem (OOBP), which is a version of the OBP that introduces the novelty of considering orders that have arrived to the warehouse once the process of retrieval of previous orders has already started. The objective function of the problem is to minimize the maximum time that an order remains in the system. This is usually known in the related literature as the turnover time. To tackle this problem we propose the use of the methodology Variable Neighborhood Search (VNS), particularly, the Basic Variable Neighborhood Search (BVNS) variant and we compare our approach with the classical approaches in the literature for other variants of the OBP.

The rest of the paper is organized as follows: in Sect. 2 we review the most outstanding proposals for the problem in the literature, and we describe in detail the methods that will be used in our experiments as a comparative framework. In Sect. 3 we propose a new version of a competitive Variable Neighborhood Search algorithm to tackle the problem. In Sect. 4 we perform the experiments in order to compare our proposal with the traditional methods for the OBP family of problems. Finally, in Sect. 5 we present our conclusions future research lines.

2   State of the Art

The Order Batching family of problems has been deeply studied by the heuristic community in the past. There are remarkable references based on different metaheuristics for most of the best-known variants of problems within the OBP literature: the classical OBP [1,11,15,16]; the Min-max OBP [7,13]; the OBSP [2,14]; and also to the OOBP tackled in this paper [10,20,22].

However, the first remarkable methods for most of the previous problems are not the metaheuristic approaches but the simpler heuristic procedures based on greedy functions [12]. Those methods were constructive procedures based on simple ideas and have been used as a baseline in many comparisons.

As far as we know, those methods have not been either used or compared in the context of the OOBP. Next, we present a brief description of the most remarkable ones that will be used later in the Sect. 4. Particularly, we consider: the First Come First Served algorithm (Sect. 2.1); the Seed algorithm (Sect. 2.2); and, the Clarke & Wright Savings algorithm (Sect. 2.3).

2.1 First Come First Served Algorithm

The First Come First Served (FCFS) algorithm is probably the simplest heuristic algorithm designed for the OBP. The algorithm receives a list of orders and returns a list of batches. First, the received list of orders is sorted according to the arrival time of each order, in such a way that the oldest order comes first. Then the list of orders is traversed one by one, assigning the next order to be processed to the next available batch. If the order fits in the current batch it is inserted in that batch. Otherwise, a new batch is created with that order, becoming this new batch the current one which will be target of the next considered order. This process is repeated with all the orders until the end of the list. Once all the orders have a batch assigned, the set of batches generated in the algorithm is returned.

2.2 Seed Algorithm

The algorithms known as "seed algorithms" are a group of methods based on a common strategy: a "seed" (in this case an order) is first chosen and assigned to a batch. Then, other available orders might be added to the same batch, as far as the capacity constraint is not violated. Therefore, for each "seed method" it will be necessary to determine how to choose the seed order, and how to choose the additional orders suitable to be assigned to a particular batch with an assigned seed. In this case, the strategy used to select a "seed order" considers the idea introduced in [18] consisting in selecting the available order with the largest number of products. Then, once the seed has been chosen, the strategy used to aggregate orders to the same batch is the one introduced in [21] consisting in selecting the order with lowest absolute difference of its Center Of Gravity (COG) to the seed. Where the COG of an order is defined as the average of the aisle numbers where the items of that order are located. The considered procedure applies a "cumulative mode" (i.e., the seed is renewed each time a new batch is created). The method, then, consist in selecting one seed order, assign it to a batch and trying to complete the batch following the criteria of the difference of COG. Once the batch is full, the method selects a new seed and so on until all the orders have been assigned to a batch.

2.3 Clark & Wright Savings Algorithm

The "Clark & Wright savings" algorithm is inspired in the idea presented in [3] in the context of vehicles routing. It is based on computing the save of time derived from collecting two orders separately versus collecting them together in the same route. The algorithm creates a square matrix with a size equals to the number of orders. Then, each row/column corresponds with one order. The crossing position of a column and a row will store the save/loss of time of collecting the two orders related, separately or together. Additionally, the diagonal of the matrix would store the time of collecting each order in isolation. Notice that this is a symmetric matrix, therefore only one half of the matrix (above or under the diagonal) is needed. For instance, the saving of collecting orders 1 and 2 would be computed as follows: $saving = t_1 + t_2 - t_{1,2}$ where t_1 and t_2 represent the time needed to collect orders 1 and 2 separately, and $t_{1,2}$ the time needed to collect them together. Then all the pairs of orders are stored in a list sorted depending on their savings, in a decreasing way. Next, the list is scanned trying first to allocate together the pairs which produce a largest saving. Notice that several situations might happen: if both orders have not been previously allocated in a batch and they fit together, they are assigned to the same batch; if one of the batches have already been allocated, then the other one will be assigned to the same batch if it fits. Otherwise the procedure will continue with the next pair; finally, if both orders involved have previously been placed in other batches the procedure will jump again to the following pair. We refer the reader to [4] for further details.

3 Algorithmic Proposal

In this section we present our algorithmic proposal to tackle the OOBP. In particular, we propose the use of the methodology Variable Neighborhood Search (VNS) [17]. VNS was originally proposed by Mladenović and Hansen in 1997 as a revolutionary idea to escape from a local optimum, based on the concept of change of the neighborhood structure. Then, the general idea behind the method is to reach local optimum by using a local search procedure and then, change the neighborhood structure (once the current solution found can not be further improved) in order to give the local search the opportunity of looking for a new local optimum in the new neighborhood.

There original idea has been notably evolved with many variants. Probably, the most remarkable ones are: Reduced VNS (RVNS) which perform a stochastic search within a neighborhood; Variable Neighborhood Descent (VND) which perform a deterministic search within the considered neighborhoods; Basic VNS (BVNS) which combines stochastic and deterministic exploration in one neighborhood; and General VNS (GVNS) which combines stochastic and deterministic exploration within a set of neighborhoods. Other well-known approaches are: Skewed VNS (SVNS); and Variable Neighborhood Decomposition Search (VNDS). For a detailed description and tutorials of all those methods we refer the reader to [8,9,17]. Other recent variants include: Variable Formulation Search

(VFS) [19], Parallel Variable Neighborhood Search [6,13] and Multi-Objective Variable Neighborhood Search [5].

In this paper we make use of the BVNS algorithm. In Algorithm 1 we present a pseudocode of this method. It receives three parameters to start the search: (i) an initial solution S generated with an external method; (ii) a value k_{max} which determines the maximum number of neighborhoods to explore; and (iii) the maximum allowed running time (t_{max}). The method explores the neighborhood of the current solution trying to obtain a better one. In order to do that, BVNS has three stages that run consecutively. The first stage is the perturbation of the current solution, performed in order to escape from the current local optimum, reaching a solution in a new neighborhood. As a second stage the method make use of a local search procedure, which is able to find a local optimum within the current neighborhood. The third stage, represented by the procedure Neighborhoodchange, determines if it is necessary to change the neighborhood to be explored, depending on whether the solution provided to the local search has been improved or not. This method updates the value of the variable k, which indicates the number of perturbations to be performed to the current solution in the Shake procedure. The value $k = 1$ indicates that an improvement has been performed, otherwise the value of k is incremented in a predefined amount (typically 1 unit).

Algorithm 1. BVNS$(S, k_{\max}, t_{\max})$

1: **repeat**
2:     $k \leftarrow 1$
3:     **while** $k \leq k_{\max}$ **do**
4:         $S' \leftarrow$ Shake(S, k)
5:         $S'' \leftarrow$ LocalSearch(S')
6:         $k \leftarrow$ NeighborhoodChange(S, S'', k)
7:     **end while**
8: **until** $t < t_{max}$
9: **return** S

A more detailed description of the method used to generate the initial solution can be found in the Sect. 3.1. Similarly, the description of the Shake and LocalSearch procedures are presented, respectively, in Sect. 3.2 and Sect. 3.3. Notice that we do not provide a detailed description of the NeighborhoodChange procedure since it follows an standard implementation.

The algorithm is executed repeatedly until the maximum allowed time is reached. In each iteration, the number of perturbations performed to the solution, before the local search, is indicated by the value of the variable k. The variable k starts at 1, indicating that the first neighborhood to be explored is the closer one. This value is increased every time that the local search does not improve the current solution, until it reaches the value of k_{max}. Then, the variable k is reset to its initial value 1 and the procedure is repeated again until the maximum allowed time is reached.

3.1   Constructive Procedure

We have used a random algorithm as a constructive method in order to provide an initial solution to the BVNS algorithm. The algorithm receives a list of orders as an input parameter. The list of orders is randomly scanned. In each iteration, an order is randomly selected and it is placed in the next available batch. When the selected order no longer fits in the current batch, a new batch is created with this order. Then, the next order will be placed in this new batch and the process is repeated until the order list is fully scanned and all the orders have a batch assigned. Once the process is finished, the procedure returns a list of batches as a solution. In Algorithm 2 we present a pseudocode of this procedure.

Algorithm 2. Constructive(L_{orders})

1: $S \leftarrow$ NewBatchList()
2: $B \leftarrow$ NewBatch()
3: **repeat**
4:     $o \leftarrow$ ChooseRandomOrder(L_{orders})
5:     $L_{orders} \leftarrow L_{orders} \setminus o$
6:     **if** Fits(B, o) **then**
7:         Add(B, o)
8:     **else**
9:         Add(S, B)
10:         $B \leftarrow$ NewBatch()
11:         Add(B, o)
12:     **end if**
13: **until** $L_{orders} = \emptyset$
14: **return** S

3.2   Shake Procedure

The perturbation procedure chosen for this problem consist in exchanging two orders from different batches. The procedure receives as input parameters an initial solution S and the parameter k that indicates the number of times the perturbation will occur. In each perturbation two random batches are selected. Then, two orders also selected at random within the selected batches are exchanged. Notice that the exchange must produce a feasible solution (i.e., it does not exceed the maximum capacity of each batch), otherwise it should be repeated. This process will be repeated as many times as the parameter k indicates. At the end of this procedure, a solution in a different neighborhood will be returned. In Algorithm 3 we present a pseudocode of this procedure.

3.3   Local Search Procedure

The local search procedure proposed to be used within the BVNS, as well as the shake procedure, is based in the one-to-one exchange move. This procedure receives an initial solution S, as an input parameter, and it returns the

---

Algorithm 3. Shake(S, k)

1: **repeat**
2:     **repeat**
3:        $B_i \leftarrow$ ChooseRandomBatch(S)
4:        $B_j \leftarrow$ ChooseRandomBatch(S)
5:     **until** $B_i \neq B_j$
6:     $o_i \leftarrow$ ChooseRandomOrder(B_i)
7:     $o_j \leftarrow$ ChooseRandomOrder(B_j)
8:     **if** Fits$(B_i \setminus o_i, o_j)$ **and** Fits$(B_j \setminus o_j, o_i)$ **then**
9:        $B_i \leftarrow B_i \setminus o_i$
10:      Add(B_i, o_j)
11:      $B_j \leftarrow B_j \setminus o_j$
12:      Add(B_j, o_i)
13:      $k \leftarrow k - 1$
14:     **end if**
15: **until** $k = 0$
16: **return** S

---

local optimum within the neighborhood of the solution. The procedure explores every order o in all the batches trying to find a feasible interchange with other order that improves the current solution. If an improve move is performed, then the procedure starts again from the new solution found, performing another whole iteration, otherwise it carries on until all candidate interchanges have been explored without improvement and returns the best solution found. In Algorithm 4 we present the pseudocode of this procedure.

Algorithm 4. LocalSearch(S)

1: **repeat**
2:    *improved* $\leftarrow$ *false*
3:    **for all** $o_i \in S$ **do**
4:       **for all** $o_j \in S$ **do**
5:         $S' \leftarrow$ Exchange(S, o_i, o_j)
6:         **if** $f(S') < f(S)$ **then**
7:           $S \leftarrow S'$
8:           *improved* $\leftarrow$ *true*
9:           **break**
10:        **end if**
11:      **end for**
12:    **end for**
13: **until** *improved* = *false*
14: **return** S

---

4 Results

We compare our proposal with the classical greedy constructive procedures presented in Sect. 2. The experiments were run an Intel (R) Core (TM) 2 Quad CPU Q6600 2.4 Ghz machine, with 4 GB DDR2 RAM memory. The operating system used was Ubuntu 18.04.1 64 bit LTS, and all the codes were developed in Java 8.

4.1 Instances

An instance to test any algorithm for the OOBP needs to consider the following aspects: the warehouse layout; the orders; and the distribution followed by the arrival of the orders.

We have selected and adapted a set of instances previously referred in the literature for the OBP to test our proposal. In particular, we have selected a subgroup of instances from those reported in [1] which have been reference instances in the OBP literature in the last few years. This data set contains instances related to four real warehouses of rectangular shape. Each warehouse has two transversal aisles, one at the front and one at the back of the warehouse and a variable number of parallel aisles. In each side of the parallel aisles there are products stored. Every warehouse has only one depot located at the front-cross aisle either at the left corner or at the center of the aisle. In the Fig. 1 we present an example of the layout of the considered warehouse. Particularly, this example warehouse has 2 crossing aisles and 5 parallel aisles, with 9 picking positions in each side of the parallel aisles, totalizing 90 picking positions. In this case, the depot is placed in the center of the front cross aisle.

The number of orders per instance varies among the following values [50, 100, 150, 200, 250]. The distribution of the products in the warehouse follows either an ABC distribution or a random one. We have selected 16 representative instances from the Warehouse 1 for our comparison. In this subset, we have selected 4 different instances for each number of orders [100, 150, 200, 250]. Notice that we have avoided the use of the smallest type of instances (i.e., the ones composed by 50 orders) since a small number of orders do not create enough congestion in the delivery of orders and, therefore, the instances become trivial for the OOBP.

Finally, we have adapted the instances by determining distribution of the delivery instant of the orders to the warehouse. We have divided each set of orders into two groups: offline/online. The first group is formed by 15 orders which will be already available at the beginning of execution. The rest of the orders will arrive to the warehouse following an uniform distribution along the time horizon of 4 h.

4.2 Comparison with the State of the Art

The BVNS algorithm has been successfully compared with three different algorithms in the state of the art. Particularly, we have selected three classical and well-known greedy constructive procedures, widely used in the OBP literature:

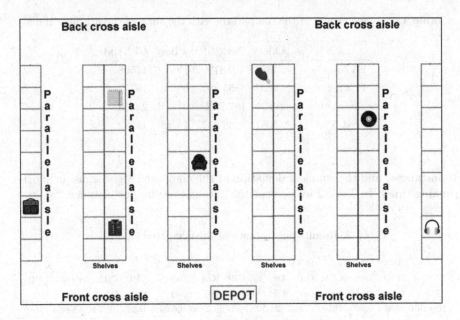

Fig. 1. Warehouse layout.

the First Come First Served (FCFS) algorithm, a variant of the Seed algorithm, and finally the Savings C&W algorithm. These three algorithms were described in the Sect. 2.

Before to perform the comparison of the BVNS with the algorithms in the state of the art, we have carried on some preliminary experiments, to empirically adjust the value of the parameter k_{max} of the BVNS. In this case, we have selected $k_{max} = 15$ for the final configuration of the BVNS. Also, the value of t_{max} was set to 10 s. Therefore, every 10 s, the algorithm starts again from a new solution constructed with the procedure described in Sect. 3.1. Notice, that every construction considers all the orders already arrived to the warehouse and not collected yet.

In Table 1 we present the average value of the objective function (O.F.), which in this case is, for each instance, the maximum time that an order remains in the system before being served; the average deviation with respect to the best solution found in the experiment (Dev.(%)); the number of best solutions found in the experiment (#Best); and the running time of the CPU in seconds (CPUt(s)). Notice, that for each instance, the minimum running time is four hours. These four hours is the time that the order dispenser will use to deliver all the orders in the instance to the system. The final execution time will depend on the time that each algorithm takes to distribute those orders into batches and on the quality of the solution.

As it is shown in Table 1 BVNS is the best algorithm of the comparison since it was able to find the largest number of best solutions found (15 out of

Table 1. Average results of the comparison with the state-of-the-art algorithms.

	O.F.	Dev.(%)	#Best	CPUt(s)
BVNS	3682	0.37%	15	17508
FCFS	5897	55.36%	0	19717
Savings C&W	10621	187.70%	0	24353
Seed	5081	38.45%	1	18232

16 instances) and the smallest deviation of the compared algorithms, in shorter running times. In Table 2 we present the detailed results per instance.

Table 2. Results per instance of the compared algorithms.

	BVNS			FCFS			Savings C&W			Seed		
	O.F. (s)	Dev (%)	CPU t(s)	O.F. (s)	Dev (%)	CPU t(s)	O.F. (s)	Dev (%)	CPU t(s)	O.F. (s)	Dev (%)	CPU t(s)
100_000	1491	0.00%	16196	1856	24.50%	16149	2399	60.94%	16522	1974	32.44%	15989
100_030	1522	0.00%	15430	1924	26.46%	15416	2276	49.54%	15416	1847	21.39%	15416
100_060	1771	0.00%	16103	1927	8.78%	15857	2841	60.39%	16055	2218	25.23%	15996
100_090	1371	0.00%	15498	1492	8.84%	15478	1945	41.87%	15597	2008	46.41%	15484
150_000	2602	0.00%	15551	4461	71.41%	17565	10843	316.67%	24491	3802	46.11%	16163
150_030	1181	0.00%	14619	1492	26.36%	14512	2263	91.60%	14916	1311	11.01%	14584
150_060	3078	0.00%	16209	5413	75.89%	18829	11488	273.28%	24984	4074	32.38%	16524
150_090	1068	0.00%	14366	1522	42.47%	14323	1432	34.10%	14285	1150	7.67%	14416
200_000	7135	0.00%	21409	10871	52.36%	25490	20998	194.30%	35617	9679	35.65%	23344
200_030	1255	5.90%	15480	1497	26.33%	15981	4624	290.19%	19136	1185	0.00%	15334
200_060	5400	0.00%	19888	8926	65.29%	23584	19491	260.92%	34003	7975	47.68%	21259
200_090	1498	0.00%	15445	2709	80.85%	17148	6829	355.82%	21218	2350	56.85%	15305
250_000	12202	0.00%	25727	19750	61.86%	33437	28913	136.96%	42601	15657	28.32%	28479
250_030	2446	0.00%	15821	5629	130.14%	19315	11356	364.31%	25067	4710	92.57%	17018
250_060	12028	0.00%	25840	18201	51.33%	32017	30147	150.64%	43962	15551	29.29%	28299
250_090	2869	0.00%	16542	6683	132.97%	20375	12096	321.65%	25784	5801	102.22%	18107

5 Conclusions

In this paper we deal with the Online Order Batching Problem, as a variant of the well-known family of problems related to the Order Batching. This variant considers that there are orders which arrive to the warehouse once the retrieving process has already started. Those orders are immediately processed and introduced in a batch in order to be collected. The problem looks for minimizing the maximum time that an order remains in the system before being served.

To tackle this problem we have proposed several heuristics within the Basic Variable Neighborhood Search framework. Particularly, we propose to start the search with a random solution and then we define a neighborhood, based on interchange moves, explored by a local search procedure which follows a first improvement strategy. The proposed method has been compared successfully with classical greedy methods in the state of the art, previously used for other variants of the OBP.

In a future research we propose the extension of our algorithm by defining new neighborhoods to be combined in a Variable Neighborhood Descent or in a General Variable Neighborhood Search procedure. Additionally, we also propose to extend the comparison performed, by considering not only the classical greedy constructive methods in the literature, but also the latest metaheuristic-based methods.

References

1. Albareda-Sambola, M., Alonso-Ayuso, A., Molina, E., De Blas, C.S.: Variable neighborhood search for order batching in a warehouse. Asia-Pac. J. Oper. Res. **26**(5), 655–683 (2009)
2. Azadnia, A.H., Taheri, S., Ghadimi, P., Mat Saman, M.Z., Wong, K.Y.: Order batching in warehouses by minimizing total tardiness: a hybrid approach of weighted association rule mining and genetic algorithms. Sci. World J. **2013**, (2013)
3. Clarke, G., Wright, J.W.: Scheduling of vehicles from a central depot to a number of delivery points. Oper. Res. **12**(4), 568–581 (1964)
4. De Koster, R., Roodbergen, K.J., Van Voorden, R.: Reduction of walking time in the distribution center of de bijenkorf. In: Speranza, M.G., Stähly, P. (eds.) New Trends in Distribution Logistics. Lecture Notes in economics and mathematical systems, vol. 480, pp. 215–234. Springer, Heidelberg (1999). https://doi.org/10.1007/978-3-642-58568-5_11
5. Duarte, A., Pantrigo, J.J., Pardo, E.G., Mladenovic, N.: Multi-objective variable neighborhood search: an application to combinatorial optimization problems. J. Glob. Optim. **63**(3), 515–536 (2015)
6. Duarte, A., Pantrigo, J.J., Pardo, E.G., Sánchez-Oro, J.: Parallel variable neighbourhood search strategies for the cutwidth minimization problem. IMA J. Manag. Math. **27**(1), 55–73 (2013)
7. Gademann, N., Velde, V.D.S.: Order batching to minimize total travel time in a parallel-aisle warehouse. IIE Trans. **37**(1), 63–75 (2005)
8. Hansen, P., Mladenović, N.: Variable neighborhood search: principles and applications. Eur. J. Oper. Res. **130**(3), 449–467 (2001)
9. Hansen, P., Mladenović, N., Moreno-Pérez, J.A.: Variable neighbourhood search: methods and applications. Ann. Oper. Res. **175**(1), 367–407 (2010)
10. Henn, S.: Algorithms for on-line order batching in an order picking warehouse. Comput. Oper. Res. **39**(11), 2549–2563 (2012)
11. Henn, S., Koch, S., Doerner, K.F., Strauss, C., Wäscher, G.: Metaheuristics for the order batching problem in manual order picking systems. Bus. Res. **3**(1), 82–105 (2010)
12. Koster, M.B.M.D., der Poort, E.S.V., Wolters, M.: Efficient orderbatching methods in warehouses. Int. J. Prod. Res. **37**(7), 1479–1504 (1999)
13. Menéndez, B., Pardo, E.G., Sánchez-Oro, J., Duarte, A.: Parallel variable neighborhood search for the min-max order batching problem. Int. Trans. Oper. Res. **24**(3), 635–662 (2017)

14. Menéndez, B., Bustillo, M., Pardo, E.G., Duarte, A.: General variable neighborhood search for the order batching and sequencing problem. Eur. J. Oper. Res. **263**(1), 82–93 (2017)

15. Menéndez, B., Pardo, E.G., Alonso-Ayuso, A., Molina, E., Duarte, A.: Variable neighborhood search strategies for the order batching problem. Comput. Oper. Res. **78**, 500–512 (2017)

16. Menéndez, B., Pardo, E.G., Duarte, A., Alonso-Ayuso, A., Molina, E.: General variable neighborhood search applied to the picking process in a warehouse. Electron. Notes Discret Math. **47**, 77–84 (2015)

17. Mladenović, N., Hansen, P.: Variable neighborhood search. Comput. Oper. Res. **24**(11), 1097–1100 (1997)

18. Pan, C.H., Liu, S.Y.: A comparative study of order batching algorithms. Omega **23**(6), 691–700 (1995)

19. Pardo, E.G., Mladenović, N., Pantrigo, J.J., Duarte, A.: Variable formulation search for the cutwidth minimization problem. Appl. Soft Comput. **13**(5), 2242–2252 (2013)

20. Pérez-Rodríguez, R., Hernández-Aguirre, A., Jöns, S.: A continuous estimation of distribution algorithm for the online order-batching problem. Int. J. Adv. Manuf. Technol. **79**(1), 569–588 (2015)

21. Rosenwein, M.B.: An application of cluster analysis to the problem of locating items within a warehouse. IIE Trans. **26**(1), 101–103 (1994)

22. Rubrico, J., Higashi, T., Tamura, H., Ota, J.: Online rescheduling of multiple picking agents for warehouse management. Robot. Comput.-Integr. Manuf. **27**(1), 62–71 (2011)

An Adaptive VNS and Skewed GVNS Approaches for School Timetabling Problems

Ulisses Rezende Teixeira[1], Marcone Jamilson Freitas Souza[2],
Sérgio Ricardo de Souza[1(✉)], and Vitor Nazário Coelho[3]

[1] Federal Center of Technological Education of Minas Gerais (CEFET-MG),
Av. Amazonas, 7675, Belo Horizonte, MG 30510-000, Brazil
ulisses.rezende@gmail.com, sergio@dppg.cefetmg.br
[2] Federal University of Ouro Preto (UFOP),
Campus Universitário, Morro do Cruzeiro, Ouro Preto, MG 35400-000, Brazil
marcone@ufop.edu.br
[3] Fluminense Federal University (UFF), Rua Miguel de Frias, 9,
Niterói, RJ 24220-900, Brazil
vncoelho@gmail.com

Abstract. The School Timetabling Problem is widely known and it
appears at the beginning of the school term of the institutions. Due to
its complexity, it is usually solved by heuristic methods. In this work, we
developed two algorithms based on the Variable Neighborhood Search
(VNS) metaheuristic. The first one, named Skewed General Variable
Neighborhood Search (SGVNS), uses Variable Neighborhood Descent
(VND) as local search method. The second one, so-called Adaptive VNS,
is based on VNS and probabilistically chooses the neighborhoods to do
local searches, with the probability being higher for the more successful
neighborhoods. The computational experiments show a good adherence
of these algorithms for solving the problem, especially comparing them
with previous works using the same metaheuristic, as well as with previous published results of the winning algorithm of the International
Timetabling Competition of 2011.

Keywords: School Timetabling · Variable Neighborhood Search ·
SGVNS · International Timetabling Competition

1 Introduction

The task of creating a school timetabling consists, in a very simplified way, in
determining for each class and time slot the school subject and its respective
teacher. This is a very hard activity and it may take a lot of hours or even
days depending on the amount of classes, time slots, and teachers [2]. This
combination of class, teacher, and subject follows some rules or constraints. The
compliance according to these constraints determines if a solution is feasible or
not and how good it is.

© Springer Nature Switzerland AG 2019
A. Sifaleras et al. (Eds.): ICVNS 2018, LNCS 11328, pp. 101–113, 2019.
https://doi.org/10.1007/978-3-030-15843-9_9

The constraints are usually divided into two groups [20]:

(i) Hard constraints: they are mandatory constraints. If they are not met, the solution is infeasible. For example, if one teacher is allocated to more than one class at the same time-slot, the solution is invalid;

(ii) Soft constraints: these are non-mandatory restrictions. They should be met only when possible but when this is not the case, the solution still remains feasible. For example, no occurrence of idle time in a timetabling of a specific teacher is expected, but the existence of it does not infeasible the solution.

Since the pioneer work of Gotlieb [8], many techniques have been used to solve timetabling problems. According to [21] and [19], this interest is due to three main points:

(i) Difficulty to find a solution: in view of the big amount of constraints, the goal of finding a feasible solution is a hard task and it may takes many days of manual work due to the amount of involved resources (classes, teachers, time slots);

(ii) Practical importance: to build a timetabling is a basic necessity of all educational institutions. A good school timetabling can impact the life of a big quantity of people, especially students and teachers. It can impact the efficiency of the classes and student's performance too;

(iii) Theoretical importance: school timetabling is a NP-hard problem [7]. Thus, it is challenging to develop efficient algorithms to solve it.

The interest of the academic community in seeking more efficient solution methods for solving the problem grew especially in the late 1990s and early 2000s. As a result, an international conference, called PATAT (Practice and Theory on Automated Timetabling), was created. This conference originated specific competitions called ITC (*International Timetabling Competition*), the most recent of them was organized in 2011.

Besides that, specialists in the School Timetabling's class of problem created a standard to represent it, called XHSTT, as well as a library specialized to manipulate their instances, called KHE.

In this paper, the school timetabling problem is approached using the VNS metaheuristic in two variances: skewed VNS and adaptive VNS. In both approaches, it is used the KHE library and the instances from ITC 2011. The results were compared with the algorithm Goal Solver, winner of ITC 2011, the last competition specialized in problems from this kind.

The paper is organized as follows: Sect. 2 presents the KHE library, the XHSTT format and the ITC 2011; Sect. 3 presents the proposed algorithms and Sect. 4 shows their results. Finally, on Sect. 5 the conclusions and some future works are presented.

2   Context

The researches in school timetabling had an impulse with the organization of specialized competitions of algorithms to solve this type of problem (such as the

ITC, described at Sect. 2.3), as well as the creation of a standard to represent and treat instances (the XHSTT, described at Sect. 2.1) and the creation of a library to handle this format (the KHE, described at Sect. 2.2).

2.1 XHSTT Standard

The XHSTT format [17], an acronym to XML for High School TimeTabling, is a format based on XML's markup language that establishes specific structures to treat resources, time-slots and their respective constraints.

This format is divided in three basic entities:

(i) Time and resource: the time entity consists of a time-slot or a set of time-slots and the resources are subdivided into three subcategories: students, teachers, and rooms;
(ii) Events: an event is the basic unit of assignment, representing a simple lesson;
(iii) Constraints: it is responsible to determine the distribution of resources in the events. It can be defined by hard or soft constraints, according to the criteria expected for a specific solution to be feasible or infeasible. Besides, it is subdivided into three subcategories: (i) basic constraints of schedule; (ii) constraints of events; and (iii) constraints of resources.

From the creation of this standard, emerged a lot of research from many countries around the world that created instances of all kind of types, some representing real cases from several countries, for example, England [26], Finland [16], Greek [25], Netherlands [6] and Brazil [22,23]. All these instances were published in a global and free public repository, to be used as benchmarking for other studies, such as the one conducted in this paper.

2.2 KHE Library

In 2006, [11] presented a library, called KHE (*Kingston High School Timetabling Engine*). This library was created exclusively for school timetabling problems, with the objective of facilitating and optimizing the management of the instances and their solutions. Completely integrated with the XHSTT standard, the main points of the use of this library are the data structures available and the possibility of using the function of generating initial solutions, called KheGeneralSolve. This routine generates an initial solution in a fast and easy fashion, even in large and complex instances. This library is available on the Internet and can be used freely for studies and researches in this area. Its creator also provides a service to evaluate solutions, called HsEval, available at http://www.it.usyd.edu.au/~jeff/cgi-bin/hseval.cgi.

2.3 ITC 2011

With this standards and variety of libraries for handling them, the PATAT members launched the third edition of an International Timetabling Competition -

ITC, dedicated to High School timetabling problems, in 2011. This was the last and most recent competition in this area. The two previous ones were organized in 2002 and 2007, with others specific themes of timetabling.

The ITC 2011 was composed by three phases:

- Phase 1: the instances were published and the competitors were responsible to generate the best solutions without restrictions of time and computational resources;
- Phase 2: the organizers were responsible for executing each algorithm under the same conditions, using instances not previously known and having a time limit of 1000 s of processing;
- Phase 3: the competitors generated the solutions in a set of hidden instances and, as in the phase one, it was not defined time and technology restrictions. Only the top five competitors of Phase 2 participated of this phase.

3   Proposed Approaches

In order to solve the School Timetabling Problem, in the current article we propose two algorithms for solving it, both of them based on the Variable Neighborhood Search (VNS) metaheuristic [13].

The first one, called Adaptive VNS, is described in Subsect. 3.1, and the second one, named SGVNS, is presented in Subsect. 3.2. Finally, in Subsect. 3.3, the types of moves used to explore the solution space of the problem are detailed.

3.1   Adaptive VNS

The proposed Adaptive VNS algorithm is a variant of the classic VNS metaheuristic, in which the used neighborhoods for local searches are chosen according to evolving probabilities.

This approach is similar to that presented in [1]. The basic principle is that the neighborhoods that generate better solutions should have probabilities higher than the other ones that are not generating good solutions at that moment. In order to avoid premature convergence of the algorithm, and avoiding getting biased to some neighborhoods, whenever a better solution is found, the probabilities are periodically reset.

The implementation follows the pseudo-code presented at Algorithm 1. Initially, in line 10, all $|N|$ neighborhoods used for local searches have the same probability of being chosen, that is, the parameter $probneighborhood(N^l)$ is set to $1/|N|$ (0.2 in our case). On the loop started at line 14 a shaking move using Kempe's chain neighborhood is applied during $k_{current}$ times. The neighborhood used to perform local search according to the current probabilities is chosen at line 18. As in timetabling problems there are many plateaus, solutions with an evaluation less than or equal to that of the current solution are accepted (line 20 of Algorithm 1). The probabilities of all neighborhoods are recalculated every *itercalc* iterations (line 40 of Algorithm 1).

Algorithm 1. Adaptive VNS

---

1 **Input:** Initial solution s_0; Maximum runtime (*MaxTime*); Maximum
   number of moves of the Kempe's Chain; Iterations for recalculating the
   probabilities (*itercalc*); set of $|N|$ neighborhoods N; Number of
   recalculating probabilities without improvement to restart probabilities
   (*IterRestart*).
2 **Output:** Best solution s.
3 **begin**
4      $s \leftarrow s_0$;
5      $s' \leftarrow s_0$;
6      $improvement \leftarrow 0$;
7      $k_{current} \leftarrow 1$;
8      $number_{itercalc} \leftarrow 1$;
9   **for** *each neighborhood N^l of N* **do**
10         $probneighborhood(N^l) \leftarrow 1/|N|$;
11   **end**
12     $iter \leftarrow 1$;
13   **while** *time $\leq$ MaxTime* **do**
14   **for** $k = 0$; $k < k_{current}$ **do**
15             $s' \leftarrow$ neighbor of s' built by applying the Kempe's Chain
                   move;
16   **end**
17         $prob \leftarrow$ random number between 0 and 1;
18         $l \leftarrow$ chosen neighborhood N^l according to the probability
               $probneighborhood(N^l)$ and random number $prob$;
19         $s' \leftarrow$ Local Search using neighborhood $N^l(s')$;
20   **if** $f(s') \leq f(s)$ **then**
21             $s \leftarrow s'$;
22             $k_{current} \leftarrow 1$; $improvement \leftarrow 1$;
23   **end**
24   **else**
25             $s' \leftarrow s$;
26   **if** $k_{current} \leq Kempe_{max}$ **then**
27                 $k_{current} \leftarrow k_{current} + 1$;
28   **end**
29   **end**
30   **if** *rest of division of iter by itercalc is* 0 **then**
31             $number_{itercalc} \leftarrow number_{itercalc} + 1$;
32   **if** *$number_{itercalc} \geq IterRestart$ and improvement = 0* **then**
33   **for** *each neighborhood N^l of N* **do**
34                     $probneighborhood(N^l) \leftarrow 1/|N|$;
35   **end**
36                 $number_{itercalc} \leftarrow 1$;
37   **end**
38   **else**
39   **for** *each neighborhood N^l of N* **do**
40   update $probneighborhood(N^l)$
41   **end**
42   **end**
43             $improvement \leftarrow 0$;
44             $iter \leftarrow iter + 1$;
45   **end**
46   **end**
47 **end**
48 **return** s

3.2   Skewed GVNS (SGVNS)

This algorithm was built by merging two variations of VNS metaheuristics: Skewed VNS (SVNS) and General VNS (GVNS).

The SVNS is a variation of VNS proposed by [13]. It uses a parameter α to accept solutions that are worse than the current solution. The concept involved is that better solutions can be far away from the current solution, so it is necessary to go through intermediate (and worse) steps to reach them.

On the other hand, the GVNS algorithm, proposed in [14], uses Variable Neighborhood Descent (VND) algorithm to perform local searches. VND [9] is a descent method that uses systematic changes of neighborhoods to explore the space solution. It returns a local optimum among all the used neighborhoods.

The proposed SGVNS uses also VND as a local search method. In addition, as in the SVNS algorithm, a parameter α is used to accept intermediate solutions that are worse than the current solution.

Algorithms that accept worse solutions can bring a problem of execution, called cycling. It occurs when the algorithm remains stuck in the same sequence of solutions. To avoid this behavior, a Tabu List was implemented in a way that a short time list stores the values of solutions already visited. In consequence, the algorithm prevents the same sequence of solutions from being generated again. This Tabu List has a length defined by a parameter and works with FIFO protocol (First In First Out) that means that when the length is achieved the first value is overwritten by the next and so on.

The pseudo-code of SGVNS is described in Algorithm 2. At line 14 it is verified if the new solution will be considered or not according to the parameter α and the list of solution values already generated. As in the Adaptive VNS algorithm, solutions with evaluation less than or equal to the current solution are accepted (line 18 of the SGVNS Algorithm).

3.3   Moves

Both algorithms use the Kempe's Chain move for shaking the current solution. This move was proposed in [10] to the graph coloring problem. It is based on the concept that some changes in the solution can generate infeasible solutions, creating conflicts and in order to remove them it is necessary to perform a sequence of other moves. These modifications in sequence applied to a determined solution are called Kempe's Chain.

When the solution does not improve, both algorithms increase the number of times that the Kempe's Chain move is executed until a limit value defined by the parameter $Kempe_{max}$. When an improved solution is found, each algorithm returns to its initial configuration and only one Kempe's Chain move is performed. This strategy has the objective to search better solutions and not get stuck in local optimums.

The SGVNS algorithm uses the classic VND algorithm to perform local searches and it returns the optimum in relation to all neighborhoods. The VND

Algorithm 2. SGVNS

---

1   **Input:** Initial solution (s_0); Maximum runtime ($MaxTime$); Maximum number of Kempe's Chain move ($Kempe_{max}$); percentage to accept worse solutions (α); Length of the Tabu List.

2   **Output:** Improved solution s found.

3   **begin**

4      $s \leftarrow s_0$;

5      $s' \leftarrow s_0$;

6      $s_{temp} \leftarrow s$;

7      $k_{current} \leftarrow 1$;

8   Insert $f(s)$ in *Tabu List*;

9   **while** *time* $\leq$ *MaxTime* **do**

10   **for** $k = 0$; $k < k_{current}$ **do**

11          $s' \leftarrow$ neighbor of s' using Kempe's Chain move;

12   **end**

13        $s' \leftarrow$ Local search using VND algorithm(s');

14   **if** $((f(s') \leq ((1 + \alpha) \times f(s)))$ *and* $(f(s') \notin$ *Tabu List*$))$ **then**

15          $s_{temp} \leftarrow s'$;

16          $k_{current} \leftarrow 1$;

17   Insert $f(s')$ in *Tabu List*;

18   **if** $(f(s') \leq f(s))$ **then**

19            $s \leftarrow s'$;

20   **end**

21   **end**

22   **else**

23          $s' \leftarrow s_{temp}$;

24   **if** $k_{current} \leq Kempe_{max}$ **then**

25            $k_{current} \leftarrow k_{current} + 1$;

26   **end**

27   **end**

28   **end**

29   **end**

30   **return** s

---

algorithm is described in the literature, so it is not presented in this article. The neighborhoods are generated with one of the moves described below:

Event Swap: this move consists in selecting two lessons and changing the time slots between them;

Event Move: this move consists in choosing one lesson and moving it to another time slot that is empty;

Event Block Swap: like to the Event Swap, it consists in swapping the time slot of two lessons. However, if the lessons have different durations, one lesson is moved to the last time slot occupied by the other lesson. That is, if one of the selected lessons has another lesson in a time slot adjacent to it, the change involves both lessons, not only the selected one. This move allows contiguous time slots to be exchanged;

Move Time: in this move, two classes are chosen and exchanged;
Change Time: this move consists in choosing one class and changing its
   resource with another resource that is current available.

On the other hand, the Adaptive VNS algorithm chooses, in a probabilistic
fashion, only one of these moves described above to perform only **one** local
search.

4   Computational Experiments

Both algorithms were implemented in C++ using the IDE Code::Blocks. All
tests were done in a notebook with Intel Core i5 processor, 4 GB RAM memory
running Windows 10.

In order to test the algorithms, instances from ITC 2011 were used. Among
the twenty-one instances presented in that event, five of them already were
in local optimum since the initial solution, so it was not necessary to work
with them. In the first phase of ITC 2011, although there were no processing
time restrictions, the computational time limit for each instance used by Goal
Solver algorithm [5] (the winner of the competition) was 1000 s. In the current experiments, the same value of computational time limit to each instance
was considered. The initial solutions were provided by the organizers and they
were made available together with the instances, in the same XHSTT file. The
three instances from Australia (AustraliaBGHS98, AustraliaSAHS96 and Aus-
traliaTES99) presented worse initial solutions compared with that generated by
KHE library. Thus, in these instances, we used the solutions generated by KHE.

4.1   Parameter Tuning

Several distinct experiments were conducted to find the best set of parameter
configurations for each algorithm. The iRace package (http://iridia.ulb.ac.be/
irace/) was used for the accomplishment of this task. This tool implements the
iterated racing procedure [12] and it is an extension of the iterated F-race (I/F-
Race) proposed by [4]. The main function of iRace is the automatic configuration
of optimization algorithms in order to determine the most appropriate parameter
settings for an optimization method. The iRace framework is implemented as an
R package [18] and builds upon the race package.

The iRace analysis was done on a budget of 3,000 runs for each algorithm
(Adaptive VNS and SGVNS). Due to the high duration of the tests, we do not
use 1,000 s as a stopping criterion in this phase. Three different times were used
as stopping criterion for each algorithm applied in each instance: 10 s, 30 s and
60 s. Tables 1 and 2 show the parameters tested by iRace for both the SGVNS
and the Adaptive VNS algorithms, respectively.

As a result, the iRace indicated the best parameters for each method, as
shown in the values depicted at Table 3. In addition to the parameter calibration,
the iRace gave us the feedback that the performance of both algorithms improved

Table 1. The parameters tested by iRace in SGVNS algorithm.

Parameter	Values				
α	0.001	0.01	0.025	0.05	0.075
$Kempe_{max}$	1	5	10	-	-
Length of the Tabu List	5	7	10	15	-

Table 2. The parameters tested by iRace in Adaptive VNS algorithm.

Parameter	Values			
Iterations to restart probabilities	5	10	15	20
$Kempe_{max}$	1	5	10	-
itercalc	100	250	500	750

from 10 s to 60 s. In this sense, the best parameters were picked from executions with 60 s, which are the ones that provide more liberty to the methods to play with exploration-exploitation concepts.

4.2 Results

The same set of neighborhoods N = {Event Swap, Event Move, Event Block Swap, Move Time and Change Time} was used for both algorithms.

Table 4 shows the best results obtained by the algorithms Goal Solver of [5], GVNS of [24] and the proposed SGVNS and Adaptive VNS algorithms. In addition, Table 5 shows the average results also obtained by these same algorithms. In both tables, the first column shows the tested instances and the second one, the value of the initial solutions provided by the organizers of ITC 2011, except for instances AustraliaBGHS98, AustraliaSAHS96 and AustraliaTES99, whose initial solutions were generated by the KHE algorithm of [11]. The following columns present the values of Goal Solver, GVNS, SGVNS e Adaptive VNS algorithms, respectively.

Each instance was executed 30 times for each algorithm. It is noteworthy that all algorithms were executed on the same machine, and the Goal Solver code was provided by its developers.

The values presented in each cell of Tables 4 and 5 are pairs x/y, where x means the sum of penalties for hard constraints not met and y the sum of penalties for soft constraints not met. In case of a tie in the penalties for hard constraints not met, the solutions that have smallest soft constraints not met are considered the best ones. A value highlighted in **bold** means that is considered to be the best result produced among all the algorithms.

Analyzing the best results as presented in Table 4, it is verified that SGVNS outperforms the other algorithms in most instances. On the other hand, the Adaptive VNS algorithm did not outperform the other algorithms in any instance, although it has produced good results as well. Considering the sixteen

Table 3. Best parameters indicated by iRace.

SGVNS	α	$Kempe_{max}$	Length of the Tabu List	Execution time
	0.025	5	10	60
Adaptive VNS	itercalc	$Kempe_{max}$	Iterations to restart probabilities	Execution time
	500	1	5	60

Table 4. Best results of the algorithms.

Instance	KHE (Initial Solution)	Goal Solver (SA + ILS)	GVNS	SGVNS	Adaptive VNS
AustraliaBGHS98	6/450	6/450	4/370	**1/401**	7/431
AustraliaSAHS96	17/55	14/50	**12/51**	13/46	17/52
AustraliaTES99	7/163	7/161	**7/151**	7/163	7/163
BrazilInstance1	0/24	0/12	**0/11**	**0/11**	0/12
BrazilInstance4	0/112	0/91	0/94	**0/90**	0/94
BrazilInstance5	0/225	0/164	0/158	**0/149**	0/165
BrazilInstance6	0/209	0/149	0/148	**0/131**	0/163
BrazilInstance7	0/330	0/264	0/249	**0/248**	0/282
EnglandStPaul	0/18,444	0/18,092	0/12,542	**0/12,466**	0/18,418
FinlandHighSchool	**0/1**	**0/1**	**0/1**	**0/1**	**0/1**
FinlandSecondarySchool	0/106	**0/86**	0/87	0/88	0/87
ItalyInstance1	0/28	0/19	**0/18**	**0/18**	**0/18**
NetherlandsGEPRO	1/566	1/566	**1/434**	1/441	1/532
NetherlandsKottenpark2003	0/1,410	0/1,409	**0/1,216**	0/1,281	0/1,372
NetherlandsKottenpark2005	0/1,078	0/1,078	0/881	**0/877**	0/1,078
SouthAfricaLewitt2009	0/58	**0/22**	0/24	0/24	0/42

instances, SGVNS algorithm reached the best results in ten ones and GVNS in seven ones. Goal algorithm, in turn, reached the best results only in three instances.

In another analysis, focusing in the average of the results, as presented in Table 5, SGVNS algorithm reached the best results in six instances and GVNS in eight ones. Goal algorithm reached the best results in five instances and the adaptive VNS in only one instance.

In order to evaluate if there were significant differences among the algorithms, the R Studio tool was used to perform the statistical analyzes of the results. For this analysis all samples were used and it was concluded that the samples did not present normal distribution applying the Shapiro-Wilk test [15]. Then, a non-

Table 5. Average results of the algorithms.

Instance	KHE (Initial Solution)	Goal Solver (SA + ILS)	GVNS	SGVNS	Adaptive VNS
AustraliaBGHS98	6/450	6/450	5/450	**3/514**	7/431
AustraliaSAHS96	17/55	**16/20**	16/30	16/91	17/53
AustraliaTES99	7/163	**7/162**	**7/162**	7/163	7/163
BrazilInstance1	0/24	0/14	**0/11**	**0/11**	0/13
BrazilInstance4	0/112	**0/98**	0/100	0/100	0/099
BrazilInstance5	0/225	0/181	0/178	**0/177**	0/188
BrazilInstance6	0/209	0/168	**0/160**	0/170	0/175
BrazilInstance7	0/330	0/280	**0/276**	0/289	0/300
EnglandStPaul	0/18,444	0/18,444	**0/14,217**	0/14,442	0/18,418
FinlandHighSchool	**0/1**	**0/1**	**0/1**	**0/1**	**0/1**
FinlandSecondarySchool	0/106	**0/89**	0/92	0/93	0/90
ItalyInstance1	0/28	0/21	0/21	**0/19**	0/24
NetherlandsGEPRO	1/566	1/566	**1/446**	1/484	1/551
NetherlandsKottenpark2003	0/1,410	0/1,409	**0/1,290**	0/1,387	0/1,377
NetherlandsKottenpark2005	0/1,078	0/1,078	**0/956**	0/1,056	1/1,078
SouthAfricaLewitt2009	0/58	0/30	0/30	**0/28**	0/48

parametric test was used, the Friedman test [15], with the goal of verifying if the algorithms had significant differences among them. The test returned a p-value of 0.0003287. Thus, considering a significance level of 95%, the algorithms had statistically significant differences among them.

From this result, we proceed the pairwise Wilcoxon test [15] to evaluate if there is statistical difference between each pair of algorithms. The results indicated that the SGVNS is statistically different from all others algorithms. The GVNS is also statistically different from Adaptive VNS. The other comparisons are statistically equivalent. It was used the BH p-value adjustment method [3].

5   Conclusions

This paper proposed two metaheuristic approaches for solving the School Timetabling problem, both based on the VNS metaheuristic. The first algorithm represents a combination between GVNS and SVNS metaheuristics and it was called SGVNS or Skewed GVNS. The second one, named Adaptive VNS, is an adaptive approach based on VNS that defines the neighborhood to perform local searches by means of probabilities, and prioritizes the neighborhoods that have obtained the best results in past iterations.

Both algorithms produced good solutions to this problem with an advantage of the SGVNS algorithm that is statistically different from all other algorithms. In turn, the Adaptive VNS is equivalent to the Goal Solver of [5]. Considering the best results, SGVNS performed equal or better than the Goal Solver and the GVNS algorithm of [24] in ten from sixteen instances.

As future work, we suggest:

- optimize the calculation of the probabilities of the Adaptive VNS algorithm;
- Evaluate both algorithms in other instances, such as those used in the second phase of ITC 2011.

Acknowledgments. We would like to thank CAPES, CNPq, FAPEMIG, CEFET-MG and UFOP for supporting the development of this work. We also thank the authors of the Goal Solver algorithm for making its source code available.

References

1. Aziz, R.A., Ayob, M., Othman, Z., Ahmad, Z., Sabar, N.R.: An adaptive guided variable neighborhood search based on honey-bee mating optimization algorithm for the course timetabling problem. Soft Comput. **21**(22), 6755–6765 (2017)
2. Bardadym, V.A.: Computer-aided school and university timetabling: the new wave. In: Burke, E., Ross, P. (eds.) PATAT 1995. LNCS, vol. 1153, pp. 22–45. Springer, Heidelberg (1996). https://doi.org/10.1007/3-540-61794-9_50
3. Benjamini, Y., Hochberg, Y.: Controlling the false discovery rate: a practical and powerful approach to multiple testing. J. R. Stat. Soc. Ser. B (Methodol.) **57**, 289–300 (1995)
4. Birattari, M., Balaprakash, P., Dorigo, M.: The ACO/F-Race algorithm for combinatorial optimization under uncertainty. In: Doerner, K.F., Gendreau, M., Greistorfer, P., Gutjahr, W., Hartl, R.F., Reimann, M. (eds.) Metaheuristics. ORSIS, vol. 39, pp. 189–203. Springer, Boston, MA (2007). https://doi.org/10.1007/978-0-387-71921-4_10
5. da Fonseca, G.H.G., Santos, H.G., Toffolo, T.Â.M., Brito, S.S., Souza, M.J.F.: GOAL solver: a hybrid local search based solver for high school timetabling. Ann. Oper. Res. **239**(1), 77–97 (2016)
6. de Haan, P., Landman, R., Post, G., Ruizenaar, H.: A case study for timetabling in a dutch secondary school. In: Burke, E.K., Rudová, H. (eds.) PATAT 2006. LNCS, vol. 3867, pp. 267–279. Springer, Heidelberg (2007). https://doi.org/10.1007/978-3-540-77345-0_17
7. Even, S., Itai, A., Shamir, A.: On the complexity of time table and multicommodity flow problems. In: Proceedings of the 16th Annual Symposium on Foundations of Computer Science, pp. 184–193 (1975)
8. Gotlieb, C.C.: The construction of class-teacher timetables. In: Proceedings of IFIP Congress, pp. 73–77 (1963)
9. Hansen, P., Mladenovic, N., Pérez, J.A.M.: Variable neighborhood search: methods and applications. 4OR: Q. J. Belg. Fr. Ital. Oper. Res. Soc. **6**, 319–360 (2008)
10. Johnson, D.S., Aragon, C.R., McGeoch, L.A., Schevon, C.: Optimization by simulated annealing: an experimental evaluation; part II, graph coloring and number partitioning. Oper. Res. **39**(3), 378–406 (1991)

11. Kingston, J.H.: Hierarchical timetable construction. In: Burke, E.K., Rudová, H. (eds.) PATAT 2006. LNCS, vol. 3867, pp. 294–307. Springer, Heidelberg (2007). https://doi.org/10.1007/978-3-540-77345-0_19
12. Lopez-Ibanez, M., Dubois-Lacoste, J., Stutzle, T., Birattari, M.: The irace package: iterated racing for automatic algorithm configuration. IRIDIA, Universite Libre de Bruxelles, Belgium, Technical report, TR/IRIDIA/2011-004 (2011)
13. Mladenović, N., Hansen, P.: Variable neighborhood search. Comput. Oper. Res. **24**(11), 1097–1100 (1997)
14. Mladenović, N., Dražić, M., Kovačevic-Vujčić, V., Čangalović, M.: General variable neighborhood search for the continuous optimization. Eur. J. Oper. Res. **191**, 753–770 (2008)
15. Montgomery, D.C.: Design and Analysis of Experiments. Wiley, Boco Raton (2017)
16. Nurmi, K., Kyngas, J.: A framework for school timetabling problem. In: Proceedings of the 3rd Multidisciplinary International Scheduling Conference: Theory and Applications, Paris, pp. 386–393 (2007)
17. Post, G., et al.: XHSTT: an XML archive for high school timetabling problems in different countries. Ann. Oper. Res. **218**, 295–301 (2014)
18. R Core Team: R: A language and environment for statistical computing. R Foundation for Statistical Computing, Vienna, Austria (2014). http://www.R-project.org/
19. Schaerf, A.: A survey of automated timetabling. Artif. Intell. Rev. **13**(2), 87–127 (1999)
20. Santos, H.G., Ochi, L.S., Souza, M.J.F.: A tabu search heuristic with efficient diversification strategies for the class/teacher timetabling problem. J. Exp. Algorithmics (JEA) **10**, 2–9 (2005)
21. Santos, H.G., Souza, M.J.F.: Timetabling in educational institutions: formulations and algorithms (in Portuguese). In: Proceedings of the XXXIX Brazilian Symposium of Operations Research, Fortaleza, Brazil, pp. 2827–2882 (2007)
22. Souza, M.J.F.: School timetabling: an approximation by metaheuristics (in Portuguese). Ph.D. thesis, Programa de Pós-graduação em Engenharia de Sistemas e Computação, Universidade Federal do Rio de Janeiro, Brazil (2000)
23. Souza, M.J.F., Maculan, N., Ochi, L.S.: A GRASP-Tabu search algorithm for solving school timetabling problems. In: METAHEURISTICS: Computer Decision-Making. Kluwer Academic Publishers, Dordrech, vol. 86, pp. 659–672 (2004)
24. Teixeira, U.R., Souza, M.J.F., de Souza, S.R.: A local search approach using GVNS for solving school timetabling problems (in Portuguese). In: Proceedings of the XXXVIII Ibero Latin American Congress on Computational Methods in Engineering (CILAMCE), Florianópolis, Brazil (2017). https://doi.org/10.20906/CPS/CILAMCE2017-1240
25. Valouxis, C., Housos, E.: Constraint programming approach for school timetabling. Comput. Oper. Res. **30**(10), 1555–1572 (2003)
26. Wright, M.: School timetabling using heuristic search. J. Oper. Res. Soc. **47**(3), 347–357 (1996)

Finding Balanced Bicliques in Bipartite Graphs Using Variable Neighborhood Search

Juan David Quintana[iD], Jesús Sánchez-Oro[✉][iD], and Abraham Duarte[iD]

Department of Computer Sciences, Universidad Rey Juan Carlos, Móstoles, Spain
{juandavid.quintana,jesus.sanchezoro,abraham.duarte}@urjc.es

Abstract. The Maximum Balanced Biclique Problem (MBBP) consists of identifying a complete bipartite graph, or biclique, of maximum size within an input bipartite graph. This combinatorial optimization problem is solvable in polynomial time when the balance constraint is removed. However, it becomes $\mathcal{NP}$–hard when the induced subgraph is required to have the same number of vertices in each layer. Biclique graphs have been proven to be useful in several real-life applications, most of them in the field of biology, and the MBBP in particular can be applied in the design of programmable logic arrays or nanoelectronic systems. Most of the approaches found in literature for this problem are heuristic algorithms based on the idea of removing vertices from the input graph until a feasible solution is obtained; and more recently in the state of the art an evolutionary algorithm (MA/SM) has been proposed. As stated in previous works it is difficult to propose an effective local search method for this problem. Therefore, we propose the use of Reduced Variable Neighborhood Search (RVNS). This methodology is based on a random exploration of the considered neighborhoods and it does not require a local search.

Keywords: Biclique · Reduced VNS · Bipartite

1 Introduction

Let $G(L, R, E)$ be a balanced bipartite graph where L and R are the two sets (or layers) of vertices of the same cardinality (i.e., $|L| = |R| = n$) and E is the set of edges. As a bipartite graph, $L \cap R = \emptyset$, and an edge can only connect a vertex $v \in L$ with a vertex $u \in R$, i.e., $\forall (v, u) \in E$, $v \in L \wedge u \in R$. Additionally, let us define a biclique $B(L'R', E')$ as an induced subgraph of G, where $L' \subseteq L$, $R' \subseteq R$, such as every vertex $v \in L'$ is connected to all the vertices $u \in R'$. In other words, B is a complete bipartite graph.

Given a balanced bipartite graph $G(L, R, E)$, this work is focused on solving the Maximum Balanced Biclique Problem (MBBP), which consists of identifying

This work has been partially founded by Ministerio de Economía y Competitividad with grant ref. TIN2015-65460-C2-2-P.

A. Sifaleras et al. (Eds.): ICVNS 2018, LNCS 11328, pp. 114–124, 2019.
https://doi.org/10.1007/978-3-030-15843-9_10

a balanced biclique $B^\star(L', R', E')$ with the largest number of vertices per layer. In other words, the objective of MBBP is maximizing the cardinality of sets L' and R'.

Figure 1 presents an example bipartite graph with 8 vertices and 13 edges and two possible solutions for the MBBP. Figure 1(b) depicts a solution $B_1(L_1, R_1, E_1)$ with two vertices in each layer. Specifically, $L_1 = \{A, B\}$, and $R_1 = \{F, G\}$. The edges involved in the induced biclique are depicted with continuous line, while those with an endpoint out of the solution are depicted with dashed line. Notice that it is not possible to insert new vertices in the solution, since the resulting induced bipartite graph will not be a balanced biclique. For instance, adding vertices E or H is not possible because they are not adjacent to vertices B and A, respectively. Furthermore, it is not possible to add new vertices just in layer L_1 since the induced biclique is not balanced (i.e., $|L_1| \neq |R_1|$).

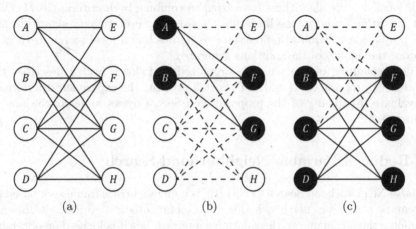

Fig. 1. (a) Bipartite graph with 8 vertices and 13 edges, (b) a feasible solution with 2 vertices in each layer (A and B in L_1, F and G in R_1), and (c) a different solution with 3 vertices in each layer (B, C and D in L_2, F, G, and H in R_2)

Figure 1(c) presents a solution $B_2(L_2, R_2, E_2)$ of better quality, since it has 3 vertices in each layer. In particular, $L_2 = \{B, C, D\}$, and $R_2 = \{F, G, H\}$. Again, no more vertices can be added without violating the balanced biclique constraint.

This problem has been proven to be $\mathcal{NP}$-hard in various works [2,9,11]. Some theoretical results proposed bounds for the maximum size that the optimal solution for the MBBP can have [6], and it has been proven to be hard to approximate within a certain factor [7].

Biclique graphs have been proven to be useful in several real-life applications, most of them in the field of biology: biclustering microarray data [4,17,18], optimization of the phylogenetic tree reconstruction [14], or identifying common gen-set associations [5], among others [3,12]. In particular, the MBBP has additional applications in a diverse set of fields: folding of programmable logic arrays in Very Large Scale Integration (VLSI) design [13], or new nanoelectronic systems design [1,15,16], among others.

Despite of the practical applications of the MBBP, not many efficient algorithms for it has been proposed, mainly due to the difficulty of the problem. However, if the MBBP does not consider that the solution must be balanced, the resulting problem is solvable in polynomial time [19], although most of the solutions obtained are totally unbalanced, making the results not adaptable for the problem under consideration.

Most of the previous approaches follow a destructive approach where the initial solution contains all the nodes and the heuristic iteratively removes nodes from L' and R' until the incumbent solution becomes feasible. The algorithms mainly differ in the order in which the vertices to be removed are selected. In particular, [15] selects the vertices with the largest degree, while [1] removes the vertex with the largest number of minimum degree nodes in the other layer. Additionally, some algorithms have tried to combine both criteria [20,21]. The best algorithm found in the literature consists of a evolutionary algorithm [22] that proposes a new mutation operator as well as a new local search method to improve the quality of the solutions generated.

The remaining of the paper is organized as follows: Sect. 2 describes the algorithmic proposal for the MBBP; Sect. 3 presents the experiments performed to evaluate the quality of the proposal; and Sect. 4 draws some conclusions on the research.

2   Reduced Variable Neighborhood Search

Variable Neighborhood Search (VNS) [10] is a metaheuristic framework based on systematic changes of neighborhoods. As a metaheuristic algorithm, it does not guarantee the optimality of the solutions obtained, but it is focused on obtaining high quality solutions in reasonable computing times. The constant evolution of VNS has resulted in several variants, among which we can highlight Basic VNS, Reduced VNS, Variable Neighborhood Descent, General VNS, Skewed VNS, Variable Neighborhood Decomposition Search, among others.

Most of the variants differ in the way of exploring the considered neighborhoods. Specifically, Variable Neighborhood Descent (VND) considers a totally deterministic exploration of the solution space, while the exploration performed by Reduced VNS (RVNS) is totally stochastic. Some other variants combine both deterministic and stochastic changes of neighborhoods (e.g., Basic VNS, General VNS).

As stated in previous works [21,22], designing a local search method for the MBBP is a very difficult task mainly due to the complexity of maintaining a feasible solution (i.e. a balanced biclique) after removing or adding vertices to a previous solution. In other words, the MBBP is not suitable for designing local search methods in order to find a local optimum with respect to a given solution. Therefore, this work proposes a Reduced VNS algorithm, which is based on a random exploration of the considered neighborhoods.

RVNS is a VNS variant useful for large instances in which local search is very time consuming, or for those problems in which it is not easy to design a local search method. Algorithm 1 presents the pseudocode of RVNS.

Algorithm 1. $RVNS(B, k_{\max}, t_{\max})$

1: **repeat**
2:     $k \leftarrow 1$
3:     **while** $k \leq k_{\max}$ **do**
4:         $B' \leftarrow Shake(B, k)$
5:         $k \leftarrow NeighborhoodChange(B, B', k)$
6:     **end while**
7: **until** $CPUTime() \leq t_{\max}$
8: **return** B

The algorithm requires three input parameters: B, the initial feasible solution, that can be randomly generated or using a more elaborated constructive procedure; $k_{\max}$, the maximum neighborhood to be explored; and $t_{\max}$, the maximum computing time in which RVNS is allowed to explore the search space. Each RVNS iteration starts from the initial neighborhood (step 2). Then, the method explores each one of the considered neighborhoods (steps 3–6) as follows. Firstly, the method generates a random solution B' in the current neighborhood k of the incumbent solution B using the *Shake* method (step 4). Then, the *NeighborhoodChange* method (step 5) is responsible for selecting the next neighborhood to be explored. Specifically, if the new solution B' is better than the incumbent one B, then it is updated $(B \leftarrow B')$, restarting the search from the initial neighborhood $(k \leftarrow 1)$. Otherwise, the search continues with the next neighborhood $(k \leftarrow k + 1)$. A RVNS iteration ends when reaching the maximum predefined neighborhood $k_{\max}$. It is worth mentioning that the maximum neighborhood considered in RVNS is usually small due to the random nature of the *Shake* method, since a large value for $k_{\max}$ would produce the same result as restarting the search from a new initial solution. The RVNS method is executed until reaching a maximum computing time $t_{\max}$.

2.1 Constructive Method

The initial solution for VNS can be generated at random or with a more elaborated constructive procedure. This work proposes a constructive procedure based on the Greedy Randomized Adaptive Search Procedure (GRASP) [8]. This methodology considers a greedy function that evaluates the importance of inserting a vertex in the solution under construction. Algorithm 2 presents the pseudocode of the constructive method proposed.

The method starts by randomly selecting a vertex from layer L (step 11), inserting it in the corresponding layer L' of the solution (step 12). Then, two candidate lists (CL) are created, one for each layer of the graph (steps 13–14).

Algorithm 2. $Construct(G = (L, R, E), \alpha)$

1: **function** UPDATELAYER(CL_1, CL_2, α)
2:     $g_{min} \leftarrow \min_{v \in CL_1} g(v)$
3:     $g_{max} \leftarrow \max_{v \in CL_1} g(v)$
4:     $\mu \leftarrow g_{min} + \alpha \cdot (g_{max} - g_{min})$
5:     $RCL \leftarrow \{v \in CL_1 \ ; \ g(v) \leq \mu\}$
6:     $v \leftarrow Random(RCL)$
7:     $CL_1 \leftarrow CL_1 \setminus \{v\}$
8:     $CL_2 \leftarrow CL_2 \setminus \{u \in CL_2 \ : \ (v, u) \notin E\}$
9:     **return** v
10: **end function**
11: $v \leftarrow Random(L)$
12: $L' \leftarrow \{v\}$
13: $CL_L \leftarrow L \setminus \{v\}$
14: $CL_R \leftarrow \{u \in R \ : \ (v, u) \in E\}$
15: **while** $CL_L \neq \emptyset$ **and** $CL_R \neq \emptyset$ **do**
16:     $v_r \leftarrow$ UPDATELAYER(CL_R, CL_L, α)
17:     $R' \leftarrow R \cup \{v_r\}$
18:     $v_l \leftarrow$ UPDATELAYER(CL_L, CL_R, α)
19:     $L' \leftarrow L \cup \{v_l\}$
20: **end while**
21: **return** $B = (L', R', \{(v, u) \ v \in L' \wedge u \in R'\})$

Notice that each candidate lists only contains those vertices from each layer that can be selected maintaining the solution feasible (i.e., the solution is a compete bipartite graph). On the one hand, in this first step, CL_L contains all the vertices from L excepting the selected vertex v. On the other hand, CL_R contains all adjacent vertices to v in R, since otherwise the solution would not be a bipartite complete graph. Then, the method iterates adding a vertex in layer R' and then in layer L', while there are still candidate vertices in both layers (steps 16–19).

The selection of the next vertex is described in function *UpdateLayer* that requires from three parameters: the candidate list from which the vertex must be selected, CL_1, the candidate list of the other layer, CL_2, and the α parameter that determines the greediness/randomness of the selection. A greedy function g that evaluates the quality of a candidate vertex v must be defined. In this work, we propose the number of adjacent vertices in the oppositive candidate list. More formally,

$$g(v, CL) = |\{u \in CL \ : \ (v, u) \in E\}|$$

The first step to select the next vertex consists of evaluating the minimum (g_{min}) and maximum (g_{max}) values for the greedy function value (steps 2–3). Then, a Restricted Candidate List (*RCL*) is constructed (step 5) with those vertices whose objective function value is larger or equal than a previously evaluated threshold μ (step 4). The values for the α parameter are in the range

0–1, where $\alpha = 0$ implies that the method is totally random, and $\alpha = 1$ transforms the algorithm in a completely greedy method. The next vertex is selected at random from the RCL (step 6), updating the corresponding candidate lists. In particular, CL_1 is updated by removing the selected vertex v from it, while CL_2 is updated by removing those vertices that are not adjacent to v, since they cannot be selected in further iterations maintaining the feasibility of the solution.

2.2 Shake

The *Shake* method is a stage inside VNS methodology designed to escape from local optima found during the search. It consists of randomly perturbing a solution with the aim of exploring a wider region of the solution space. This phase of the VNS methodology is focused on diversifying the search.

Given a neighborhood k, the *Shake* method proposed in this work removes k elements at random from each layer. The resulting solution is feasible but the value of the objective function is always smaller, since it reduces the number of vertices selected.

Considering the constraints of the problem, if a vertex v is included in the solution, then all the vertices of the opposite layer that are not adjacent to v cannot be included in future iterations, since the resulting solution would not be a biclique. However, the removal of some vertices in the *Shake* method can eventually allow new vertices to be included in the solution (i.e., those that were not adjacent to any of the vertices removed).

Therefore, we propose a reconstruction stage that is executed after each *Shake* method. In particular, the reconstruction phase tries to add new vertices to the solution, from those that were not feasible to add before executing the *Shake* method.

The reconstruction stage always improves the quality of the solution or, at least, maintains the quality of the solution produced by the *Shake* method. Notice that the reconstructed solution outperforms the initial one if and only if reconstruction stage is able to insert more than k vertices in each layer.

The random nature of this procedure makes it difficult to improve the quality of the solution. In order to mitigate this effect we consider four variants for the *Shake* method. These differ in how the destruction and reconstruction phases are performed. Each stage can be either random (R) or greedy (G), which leaves us with four variants shown in Table 1. For instance, the *Shake* variant GR firstly destroys the solution with a greedy selection of vertices and then reconstructs it randomly.

The reconstruction stage follows the same idea that the constructive method described in Sect. 2.1, with the parameters $\alpha = 0$ for the random construction (R) and $\alpha = 1$ for the greedy one (G). For the destruction stage we use a template similar to the constructive method, where we evaluate all candidates with a heuristic function, in this case those vertices already included in the solution, and then choose one of the more promising vertices. We need to define a new heuristic function g' that allows us to score the candidates for removal. Without

Table 1. Summary of the four variants considered for the *Shake* procedure.

Variant	Destruction	Reconstruction
RR	Random	Random
RG	Random	Greedy
GR	Greedy	Random
GG	Greedy	Greedy

loss of generality, for a given vertex $v \in S$ located in layer L, we calculate $g'(v)$ as the number of vertices in the opposite layer R not connected to v. More formally,

$$g'(v, S) = |\{u \in R \; : \; (u, v) \notin E\}|$$

3   Computational Results

In this section we present two sets of experiments: the preliminary experiments, performed to tune the parameters of the proposed algorithm; and the final experiment designed to test the quality of our proposal and compare it with the current state of the art. All the algorithms have been coded in Java 8 and were executed on a computer with an Intel i7 (7660U) CPU with 2.5 GHz and 8 GB RAM.

We use the same data set presented in [22], which consist of 30 instances with sizes $n = \{250, 500\}$ and different densities. This data set is used to compare the performance of our algorithm with the current state of the art.

In these experiments we report: the average size of the largest balanced biclique obtained, Avg. Size; the average execution time per instance, Avg. Time (s); the average percentage deviation to the best solution found in the experiment, %Dev.; and the number of times an algorithm reaches the best solution found for a given instance in the current experiment, # Best.

3.1   Preliminary Experiments

The following experiments are designed to select the best variant for the proposed algorithm. A small group of 6 representative instances, selected from the original 30, was used in these experiments to avoid overfitting the parameters to the data set in the final experiment.

The first experiment is designed for testing the effect of the α parameter in the constructive procedure, considering $\alpha = \{0.25, 0.50, 0.75, RND\}$. The *RND* value means that we will use a different α-value, selected randomly, in each iteration. The results in Table 2 show us that the best performance is achieved when alpha is selected randomly in each iteration. The procedure obtains on average a balanced biclique of 52.67 vertices and finds the best solutions among this experiment for all 6 instances. The results obtained shows that considering small values of α (i.e., increasing the randomness of the method) always results

Table 2. Comparison of the constructive method when considering different values for α.

α	Avg. size	Avg. time (s)	%Dev.	#Best
0.25	46.17	375.00	11.67	0
0.50	48.33	375.00	8.16	0
0.75	51.83	375.00	1.40	2
RND	**52.67**	**375.00**	**0.00**	**6**

in worse quality solutions. However, the *RND* value allows us to diversify the search by considering both small and large values of α, thus obtaining the best results in terms of average objective function value, average deviation, and total number of best solutions found. Therefore, in the following experiments we will use this configuration for the α parameter in our proposal.

Table 3. Comparison of the RVNS algorithm for a fixed neighborhood $K_{max} = 50$ and different variations of the shake procedure.

Shake	Avg. size	Avg. time (s)	%Dev.	#Best
RR	48.83	375.00	9.59	0
RG	**54.17**	375.00	0.00	6
GR	49.50	375.00	8.48	0
GG	53.17	375.00	2.07	0

The next experiment is designed to select the best variant for the shake procedure. We assume a $k_{max} = 0.5 \cdot n$ and consider the four variants of the shake procedure according the type of destruction, random (R) or greedy (G), and the type of reconstruction, random (R) or greedy (G), as presented in Sect. 2.2. We can see in Table 3 the results for this experiment and how the variant RG, random removal with greedy reconstruction, has the best performance. It achieves an average size of 54.14 and find better solutions than all other variants in all 6 instances. Notice that the best results are obtained when considering a greedy reconstruction, but the inclusion of the random destruction is able to reach better quality solutions than the greedy destruction.

In the last preliminary experiment we want to measure the impact of the maximum neighborhood explored in our algorithm. We use the best configuration of the previous experiments and test different neighborhoods $k_{max} = \{0.10, 0.20, 0.30, 0.40, 0.50\}$ for our *RVNS* framework. It is important to remark that a neighborhood k removes $k \cdot n$ vertices of the solution. In this experiment we can see that expanding the size of the neighborhood generally allows to reach better solutions as the %Dev. decreases. However, this improvement stagnates after $k_{max} = 0.40$ where the algorithm reaches its maximum performance (Table 4).

Table 4. Comparison of the RVNS algorithm when considering different values for K_{max}.

K_{max}	Avg. cost	Avg. time (s)	%Dev.	#Best
0.10	53.50	375.00	1.45	2
0.20	53.50	375.00	1.17	2
0.30	53.83	375.00	0.62	4
0.40	**54.17**	375.00	0.00	6
0.50	54.17	375.00	0.00	6

Analyzing the preliminary experimentation, the best algorithm is configure with $\alpha = RND$ for the constructive procedure, the shake variant RG which considers random destruction and greedy reconstruction, and a maximum neighborhood of $k_{\mathrm{max}} = 0.40$.

3.2 Final Experiment

In the last experiment we compare our proposal with the best algorithm found in the state of the art [22] using the same set of 30 instances. In particular, it consists of a memetic algorithm that considers a local search based on structure mutation. RVNS is executed iteratively until reaching a time limit in seconds equal to three times the size of the current instance. Table 5 shows the results obtained when comparing the best variant of RVNS with the memetic algorithm (EA/SM). As it can be derived from the table, RVNS is able to find (on average) bicliques of just one node less than the bicliques obtained by the memetic algorithm. However, it has an execution time that is roughly half of the memetic algorithm.

Table 5. Comparison of the RVNS algorithm with the best in the state of art.

	Avg. size	Avg. time (s)	%Dev.	#Best
EA/SM	55.10	2075.11	0.04	29
RVNS	54.33	1125.00	1.71	10

4 Conclusions

This work analyzes the performance of Reduced VNS for generating high quality solutions for the Maximum Balanced Biclique Problem efficiently. Specifically, we propose an intensified shaking stage which is conformed by a destruction and reconstruction phase. The experiments performed show the relevance of performing these phases in a random or greedy manner. The results obtained show the possibilities of the RVNS proposal, obtaining, on average, solutions that are really close to the best ones found in the state of the art. Furthermore, the absence of a local search in the proposed algorithm allows it to require half of the computing time of the best algorithm found in the literature.

References

1. Al-Yamani, A.A., Ramsundar, S., Pradhan, D.K.: A defect tolerance scheme for nanotechnology circuits. IEEE Trans. Circuits Syst. **54**–I(11), 2402–2409 (2007)
2. Alon, N., Duke, R.A., Lefmann, H., Rödl, V., Yuster, R.: The algorithmic aspects of the regularity lemma. J. Algorithms **16**(1), 80–109 (1994)
3. Baker, E.J., et al.: Ontological discovery environment: a system for integrating gene-phenotype associations. Genomics **94**(6), 377–387 (2009)
4. Cheng, Y., Church, G.M.: Biclustering of expression data. In: Proceedings of the 8th ISMB, pp. 93–103. AAAI Press (2000)
5. Chesler, E.J., Langston, M.A.: Combinatorial genetic regulatory network analysis tools for high throughput transcriptomic data. In: Eskin, E., Ideker, T., Raphael, B., Workman, C. (eds.) RRG/RSB -2005. LNCS, vol. 4023, pp. 150–165. Springer, Heidelberg (2007)
6. Dawande, M., Keskinocak, P., Swaminathan, J.M., Tayur, S.: On bipartite and multipartite clique problems. J. Algorithms **41**(2), 388–403 (2001)
7. Feige, U., Kogan, S.: Hardness of approximation of the balanced complete bipartite subgraph problem. Technical report (2004)
8. Feo, T.A., Resende, M.G.: Greedy randomized adaptive search procedures. J. Global Optim. **6**(2), 109–133 (1995)
9. Garey, M.R., Johnson, D.S.: Computers and Intractability: A Guide to the Theory of NP-Completeness. Freeman, New York (1979)
10. Hansen, P., Mladenović, N.: Variable Neighborhood Search, pp. 313–337. Springer, Boston (2014)
11. Johnson, D.S.: The NP-completeness column: an ongoing guide. J. Algorithms **13**(3), 502–524 (1992)
12. Mushlin, R.A., Kershenbaum, A., Gallagher, S.T., Rebbeck, T.R.: A graph-theoretical approach for pattern discovery in epidemiological research. IBM Syst. J. **46**(1), 135–150 (2007)
13. Ravi, S.S., Lloyd, E.L.: The complexity of near-optimal programmable logic array folding. SIAM J. Comput. **17**(4), 696–710 (1988)
14. Sanderson, M.J., Driskell, A.C., Ree, R.H., Eulenstein, O., Langley, S.: Obtaining maximal concatenated phylogenetic data sets from large sequence databases. Mol. Biol. Evol. **20**(7), 1036–1042 (2003)
15. Tahoori, M.B.: Application-independent defect tolerance of reconfigurable nanoarchitectures. JETC **2**(3), 197–218 (2006)
16. Tahoori, M.B.: Low-overhead defect tolerance in crossbar nanoarchitectures. JETC **5**(2), 11 (2009)
17. Tanay, A., Sharan, R., Shamir, R.: Discovering statistically significant biclusters in gene expression data. In: ISMB, pp. 136–144 (2002)
18. Wang, H., Wang, W., Yang, J., Yu, P.S.: Clustering by pattern similarity in large data sets. In: Franklin, M.J., Moon, B., Ailamaki, A. (eds.) SIGMOD Conference, pp. 394–405. ACM (2002)

19. Yannakakis, M.: Node-deletion problems on bipartite graphs. SIAM J. Comput. **10**(2), 310–327 (1981)
20. Yuan, B., Li, B.: A low time complexity defect-tolerance algorithm for nanoelectronic crossbar. In: International Conference on Information Science and Technology, pp. 143–148 (2011)
21. Yuan, B., Li, B.: A fast extraction algorithm for defect-free subcrossbar in nanoelectronic crossbar. ACM J. Emerg. Technol. Comput. Syst. (JETC) **10**(3), 25 (2014)
22. Yuan, B., Li, B., Chen, H., Yao, X.: A new evolutionary algorithm with structure mutation for the maximum balanced biclique problem. IEEE Trans. Cybern. **45**(5), 1040–1053 (2015)

General Variable Neighborhood Search for Scheduling Heterogeneous Vehicles in Agriculture

Ana Anokić[1], Zorica Stanimirović[2], Đorđe Stakić[3],
and Tatjana Davidović[4(✉)]

[1] Branka Pešića 27, 11080 Zemun, Serbia
anokicana@gmail.com
[2] Faculty of Mathematics, University of Belgrade,
Studentski trg 16/IV, 11000 Belgrade, Serbia
zoricast@matf.bg.ac.rs
[3] Faculty of Economics, University of Belgrade,
Kamenička 6, 11000 Belgrade, Serbia
djordjes@ekof.bg.ac.rs
[4] Mathematical Institute of the Serbian Academy of Science and Arts,
Kneza Mihaila 36, 11000 Belgrade, Serbia
tanjad@mi.sanu.ac.rs

Abstract. A new variant of Vehicle Scheduling Problem (VSP), denoted as *Vehicle Scheduling Problem with Heterogeneous Vehicles* (VSP-HV), which arises from optimizing the sugar beet transportation in a sugar factory in Serbia is introduced. The objective of the considered VSP-HV is to minimize the time required for daily transportation of sugar beet by heterogeneous vehicles under problem-specific constraints. General Variable Neighborhood Search (GVNS) is designed as a solution method for the considered problem. A computational study is conducted on the set of real-life instances, as well as on the set of generated instances of larger dimensions. A Mixed Integer Quadratically Constraint Programming (MIQCP) model is developed and used within commercial Lingo 17 solver to obtain optimal or feasible solutions for small-size real-life problem instances. Experimental results show that the proposed GVNS quickly reaches all known optimal solutions or improves the upper bounds of feasible solutions on small-size instances. On larger problem instances, for which Lingo 17 could not find feasible solutions, GVNS provided its best solutions for limited CPU time.

Keywords: Vehicle scheduling problem · Single depot · Heterogeneous vehicles · Variable neighborhood search · Transportation in agriculture

A. Anokić—Independent Researcher.
This research was partially supported by Serbian Ministry of Education, Science and Technological Development under the grants nos. 174010 and 174033.

A. Sifaleras et al. (Eds.): ICVNS 2018, LNCS 11328, pp. 125–140, 2019.
https://doi.org/10.1007/978-3-030-15843-9_11

1 Introduction

Vehicle Scheduling Problem (VSP) is a classical problem in operations research, appearing in applications such as public transport systems and transportation of different types of goods. It also arises as a subproblem of more complex scheduling and management problems when optimizing the performance of large systems. The goal of VSP is to find the set of trips that each vehicle will make during the considered time period in order to optimize a given objective function [9]. In the literature, there are many variants of VSP proposed up to now, which differ by the objective function used, constraints involved, and specific restrictions that arise from the observed situation in practice (for example, differences across transport agencies, type of goods to be transported, policies of different agencies, time or financial limits imposed, etc). In general, vehicle scheduling problems are very hard to solve and most of them belong to the class of NP-hard problems. A survey on vehicle scheduling problems, their classification and solution methods can be found in [4, 7, 18].

This study is motivated by previous research on optimization of sugar beet transportation in sugar industry in Serbia [1, 2]. Having in mind that sugar beet has very low price on the market, major costs in sugar production refer to the costs of transport. In order to keep the farmers interested in producing sugar beet, large sugar companies usually take over all transportation costs. More precisely, they rent vehicles and hire workers to perform loading of goods at agreed locations, transportation to the factory, and unloading at the factory area. Decreasing the costs of transporting of sugar beet to the factory may significantly increase the profit of a sugar company. Therefore, an efficient scheduling of vehicles that includes all problem specific constraints with the minimum expenditure of time and money is required.

Instead of assuming that vehicles used for transportation are homogenous [1, 2], we consider that a factory rents vehicles of different types, which implies different capacities of vehicles. Therefore, the variant of VSP considered in this study represents a generalization of the problem from [1, 2] and is denoted as *Vehicle Scheduling Problem with Heterogenous Vehicles* - VSP-HV.

Majority of VSP applications are related to public transportation, transportation of fuel, valuable items, etc. Only few papers in the literature focus on the transport of agricultural goods in sugar industry. Studies [13, 16, 19] deal with optimizing of sugar cane transportation and propose different variants of vehicle scheduling and vehicle routing problems, as well as adequate solution approaches. However, these problems are different from the variant of VSP considered in this study, due to the differences between sugar cane and sugar beet, the type of vehicles used for transport, and the available resources of the sugar company.

We formulated VSP-HV for sugar beet transportation as a Mixed Integer Quadratically Constrained Program (MIQCP) and used it within the framework of the commercial Lingo 17 solver in order to solve real-life and generated problem instances. As Lingo 17 returned optimal or feasible solutions only for small-size instances, we developed a variant of Variable Neighborhood Search (VNS) metaheuristic as solution approach to the considered VSP-HV.

VNS is a metaheuristic method proposed by Mladenović and Hansen [17] based on the use of different neighborhood structures while exploring the solution space. VNS and its variants have been successfully applied to a wide variety of NP-hard problems of combinatorial and continuous optimization, see [10,11]. In the literature, VNS-based methods have been proposed as solution approaches to vehicle scheduling and vehicle routing problems, such as: inventory routing and scheduling problems in supply chains [14], location routing scheduling problem [15], dynamic rich vehicle routing problem with time windows [3], vehicle routing problem with multiple trips [8], heterogeneous fleet vehicle routing problem for transportation of hazardous materials [6], periodic vehicle routing problem [12], vehicle routing problem with clustered backhauls and 3D loading constraints [5], etc.

Studies [1,2] that deal with VSP for sugar beet transportation with homogenous vehicles and limited resources at collection centers and at the factory area also propose the application of VNS-based methods. In [2], basic variant of VNS (BVNS) is proposed, while the study [1] presents an improved variant of BVNS from [2] and introduces Skewed VNS (SVNS) as another solution approach for the same problem. Computational results presented in [1,2] show that VNS-based methods represent promising solution approaches to VSP for sugar beet transportation. Therefore, in this study, we also propose VNS-based method, a General Variable Neighborhood Search (GVNS) for VSP-HV. The assumption that vehicles have different capacities leads to differences in the proposed GVNS implementation when compared to VNS approaches from [1,2] for VSP with homogenous vehicles. For example, we have to take into account the differences in loading and unloading times among nonhomogenous vehicles, moves within the neighborhoods that will preserve the feasibility of solutions; the changes in the amount of goods arrived to the factory when two vehicles of different types exchange their tours, etc.

The rest of the paper is organized as follows. Problem description is presented in Sect. 2. The proposed GVNS heuristic is explained in details in Sect. 3. Computational results are presented and analyzed in Sect. 4. Finally, in Sect. 5, some conclusions and directions for future work are given.

2   Problem Description

Within the sugar beet harvesting season, the transportation plan is made on a daily basis. The ultimate request is to satisfy the daily factory needs, concerning the amount of sugar beet transported to the factory. As the starting process of factory machines is very expensive, we need to assure that they work continuously during the whole season.

In the considered problem, it is assumed that rented vehicles are not homogeneous regarding their capacities. On the other hand, as the maximal speed of heavy vehicles is limited, we suppose that all vehicles have the same average speed. The factory area is starting and finishing point for each vehicle. After departing from the factory, a vehicle drives to a location with collected sugar beet, where it is being loaded, and then drives back to the factory area (Fig. 1). Note that a vehicle visits only one location per tour, because the company wants to prevent the possibility of obtaining a low quality mixture of sugar beet from different locations. On the other hand, the quantities of sugar beet collected at each location exceed vehicle's capacity, and therefore, each location has to be visited several times in order to be emptied. A vehicle departs from a location fully loaded, except maybe in the last visit to a location. After arrival at the factory area, a vehicle needs some time for unloading and analyzing the samples before it starts the next tour. However, the sugar beet is time sensitive agricul-ture material, and therefore, the collected goods should not stand in the open for too long, otherwise, the quality will be lost. In the case that at some location, the collected sugar beet is kept longer than a predetermined number of days, this location is considered as urgent and must be emptied during the day. Therefore, the objective is to empty all urgent locations and to satisfy the daily factory needs while minimizing the transportation time. In addition, it is assumed that two vehicles cannot be loaded at the same location in the same time, as there is usually only one loading machine available. For this reason, the difference in arrival times of two different vehicles at a location must not be smaller than the duration of loading of the vehicle that arrived first to this location. It is also assumed that the factory has enough labor, machines, and space for handling a limited number of vehicles. Therefore, the queues of vehicles are not allowed at the factory area and at each location.

Fig. 1. Sugar factory (left) and loading of vehicle (right) (Photos by courtesy of company Sunoko www.sunoko.rs)

A vehicle's schedule is defined by the array of locations to be visited during the working day and the corresponding departure times from the factory. The goal of the problem is to find a feasible schedule (the one in which all urgent locations are emptied and daily factory needs are satisfied) for a given set of vehicles such that the required working time, i.e., the moment of time when all vehicles finish their tours is minimized.

The considered VSP-HV represents a generalization of the VSP proposed in [1], which is proved to be NP hard in the same study. Having in mind that VSP-HV includes all problem specific constraints of VSP from [1], except the assumption of heterogeneous instead of homogeneous vehicles, it can be concluded that the considered VSP-HV is also NP-hard optimization problem.

We have developed a Mixed Integer Quadratically Constraint Programming (MIQCP) model for the considered VSP-HV. As the presentation of the proposed MIQCP model is too large for this paper, the complete formulation can be found at http://www.mi.sanu.ac.rs/~tanjad/MIQCP_VSP-HV.pdf.

In our MIQCP formulation, we have used the concept of virtual tours in order to equalize the number of tours for all vehicles [1,2]. It is assumed that during a virtual tour a vehicle stays at the factory area, meaning that the duration of a virtual tour is equal to zero. Therefore, virtual tours do not affect the objective function value and the problem constraints, as they are simply discarded from a solution during the objective function calculation.

Example 1. Let us consider a problem instance, denoted as $E_{4,4,4}$, which includes 4 locations, 4 vehicles and maximally 4 tours during the working day with the following data:

- the quantities at locations: 50, 150, 70, 150 tons;
- daily factory needs: 310 tons;
- the number of days that the goods are kept in the open: 4, 8, 5, 9 days;
- the maximal number of days that the goods can stay in the open without losing quality: 7 days;
- the distances of locations from the factory: 60, 30, 40, 50 km;
- capacity of vehicles: 26, 20, 13, 25 tons;
- vehicle's loading times: 0.11, 0.09, 0.06, 0.11 h;
- the times that vehicles spend in factory area between two tours: 0.17, 0.16, 0.14, 0.17 h;
- the average speed of a vehicle: 35 km/h;
- the maximal number of vehicles that can be unloaded in the same time at the factory area: 3;
- starting time: 6 h;
- the end of working day: 24 h.

For this instance, the optimal solution can be represented by the following two matrices:

$$
S_{opt} = \begin{bmatrix} 2\,4\,4\,2 \\ 4\,4\,4\,0 \\ 3\,2\,2\,4 \\ 4\,2\,2\,2 \end{bmatrix}, \quad
T_{opt} = \begin{bmatrix} 6.000\ 7.994\ 11.131\ 14.269 \\ 6.110\ 9.217\ 12.324\ 15.431 \\ 6.000\ 8.486\ 10.400\ 12.314 \\ 6.000\ 9.137\ 11.131\ 13.126 \end{bmatrix}.
$$

Each row of the matrix S_{opt} corresponds to one vehicle and contains indices of locations visited by this vehicle. The last tour of the second vehicle, denoted by 0, is virtual. According to the presented data, it can be concluded that the urgent locations 2 and 4 are emptied, as the sum of capacity of vehicles that visit

each of these locations multiplied by the number of visits exceeds the quantities
at these locations. The transported amounts from locations 1, 2, 3, and 4 are:
0, 150, 13, and 150 tons, respectively. Therefore, the total amount of 313 tons
is delivered to the factory, meaning that the daily factory needs of 310 tons are
satisfied. Matrix T_{opt} contains the departure times for all vehicles in their tours,
i.e., each row corresponds to a vehicle and column corresponds to a tour. The
finishing times for vehicles are calculated based on the values in the last column
of matrix T_{opt} and the location's serving times. In this example, the matrix of
serving times P is:

$$P = \begin{bmatrix} 3.7086 & 1.9943 & 2.5657 & 3.1371 \\ 3.6786 & 1.9643 & 2.5357 & 3.1071 \\ 3.6286 & 1.9143 & 2.857 & 3.0571 \\ 3.7086 & 1.9943 & 2.5657 & 3.1371 \end{bmatrix},$$

where rows correspond to vehicles, while columns are related to locations.

The finishing time of a vehicle is calculated as the departure time in its last
tour increased by the time needed for serving visited location. As the second
vehicle has a virtual tour as the last one, its finishing time is equal to the
departure time in its last tour. Therefore, the finishing times for vehicles in
this example are: 16.263, 15.431, 15.371, and 15.120 respectively. The maximum
among these values is the objective function value of 16.263 (approximately 16 h
15 min 47 s).

3   General Variable Neighborhood Search for Vehicle Scheduling Problem with Heterogeneous Vehicles

VNS is a metaheuristic method based on systematic change of neighborhoods
during the search [10,11]. It consists of three main VNS steps: *Shaking*, *Local
Search* and *Move or Not*, that are repeated within VNS loop until a termination criterion is satisfied. In order to apply VNS metaheuristic to the specific
optimization problem, main elements of VNS implementation (solution representation, objective function calculation, neighborhood structures, the strategy
for generating initial solution, and termination criterion) must be adapted to the
considered problem.

In this section, we present GVNS for VSP-HV. GVNS is a variant of VNS
method that uses *Variable Neighborhood Descent* (VND), a deterministic variant
of VNS, instead of classical Local Search step [11]. In the next subsections, the
elements of the proposed GVNS are explained in details.

3.1   Solution Representation

In our GVNS implementation, solution is represented by two matrices $S = [v_1, v_2, \ldots, v_m]^T$ and $T = [t_1, t_2, \ldots, t_m]^T$. S is an integer matrix of dimension
$m \times r_{max}$, where m denotes the number of available vehicles and r_{max} represents
the maximal number of tours. Each row $v_i = [v_{i,1}, v_{i,2}, \ldots, v_{i,r_{max}}]$ corresponds

to one vehicle and contains a list of locations that the observed vehicle visits during the working day. More precisely, $v_{i,r} = j$ means that vehicle i visits location j in its tour r. A row $t_i = [t_{i,1}, t_{i,2}, \ldots, t_{i,r_{max}}]$ of the real valued matrix T, that has the same dimension as the matrix S, contains the departure times of vehicle i in its tours $r = 1, 2, \ldots, r_{max}$.

Although the maximal number of tours of a vehicle during the working day is set to r_{max}, it may happen that some vehicles make less than r_{max} tours, having in mind that the daily transport finishes when the required amount of goods is transported to the factory and all urgent locations are emptied. In the case when a tour r of a vehicle i is virtual, $s_{i,r} = 0$ holds. For simplicity, virtual tours, if they exist, are located at the end of each row of matrix S. The departure times that correspond to virtual tours of the same vehicle are mutually equal and represent the finishing time for the observed vehicle.

3.2   Objective Function Calculation

In order to calculate the objective function value more efficiently, we perform a preprocessing phase to compute serving time for each location by each vehicle. The serving time for location j by vehicle i is a real value that depends on the distance of the considered location from the factory and the type of vehicle and presents the duration of driving from the factory to location and back with the corresponding loading and unloading times. All calculated serving times are stored in matrix P.

Based on the values in matrices S and P, the procedure for objective function calculation computes departure time for each vehicle in each tour as follows. The departure time in the first tour of each vehicle is set to the value that corresponds to the beginning of a working day. The departure time in each of the remaining tours of a vehicle is computed as the departure time in its previous tour increased by the serving time of the location visited in the previous tour by the considered vehicle.

The obtained matrix T is further transformed in order to satisfy two problem specific constraints: two vehicles cannot be loaded in the same time at any location and only p vehicles, loaded with sugar beet, can be served at the factory area in the same time. For this reason, the procedure for objective function calculation iteratively corrects the elements $t_{i,r}$ as follows. In the first step, the procedure checks if there is a conflict at some location, i.e., if one or more vehicles arrive to a location during the loading of some other vehicle. For all those vehicles, (if there is any) the departure times in their current tours and all subsequent ones are increased by the minimal value that resolves the conflict. This step is performed for all vehicles and all tours. The second step refers to resolving conflicts at the factory area. If there is more than p vehicles at the factory area loaded with sugar beet in the same time, the departure times for the p vehicles that have the lowest values of their arrival times to the factory area stay unchanged, while departure times for the remaining vehicles in their current tours are increased by the smallest possible value that resolves this conflict. Once the departure time for a vehicle in a tour is increased, the departure times

for all subsequent tours of the same vehicle are increased by the same value and
the second step is completed. The second step is performed for all vehicles and
all tours. As a result of the second step, new conflicts at locations may occur.
Therefore, the described steps alternate until all conflicts are resolved and the
matrix of departure times that corresponds to a feasible solution is determined.
For each vehicle, the finishing time is calculated as the departure time in its
last non-virtual tour increased by the time needed to serve location visited in
this tour. Finally, the maximum of finishing times among all vehicles has to be
determined as it represents the objective function value.

3.3 Generating Initial Solution

In order to generate feasible initial solution of the considered problem, the following strategy is used. First, the locations are sorted according to two criteria:
urgency and their distances from the factory. Priority is given to urgent locations, which are sorted in nondecreasing order according to their distances from
the factory, followed by non-urgent locations sorted in the same way. The matrix
of initial solution S_{init} is generated by taking one by one location from the sorted
list and filing the columns of matrix S_{init} starting from the first one, then the
second, third etc. The index of each location appears several times as the element of matrix S_{init}, depending on the number of visits required to empty the
observed location. When choosing a vehicle that will serve a location in the considered tour, the priority is given to the vehicles of higher capacities. As the
matrix S_{init} is filled column by column, at the beginning we define first tours of
all vehicles, then the second ones, and the rest of them.

A vehicle can start its tour to a location if this tour can be served within
a working time, otherwise a tour to that location will be assigned to the next
vehicle. When all urgent locations are served and the factory needs regarding
the amount of delivered goods are satisfied, all remaining tours become virtual.
The matrix of departure times and objective value for initial solution generated
in this way are obtained using the procedure described in Subsect. 3.2.

Example 2. For instance $E_{4,4,4}$ from Example 1, the list of locations sorted by
priority is $(2, 4, 3, 1)$. Locations 2 and 4 are urgent, as the goods at these two
locations are kept in the open for more than 7 days. Note that location 2 has
higher priority than location 4, as it is closer to the factory. Urgent locations
2 and 4 are followed by non-urgent locations 3 and 1, which are also sorted in
non-decreasing order according to their distances from the factory. Matrix S_{init}
and the corresponding matrix T_{init} are:

$$S_{init} = \begin{bmatrix} 2\,2\,4\,4 \\ 2\,2\,4\,4 \\ 2\,4\,4\,3 \\ 2\,2\,4\,4 \end{bmatrix}, \quad T_{init} = \begin{bmatrix} 6.260\ 8.254\ 10.249\ 13.386 \\ 6.170\ 8.134\ 10.099\ 13.206 \\ 6.000\ 7.914\ 10.971\ 14.029 \\ 6.060\ 8.054\ 10.049\ 13.186 \end{bmatrix}.$$

As location 2 with 150 tons has the highest priority, it must be visited in
the first tours of all vehicles, as well as in three additional tours in the second

column of matrix S_{init}. The next location 4 with the same amount of goods is visited one more time compared to location 2, due to the smaller capacity of remaining available vehicles. Finally, one more tour to the location 3 is needed to complete the daily factory needs.

Starting from the generated matrix S_{init}, the elements of the corresponding matrix T are calculated using the procedure described in Subsect. 3.2. From the first column of matrix T, it can be seen that the departure times of the first tours of vehicles 1, 2, and 4 are increased compared to the starting time of 6.000 h. This is the result of the transformation applied to matrix T in order to provide feasibility of the solution. Namely, the departure times for vehicles 2 and 4 are increased in order to avoid conflicts of vehicles at a location, while the departure time of vehicle 1 is further increased to resolve the conflict at the factory area. Finally, the finishing times of vehicles are: 16.523, 16.313, 16.514, and 16.323, respectively. Therefore, the objective function value of the generated solution is 16.523 h (approximately 16 h 31 min 23 s).

3.4 Neighborhood Structures

Four neighborhood structures are used in our GVNS implementation. Neighborhood N_1 of solution S consist of all neighbors obtained when a pair of vehicle exchanges a pair of their non-virtual tours. Figure 2 illustrates a move within neighborhood N_1, using data from the Example 1. The pair of tours exchanged between the first and the second vehicle are bolded in Fig. 2. A N_2-neighbor of solution S is obtained when a location in a non-virtual or the first virtual tour is replaced with another location that is not emptied. An example for each one of these two moves is presented on the left and the right side of Fig. 3, respectively. Neighborhood N_3 consists of all solutions obtained from S by exchanging locations in a pair of non-virtual tours of the same vehicle (see Fig. 4). Finally, a N_4-neighbor of solution S is defined by the following move: the last non-virtual tour of a vehicle is replaced with a virtual tour (see Fig. 5).

$$S = \begin{bmatrix} 2\ 2\ 4\ 4 \\ 2\ 2\ 4\ 4 \\ 2\ 4\ 4\ 3 \\ 2\ 2\ 4\ 4 \end{bmatrix} \rightarrow S' = \begin{bmatrix} 2\ 2\ 4\ 2 \\ 2\ 4\ 4\ 4 \\ 2\ 4\ 4\ 3 \\ 2\ 2\ 4\ 4 \end{bmatrix}$$

Fig. 2. A move in neighborhood N_1

$$S = \begin{bmatrix} 2\ 2\ 4\ 2 \\ 4\ 4\ 4\ 0 \\ 3\ 2\ 2\ 4 \\ 4\ 2\ 2\ 2 \end{bmatrix} \rightarrow S' = \begin{bmatrix} 2\ 2\ 4\ 2 \\ 4\ 4\ 4\ 0 \\ 1\ 2\ 2\ 4 \\ 4\ 2\ 2\ 2 \end{bmatrix} \quad \text{or} \quad S = \begin{bmatrix} 2\ 2\ 4\ 2 \\ 4\ 4\ 4\ 0 \\ 3\ 2\ 2\ 4 \\ 4\ 2\ 2\ 2 \end{bmatrix} \rightarrow S' = \begin{bmatrix} 2\ 2\ 4\ 2 \\ 4\ 4\ 4\ 1 \\ 3\ 2\ 2\ 4 \\ 4\ 2\ 2\ 2 \end{bmatrix}$$

Fig. 3. A move in neighborhood N_2

$$S = \begin{bmatrix} 2\ 2\ 4\ 4 \\ 2\ 2\ 4\ 4 \\ 2\ 4\ 4\ 3 \\ 2\ 2\ 4\ 4 \end{bmatrix} \rightarrow S' = \begin{bmatrix} 2\ 2\ 4\ 4 \\ 2\ 2\ 4\ 4 \\ 2\ 4\ 4\ 3 \\ 4\ 2\ 2\ 4 \end{bmatrix}$$

Fig. 4. A move in neighborhood N_3

$$S = \begin{bmatrix} 2\ 2\ 4\ 4 \\ 2\ 2\ 4\ 4 \\ 2\ 4\ 4\ 3 \\ 2\ 2\ 4\ 4 \end{bmatrix} \rightarrow S' = \begin{bmatrix} 2\ 2\ 4\ 4 \\ 2\ 2\ 4\ 4 \\ 2\ 4\ 4\ 0 \\ 2\ 2\ 4\ 4 \end{bmatrix}$$

Fig. 5. The neighborhood N_4

During the search within neighborhoods N_2 and N_4, we allow the algorithm to remove urgent locations from a solution by replacing them with other locations or a virtual tour, respectively. This was not a case in the VNS based implementations from [1,2] that deal with scheduling of homogeneous vehicles, where the number of visits to urgent locations in the solution remains constant during the search. If vehicles have different capacities, the number of visits to an urgent location depends on the capacity of available vehicles. Note that the moves within neighborhoods N_2 and N_4 may violate feasibility of a solution but in the case of neighborhood N_2, another move within the same neighborhood can produce feasible solution again. The infeasible moves within N_4 are discarded.

3.5   GVNS Implementation

The main steps of the proposed GVNS method are presented in Algorithm 1. After generating an initial solution, GVNS continues by repeating the three main steps (Shaking, VND, and Move or Not together with the neighborhood change step) until a stopping criterion is met. In our implementation, algorithm stops when maximum CPU running time (t_{max}) is reached. Shaking uses two neighborhood structures N_1 and N_2. First, neighborhood N_1 is explored in different sizes from $k = 1$ to $k = k_{max}$. A N_1-neighbor of solution S of size k is obtained by performing k times a random move that defines this neighborhood. When k reaches the maximum value k_{max}, Shaking switches from neighborhood N_1 to N_2. This neighborhood change is controlled by indicator ind. If Shaking in neighborhood N_2 produced an infeasible solution, a move is repeated in the same neighborhood until a feasible solution is found.

Feasible solution obtained by Shaking is passed to the local search phase - VND for a potential improvement. The local minimum S'' returned by VND is compared with the incumbent S. If S'' is better than S regarding the objective function value, S is replaced with S'' and the algorithm starts from the first neighborhood N_1 ($k = 1$). Otherwise, k is increased by 1 or the neighborhood is changed to N_2 in the case when $k = k_{max}$. Note that Shaking in N_2 consists of a single move defining that neighborhood, i.e., k is not used while Shaking in N_2.

Algorithm 1. The proposed GVNS for VSP-HV

---

procedure GVNS($Problem\ Data, k_{max}, t_{max}$)

  Generate initial solution S;

  repeat

    $ind \leftarrow 1$;

    $k \leftarrow 1$;

    while $(ind \leq 2)$ **do**

      if $k \leq k_{max}$ **then**

        $S' \leftarrow ShakeN_1(S, k)$:   //Shaking

      else

        $S' \leftarrow ShakeN_2(S)$:

      if S' is feasible **then**

        $S'' \leftarrow VND(S')$:   //VND

        if S'' is better than S **then**   //Move or Not

          $S \leftarrow S''$;

          $k \leftarrow 1$;

          $ind \leftarrow 1$;

        else $k \leftarrow k + 1$;

        if $k > k_{max}$ **then** $ind \leftarrow ind + 1$;

  until $SessionTime \geq t_{max}$

The structure of the proposed VND is presented in Algorithm 2. This step includes deterministic search of two neighborhoods N_3 and N_4 using *Best improvement* strategy. The search alternates between neighborhoods as long as an improvement is obtained.

Algorithm 2. VND

---

procedure VND($Problem\ Data, S'$)

  $S'' \leftarrow S'$;

  while (Improvement) **do**

    $ind \leftarrow 1$;

    while $(ind \leq 2)$ **do**   //Local Search

      if $ind = 1$ **then**

      Find the best neighbor $S'' \in N_3(S')$;

      else

      Find the best neighbor $S'' \in N_4(S')$;

      if $f(S'') < f(S')$ **then**   //Move or Not

        $S' \leftarrow S''$;

        $ind \leftarrow 1$;

      else $ind \leftarrow ind + 1$;

  return (S');

---

4   Computational Study

Experimental results obtained by Lingo 17 commercial solver were carried out on Intel Core i7-4578U processor on 3.00 GHz with 16 GB RAM memory under Mac operating system. The designed GVNS was run using Intel Core i7-2600 processor on 3.40 GHz with 12 GB RAM memory under Linux operating system. According to the PassMark single thread CPU Benchmark (https://www.cpubenchmark.net/singleThread.html) and single core Mac Benchmarks (https://browser.geekbench.com/mac-benchmarks), ratings of the Intel Core i7-4578U and the Intel Core i7-2600 processors are 3757 and 1942, respectively. In order to ensure fair comparison of the results obtained on these two platforms, Lingo running times should be multiplied by 1.9346.

Two sets of instances were used: the set of real-life problem instances that includes 30 instances up to 15 locations, 20 vehicles, and maximum 15 tours, and the set of 10 generated problem instances with up to 50 locations, 50 vehicles, and 20 tours. All instances used in our computational study are available at http://www.mi.sanu.ac.rs/~tanjad/VSP-HV-Instances.zip.

In our computational study, we imposed the time limit of 10 h on Lingo 17 solver. Within this time limit, Lingo 17 produced optimal or feasible solutions only for small-size problem instances. Due to the stochastic nature, the designed GVNS algorithm was run 30 times on each problem instance. The value of parameter t_{max} used in stopping criterion was set to $t_{max} = 1$ s for small-size instances solved to optimality, $t_{max} = 10$ s for small-size instances for which Lingo produced only feasible solutions, $t_{max} = 20$ s for medium-size real-life instances, and $t_{max} = 200$ s for large-size generated problem instances. Parameter tuning tests were carried out on the subset of real-life and generated instances in order to chose adequate value for parameter k_{max}. Eight different expressions for k_{max} as a linear function of maximal number of tours (r_{max}) were examined. Based on the obtained results, we have chosen $k_{max} = (r_{max} + 1)/2 + 1$, as it produced solutions of highest quality on the considered subset of instances.

Experimental results on small-size real-life instances, previously solved to optimality by Lingo solver are presented in Table 1. Instance's name ($E_{n,m,r_{max}}$) is indicated in the first column. We used n to denote the number of locations, m stands for the number of vehicles, and r_{max} indicates the maximal number of tours during the working day. The next two columns show the results provided by Lingo solver and contain optimal solution (*opt. sol.*), with the corresponding Intel Core i7-4578U running times $t(s)$ in seconds. GVNS results are presented in the last three columns of Table 1: the best objective function value (*best*), the average Intel Core i7-2600 running time $t(s)$ that GVNS needed to reach best/optimal solution, and the average percentage gap (*gap(%)*) of GVNS solution from the best/optimal one, calculated over 30 consecutive runs. We used the mark *opt* to denote the case when GVNS reached optimal solution in at least one of 30 runs. The last row (*Average*) contains the average values of data presented in each column.

Based on the results presented in Table 1, it can be seen that GVNS reached all optimal solutions almost immediately (i.e., for 0.007 s on average), while Lingo average running time is 3694.912 s. The exceptions are instances $E_{4,2,3}$, $E_{5,2,4}$, and $E_{6,2,4}$ on which GVNS provided lower quality solutions in some executions. For this reason, the average percentage gap is 0.345%. The results on all instances

Table 1. Computational results on small-size real-life instances solved to optimality by Lingo 17 solver

Instance	Lingo 17		GVNS		
$E_{n,m,r_{max}}$	opt. sol.	t(s)	best	t(s)	gap(%)
$E_{3,2,4}$	**15.491**	325.87	opt	0.000	0.000
$E_{3,3,2}$	**10.380**	19.96	opt	0.000	0.000
$E_{3,3,3}$	**13.607**	2330.48	opt	0.000	0.000
$E_{3,4,2}$	**10.610**	30009.75	opt	0.000	0.000
$E_{4,2,3}$	**10.976**	23.23	opt	0.000	3.573
$E_{4,2,4}$	**11.89**	39.37	opt	0.000	0.000
$E_{4,3,2}$	**10.686**	84.37	opt	0.000	0.000
$E_{4,3,3}$	**11.713**	8119.9	opt	0.121	0.000
$E_{5,2,3}$	**9.807**	37.670	opt	0.000	0.000
$E_{5,2,4}$	**12.299**	359.94	opt	0.000	0.014
$E_{5,3,2}$	**10.806**	76.870	opt	0.000	0.000
$E_{5,3,3}$	**13.637**	4338.98	opt	0.000	0.000
$E_{5,3,4}$	**16.263**	15397.58	opt	0.000	0.000
$E_{6,2,3}$	**11.411**	56.800	opt	0.000	0.000
$E_{6,2,4}$	**12.469**	722.190	opt	0.000	2.278
$E_{6,3,2}$	**9.713**	232.43	opt	0.000	0.000
$E_{6,4,2}$	**9.417**	638.11	opt	0.000	0.000
Average	**11.834**	3694.912	opt	**0.007**	0.345

unsolved to optimality by Lingo 17 within 10 h are presented in Table 2. The first column and the last three columns are organized in the same way as in Table 1. In cases when Lingo produced only feasible solutions, we presented the upper (UB) and lower bound (LB) in the second and third column, respectively. Average values are calculated and presented in the last row (Average), but only for columns that contain data for each instance.

As it can be seen from Table 2, Lingo succeeded in obtaining feasible solutions only for seven real-life problem instances, but no optimal solution is provided within time limit of 10 h. GVNS improved upper bounds for six of these seven instances ($E_{3,3,5}$, $E_{4,4,3}$, $E_{4,4,4}$, $E_{5,3,4}$, $E_{5,5,2}$, and $E_{7,3,3}$) and in the case of instance $E_{3,3,4}$ GVNS returned solution that coincides with the feasible one provided by Lingo. For the remaining medium-size real-life and generated problem instances, Lingo failed to provide even a feasible solution within 10 h. On the other hand, the proposed GVNS produced its best solutions on all instances presented in Table 2 within 70.345 s on average. The average percentage gap of GVNS solutions from the best ones is 1.301%.

Table 2. Computational results on medium-size real-life and generated problem instances unsolved to optimality by Lingo 17 solver

Instance	Lingo 17		GVNS		
$E_{n,m,r_{max}}$	UB	LB	best	$t(s)$	gap(%)
$E_{3,3,4}$	**15.160**	13.778	**15.160**	0.000	0.000
$E_{3,3,5}$	17.750	15.586	**15.586**	0.000	0.000
$E_{4,4,3}$	14.358	13.126	**14.179**	0.001	0.000
$E_{4,4,4}$	18.498	15.519	**16.263**	0.250	0.535
$E_{5,4,3}$	12.966	12.946	**12.946**	0.012	0.000
$E_{5,5,2}$	11.783	10.155	**10.846**	0.000	0.000
$E_{7,3,3}$	13.216	10.854	**11.727**	0.000	0.000
$E_{4,5,7}$	/	/	**14.297**	3.864	0.297
$E_{5,5,5}$	/	/	**18.679**	2.675	0.737
$E_{5,5,10}$	/	/	**17.329**	4.839	0.425
$E_{8,6,5}$	/	/	**13.930**	3.210	1.227
$E_{10,10,10}$	/	/	**15.369**	16.447	1.105
$E_{15,20,15}$	/	/	**30.090**	17.259	2.259
$E^r_{10,20,10}$	/	/	**24.707**	147.870	1.618
$E^r_{10,50,10}$	/	/	**40.966**	151.276	3.031
$E^r_{20,25,15}$	/	/	**33.234**	181.332	4.750
$E^r_{20,30,8}$	/	/	**23.716**	157.454	2.312
$E^r_{30,10,8}$	/	/	**18.130**	96.695	0.796
$E^r_{30,15,10}$	/	/	**23.914**	169.929	1.542
$E^r_{40,20,12}$	/	/	**27.451**	153.546	2.695
$E^r_{40,25,20}$	/	/	**52.391**	173.001	1.684
$E^r_{50,15,10}$	/	/	**24.971**	166.299	3.422
$E^r_{50,25,15}$	/	/	**36.403**	171.974	1.498
Average	/	/	**22.268**	70.345	1.301

5   Conclusion

This study considers a generalization of vehicle scheduling problem for sugar beet transportation analyzed in our previous work. Instead of homogeneous vehicles, we assume that vehicles of different capacities are used, which implies different loading and unloading times. For the obtained problem extension, we developed a MIQCP model by introducing necessary changes in several constraints. The MIQCP model was used within the framework of Lingo 17 commercial solver, which provided optimal or feasible solutions only for small-size problem instances. Therefore, we designed GVNS metaheuristic approach that confirmed or improved solutions obtained by Lingo and provided solutions for medium-size real-life and generated problem instances in short running times. Low values of average gaps of GVNS solutions from the optimal or the best known ones indicate the stability of the proposed GVNS approach. As a future work, it would be challenging to combine our GVNS with exact or metaheuristic approach in order to improve the results. In addition, the problem can be further extended by involving multidepots, time windows or some additional constraints from practice. Finally, our GVNS can be adapted for solving similar vehicle scheduling or routing problems.

References

1. Anokić, A., Stanimirović, Z., Davidović, T., Stakić, D.: Variable neighborhood search based approaches to a vehicle scheduling problem in agriculture. Int. Trans. Oper. Res. https://doi.org/10.1111/itor.12480 (2017)
2. Anokić, A., Stanimirović, Z., Davidović, T., Stakić, D.: Variable neighborhood search for vehicle scheduling problem considering the transport of agricultural raw materials. Electron. Notes Discrete Math. **58**, 137–142 (2017)
3. de Armas, J., Melián-Batista, B.: Variable neighborhood search for a dynamic rich vehicle routing problem with time windows. Comput. Ind. Eng. **85**, 120–131 (2015)
4. Bodin, L., Golden, B.: Classification in vehicle routing and scheduling. Networks **11**(2), 97–108 (1981)
5. Bortfeldt, A., Hahn, T., Männel, D., Mönch, L.: Hybrid algorithms for the vehicle routing problem with clustered backhauls and 3D loading constraints. Eur. J. Oper. Res. **243**(1), 82–96 (2015)
6. Bula, G., Prodhon, C., Gonzalez, F.A., Afsar, H., Velasco, N.: Variable neighborhood search to solve the vehicle routing problem for hazardous materials transportation. J. Hazard. Mater. **324**(Part B), 472–480 (2016)
7. Bunte, S., Kliewer, N.: An overview on vehicle scheduling models. Public Transp. **1**(4), 299–317 (2009)
8. Cheikh, M., Ratli, M., Mkaouar, O., Jarboui, B.: A variable neighborhood search algorithm for the vehicle routing problem with multiple trips. Electron. Notes Discrete Math. **47**, 277–284 (2015)
9. Dantzig, G., Fulkerson, D.: Minimizing the number of tankers to meet a fixed schedule. Naval Res. Logistic Q. **1**(3), 217–222 (1954)
10. Hansen, P., Mladenović, N.: Variable neighborhood search. In: Burke, E.K., Graham, R.D. (eds.) Search Methodologies: Introductory Tutorials in Optimization and Decision Support Techniques, pp. 313–337. Springer, New York (2014). https://doi.org/10.1007/978-1-4614-6940-7_12
11. Hansen, P., Mladenović, N., Pérez, J.A.: Variable neighbourhood search: methods and applications. Ann. Oper. Res. **175**(1), 367–407 (2010)
12. Hemmelmayr, V.C., Doerner, K.F., Hartl, R.F.: A variable neighborhood search heuristic for periodic routing problems. Eur. J. Oper. Res. **195**(3), 791–802 (2009)
13. Higgins, A.: Scheduling of road vehicles in sugarcane transport: a case study at an Australian sugar mill. Eur. J. Oper. Res. **170**(3), 987–1000 (2006)
14. Liu, S., Chen, A.: Variable neighborhood search for the inventory routing and scheduling problem in a supply chain. Expert Syst. Appl. **39**(4), 4149–4159 (2012)
15. Macedo, R., et al.: Skewed general variable neighborhood search for the location routing scheduling problem. Comput. Oper. Res. **61**, 143–152 (2015)

16. Milan, E., Fernandez, S., Aragones, L.: Sugar cane transportation in cube, a case study. Eur. J. Oper. Res. **174**(1), 374–386 (2006)
17. Mladenović, N., Hansen, P.: Variable neighborhood search. Comput. Oper. Res. **24**(11), 1097–1100 (1997)
18. Raff, S.: Routing and scheduling of vehicles and crews: the state of the art. Comput. Oper. Res. **10**(2), 63–211 (1983)
19. Thuankaewsing, S., Khamjan, S., Piewthongngam, K., Pathumnakul, S.: Harvest scheduling algorithm to equalize supplier benefits: a case study from the Thai sugar cane industry. Comput. Electron. Agric. **110**, 42–55 (2015)

Detecting Weak Points in Networks Using Variable Neighborhood Search

Sergio Pérez-Peló[ID], Jesús Sánchez-Oro[✉][ID], and Abraham Duarte[ID]

Department of Computer Sciences, Universidad Rey Juan Carlos, Móstoles, Spain
{sergio.perez.pelo,jesus.sanchezoro,abraham.duarte}@urjc.es

Abstract. Recent advances in networks technology require from advanced technologies for monitoring and controlling weaknesses in networks. Networks are naturally dynamic systems to which a wide variety of devices are continuously connecting and disconnecting. This dynamic nature force us to maintain a constant analysis looking for weak points that can eventually disconnect the network. The detection of weak points is devoted to find which nodes must be reinforced in order to increase the safety of the network. This work tackles the α separator problem, which aims to find a minimum set of nodes that disconnect the network in subnetworks of size smaller than a given threshold. A Variable Neighborhood Search algorithm is proposed for finding the minimum α separator in different network topologies, comparing the obtained results with the best algorithm found in the state of the art.

Keywords: Alpha-separator · Reduced VNS · Betweenness

1 Introduction

Nowadays cybersecurity is becoming one of the most relevant fields for any kind of users: from companies and institutions to individual users. The increase in the number of attacks to different networks in the last years, as well as the relevance of the privacy in the Internet, have created the necessity of having more secure, reliable and robust networks. A cyberattack to a company that causes loss of personal information of their clients can result in important economic and social damage [1]. Furthermore, Denial of Service (DoS) and Distributed Denial of Service (DDoS) attacks are becoming more common since a successful attack can result in disabling a service of an Internet provider, for instance. Even more, if several services are dependent on the service under attack, a cascade failure can occur, affecting to a large number of clients [3].

It is important to identify which are the most relevant nodes in a network. This is a matter of interest for both actors in a cyberattack: the attacker and the defender. The former is interested in disabling these nodes in order to make

This work has been partially founded by Ministerio de Economía y Competitividad with grant ref. TIN2015-65460-C2-2-P.

A. Sifaleras et al. (Eds.): ICVNS 2018, LNCS 11328, pp. 141–151, 2019.
https://doi.org/10.1007/978-3-030-15843-9_12

the network more vulnerable while the latter is focused on reinforcing these important nodes with more robust security measures. On the one hand, the attacker is interested in causing the maximum damage to the network while consuming the minimum amount of resources. On the other hand, the defender wants to reinforce the network minimizing the increase in the maintenance and security costs. Therefore, it is interesting for both parts to identify which are the weak points in a network.

We define a network as a graph $G = (V, E)$, where V is the set of vertices, $|V| = n$, and E is the set of edges, $|E| = m$. A vertex $v \in V$ represents a node of the network while an edge $(v, u) \in E$, with $v, u \in V$ indicates that there is a connection in the network between vertices v and u. Let us also define a separator of a network as a set of vertices $S \subseteq V$ whose removal cause the partition of the network into two or more connected components. More formally,

$$V \setminus S = C_1 \cup C_2 \ldots \cup C_p$$
$$\forall\, (u, v) \in E^\star \,\exists\, C_i \,:\, u, v \in C_i$$

where $E^\star = \{(u, v) \in E \,:\, u, v \notin S\}$.

This work is focused on finding a minimum α-separator $S^\star$ which is able to split a network G into connected components of sizes smaller than $\alpha \cdot n$. In mathematical terms,

$$S^\star \leftarrow \arg\min_{S \in \mathbb{S}} |S| \quad : \quad \max_{C_i \in V \setminus S} |C_i| \leq \alpha \cdot n$$

It is worth mentioning that the number of resulting connected components is not relevant in this problem neither as a constraint nor for evaluating the objective function value. The actual constraint of the α-separator problem (α-SP) is that the number of vertices in each connected component must be lower or equal than $\alpha \cdot n$, where α is an input value. This problem is $\mathcal{NP}$-hard for general networks topologies when considering $\alpha \leq \frac{2}{3}$ [6]. Some polynomial-time algorithms have been proposed when the topology of the network is a tree or a cycle [15]. However, these algorithms require to have previous knowledge on the topology of the network, which is not usual in real-life problems.

Figure 1 shows an example of a network and two feasible solutions for the α-SP. The network depicted in Fig. 1(a) is conformed with 9 vertices and 10 edges connecting those vertices. We consider $\alpha = \frac{2}{3}$ for this example, so the connected components of the network must contain $\lceil \frac{2}{3} \cdot 9 \rceil = 6$ vertices at most. Figure 1(b) shows a feasible solution $S_1 = \{B, C, E, I\}$ which divides the network into two connected components $C_1 = \{A, D\}$, and $C_2 = \{F, G, H\}$, whose number of vertices (2 and 3, respectively), are smaller than 6. The second solution, $S_2 = \{A, B\}$, depicted in Fig. 1(c), divides the network in three connected components: $C_1 = \{E\}$, $C_2 = \{D\}$, and $C_3 = \{C, G, H, I\}$ (all of them with sizes smaller than 6). Notice that S_2 is better than S_1 in terms of objective function value since it is able to disconnect the network by removing just 2 vertices, while S_1 requires to remove 4 vertices in order to disconnect the network. Notice that neither the size nor the number of connected components affect to the quality of the solution.

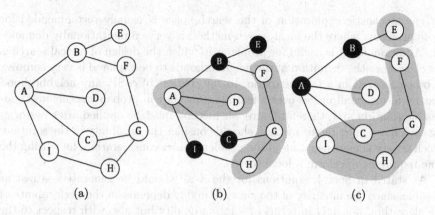

Fig. 1. (a) Example of a graph derived from a network, (b) a feasible solution with 4 nodes in the separator (B,C,E, and I), and (c) a better solution with 2 nodes in the separator (A and B)

This problem has been tackled for both exact and heuristic perspectives. Specifically, polynomial-time algorithms have been presented for special topologies as trees or cycles, as well as a greedy algorithm with approximation ratio of $\alpha \cdot n + 1$ [15]. Additionally, a heuristic algorithm for studying the node separators in the Internet Autonomous Systems was proposed [17]. Depending on the α value, the α-separator problem can be related to different well-known problems. In particular, when $\alpha = \frac{1}{n}$, it is equivalent to the minimum vertex cover problem, and when $\alpha = \frac{2}{n}$ it is analogous to the minimum dissociation set problem. Therefore, the α-separator problem is a generalization of these problems, which are also $\mathcal{NP}$-hard [9]. As far as we know, the best previous heuristic consists of a random walk algorithm which is based on a Markov Chain Monte Carlo method [13].

2   Algorithmic Proposal

Variable Neighborhood Search (VNS) [11] is a metaheuristic framework based on systematic changes of neighborhoods. As a metaheuristic algorithm, it does not guarantee the optimality of the solutions obtained, but it is focused on obtaining high quality solutions in reasonable computing times. There are several variants of VNS, which are classified taking into account fundamentally the exploration of the considered neighborhoods. Typically, the neighborhood structures are analyzed using three different criterion: stochastic (Reduced VNS, RVNS), deterministic (Variable Neighborhood Descent, VND), or a combination of both deterministic and stochastic (Basic VNS, BVNS). Furthermore, several additional variants have been proposed in the last years: General VNS (GVNS), Variable Neighborhood Decomposition Search (VNDS), Skewed VNS (SVNS), Variable Formulation Search (VFS), among others.

The stochastic exploration of the search space is usually recommended for those problems where the local search method is very computationally demanding. Additionally, it is useful for problems in which the design of a local search is not clear since the definition of the neighborhoods to be explored is very complex to be considered in a fast heuristic. In the context of α-SP, any neighborhood that performs small moves over a feasible solution will probably result in a non-feasible solution and, therefore, a repair method must be applied after performing each move. The repair method should consider the feasibility of the solution, which is very time consuming in the problem under consideration, increasing the time required to perform a local search.

As stated in Sect. 1, solutions for the α-SP should be generated as fast as possible, since the integrity of the network highly depends on the performance of the algorithm, not only in terms of solution quality but also with respect to the computing time required to produce a solution. RVNS is usually compared with a Monte-Carlo method, but being RVNS more systematic [14]. Indeed, RVNS has been able to obtain results competitive with the Fast Interchange [18] method when applied to the p-Median problem [10]. This work presents a Reduced VNS algorithm in order to generate high quality solutions in short computing time. Algorithm 1 presents the general framework of RVNS.

Algorithm 1. $RVNS(S, k_{\max})$

1: **repeat**
2:    $k \leftarrow 1$
3:    **while** $k \leq k_{\max}$ **do**
4:        $S' \leftarrow Shake(S, k)$
5:        $k \leftarrow NeighborhoodChange(S, S', k)$
6:    **end while**
7: **until** $StoppingCriterion$
8: **return** S

RVNS starts from an initial solution S, which can be generated using a random procedure or any other complex heuristic or metaheuristic. Additionally, the maximum neighborhood $k_{\max}$ to be explored must be indicated. As stated in previous works [12], the maximum neighborhood to be considered in the RVNS algorithm is usually small, to avoid exploring completely different solutions in each iteration. Finally, the third parameter of the algorithm indicates the stopping criterion. For this work, we consider a maximum number λ of RVNS iterations.

In each iteration, the algorithm starts from the first considered neighborhood $k = 1$ (step 2). Then, the algorithm iterates until reaching the maximum neighborhood $k_{\max}$ (steps 3–6). In each iteration, a random solution S' in the current neighborhood k is generated (step 4). Then, the algorithm selects the next neighborhood to be explored (step 5). In particular, if the objective function value of S' is better than the one of S, an improvement is found, updating the best solution found ($S \leftarrow S'$) and restarting the search from the first neighborhood ($k = 1$). Otherwise, the search continues with the next neighborhood

($k \leftarrow k + 1$). The method ends when performing λ iterations of RVNS (steps 1–7), returning the best solution found during the search (step 8).

2.1 Constructive Method

The main objective of the α-SP is to identify the most important nodes in a network trying to minimize the size of the separator. To achieve this goal, we can leverage several characteristics of a node of a graph, such as its degree (number of edges of a node), its position in the network, or any centrality measure, among others, as a selection criterion to find the most important vertices in a graph.

This work proposes a greedy constructive procedure for generating the initial solution. In particular, we propose a greedy function that evaluates the relevance of a vertex in a graph using a centrality metric known as betweenness centrality [2], an extended metric in the context of finding relevant users in social networks.

Betweenness centrality considers that a node is important within a network if it acts as a flow of information in the graph. In order to look for relevant nodes with respect to this metric, it is necessary to evaluate all the paths between any pair of nodes of the graph, being the relevance of a vertex directly related to the number of paths in which it appears. The rationale behind this it that if a vertex v participates in several paths of the network, then any information transmitted through it will eventually traverse v.

The betweenness centrality of a node $v \in V$, named as $b(v)$, is evaluated as the number of paths between any pair of nodes s and t in which v is included, $\sigma(s,t|v)$, divided by all the paths that connect s and t, $\sigma(s,t)$. More formally,

$$b(v) \leftarrow \sum_{s,t \in V \setminus \{v\}} \frac{\sigma\left(s,t|v\right)}{\sigma\left(s,t\right)}$$

It is worth mentioning that the evaluation of the betweenness is a computationally demanding process. Specifically, the betweenness of a node v requires from the evaluation of all shortest paths between every pair of vertices $s, t \in V$.

In order to reduce the complexity of this evaluation, we consider an approximation of this metric by evaluating a number of shortest paths between every pair of nodes using a fast algorithm for finding shortest paths without loops in networks based on Yen's algorithm [19].

We have selected this criterion because it seems logical to think that, the more information circulates through a node in a graph, the more important this node will be within the network. Therefore, it will be easier to disconnect the network if priority is given to eliminate the nodes with a higher value of betweenness centrality.

A totally greedy algorithm will always produce the same initial solution, since it is focused on intensifying the search. However, several works [4,16] have shown that introducing some randomness in the search, thus increasing the diversification, results in better initial solutions. We propose the use of Greedy Randomized Adaptive Search Procedure (GRASP) methodology in order generate more diverse initial solutions that will eventually lead the algorithm to obtain better results.

GRASP methodology was originally proposed in 1989 [7] but it was not formally defined until 1994 [8]. Traditional GRASP algorithms consists of two well-differenced phases: construction and local search. In this work we just consider the construction stage, since a local search for the α-SP is rather computationally demanding, resulting in large computing times.

The constructive procedure proposed starts by randomly selecting the first node v to be removed from the graph. Then, a candidate list is constructed with all the vertices $u \in V \setminus \{v\}$. In each iteration, a vertex is selected from the candidate list and removed from the graph. The selection of the next vertex is performed as follows. Firstly, a restricted candidate list is created with the most promising nodes of the candidate list. For this problem, we evaluate the betweenness of all the candidate vertices. Let us consider that g_{max} and g_{min} are the maximum and minimum values for this metric, respectively. Then, a threshold μ is evaluated as:

$$\mu = g_{max} - \beta * (g_{max} - g_{min})$$

The restricted candidate list contains all the candidate vertices whose betweenness value is larger or equal than the threshold μ. Notice that $\beta \in [0, 1]$ is a parameter of the constructive method that controls its degree of randomness. On the one hand, $\beta = 0$ results in $\mu = g_{max}$, which considers a totally greedy algorithm. On the other hand, $\beta = 1$ results in $\mu = g_{min}$, which considers a completely random procedure. Section 3 will evaluates different values for this parameter, discussing how it affects to the initial solution quality.

We have made a series of preliminary experiments in order to test what is the α parameter that works better with our greedy criterion. Concretely, we have tested our algorithm using 0.25, 0.5, 0.75 and RND values. We present the results of these experiments in Sect. 3.

2.2 Perturbing Solutions for the α-SP

Shake method is responsible for finding new solutions in the neighborhood under exploration in the RVNS framework. First of all, it is important to define the neighborhoods considered in this work. We define the neighborhood $N_1(S)$ of a solution S as the insertion of a new node in the solution. In mathematical terms,

$$N_1(S) = \{S' \leftarrow S \cup \{v\} \: : \: v \in V \setminus S\}$$

Analogously, we define the neighborhood $N_k(S)$ as the insertion of k new nodes in the solution S. More formally,

$$N_k(S) = \{S' \leftarrow S \cup T \: : \: \forall v \in T, \: (v \in V \setminus S) \wedge (|T| = k)\}$$

Notice that any solution S' obtained in the neighborhood N_k of solution S presents more vertices included in it. If the quality of a solution is given by the number of vertices included in it and α-SP is a minimization problem, then any solution $S' \in N_k(S)$ is worse than S in terms of objective function value. However, the main advantage of exploring this neighborhood is that every solution is always feasible, being unnecessary to check the constraint of the problem, which is one of the most time consuming parts of the algorithm.

Since the *Shake* procedure always deteriorates the quality of the solution perturbed, it is necessary to define a post-processing method that tries to improve its quality. For this purpose, we define a refining process that consists of removing all the vertices that are unnecessarily included in the perturbed solution S'. Specifically, the method randomly traverses the set of vertices that were originally in the solution (i.e., $S' \setminus T$). If the solution becomes unfeasible after removing a vertex, it is included again in S'. After trying to remove all vertices in $S' \setminus T$, the method repeats this instructions for all the vertices included in T. Following this procedure, if the number of removed nodes from the solution during the refining process is larger or equal than k, an improvement has been found, restarting the search from the first neighborhood. Otherwise, the search continues in the next neighborhood. The method stops when reaching the maximum predefined one, k, returning the best solution found during the search.

3   Computational Results

This Section is devoted to analyze the performance of the proposed algorithms and compare the obtained results with the best previous method found in the state of the art. The algorithms have been developed in Java 9 and all the experiments have been conducted on an Intel Core 2 Duo 2.66 GHz with 4 GB RAM.

The set of instances used in this experimentation has been generated using the same graph generator proposed in the best previous work [13]. Specifically, graphs are generated by using the Erdös-Réyi model [5], in which a new node is linked to the nodes already in the graph with the same probability. We have generated a set of 50 instances whose number of vertices ranges from 100 to 200 and the number of edges ranges from 200 to 2000.

We have divided the experiments into two different parts: preliminary and final experimentation. The former is designed for finding the best parameters of the GRASP constructive procedure and RVNS algorithm, while the latter is devoted to analyze the performance of the best variant when compared with the best previous method found in the state of the art.

All the experiments report the following metrics: Avg., the average objective function value; Time (s), the average computing time in seconds; Dev (%), the average deviation with respect to the best solution found in the experiment; and # Best, the number of times that an algorithm reaches the best solution of the experiment.

The preliminary experiments consider a subset of 20 representative instances out of 50 in order to avoid overfitting, while the final experimentation considers the total set of 50 instances.

The first experiment is designed for evaluating the effect of the β parameter in the quality of the initial solutions generated. We have considered the following values for $\beta = \{0.25, 0.50, 0.75, RND\}$, where RND indicates that a random value is selected in each iteration of the construction phase. Table 1 shows the results obtained in this experiment. It is worth mentioning that 100 solutions have been generated for each instance, returning the best solution found and computing the accumulated time required for constructing all of them.

Table 1. Performance of GRASP constructive procedure with different values for the β parameter.

β	Avg.	Time (s)	Dev (%)	#Best
0.25	59.80	592.43	0.59	16
0.5	60.45	590.30	2.08	12
0.75	62.25	590.79	4.26	9
-1.00	59.85	612.49	0.94	13

Analyzing Table 1 we can clearly see that the best results are obtained when considering $\beta = 0.25$, closely followed by $\beta = RND$. Specifically, $\beta = 0.25$ is able to find 16 out of 20 best solutions, and the average deviation of 0.59% indicates that in those instances in which it is not able to reach the best value, the constructive method obtains a high quality solution really close to the best one. The value of $\beta = 0.25$ indicates that the best results are obtained when introducing a small random part in the constructive procedure, and increasing the randomness of the method results in worse solutions. The worst results are obtained with the largest β value, 0.75, obtaining just 9 out of 20 best solutions with an average deviation of 4.26%. This behavior also confirms that the betweenness metric is a good selection as a greedy function value for the constructive procedure.

One of the main advantages of VNS is the reduced number of parameters that must be tuned in order to obtain high quality solutions. In particular, the RVNS algorithm proposed requires from just one parameter, k_{max}, that corresponds to the largest neighborhood to be explored. The preliminary experiment then considers the following values for $k_{max} = \{0.05, 0.10, 0.25, 0.50\}$. We do not consider larger values of k_{max} since a solution in such a large neighborhood will be completely different from the original one. It is worth mentioning that the number of vertices that will be included in the neighbor solution is evaluated as $k \cdot |S|$, being $|S|$ the number of vertices of the initial solution. Table 2 shows the performance of the different values for k_{max}.

The experiment clearly shows that the best value for k_{max} is 0.25, finding all the best solutions, while the remaining values are able to obtain just one best solution out of 20. This results are in line with those presented by Mladenovic et al. [14], which recommends considering small values of k_{max} in the context of RVNS. This is mainly because large values for k_{max} explores solutions that are not close in the search space, resulting in a completely random search, which is against the VNS methodology.

Table 2. Performance of the RVNS when considering different values for the k_{max} parameter

k_{max}	Avg.	Time (s)	Dev (%)	#Best
0.05	49.00	303.09	15.40	1
0.10	48.40	314.81	14.16	1
0.25	43.45	310.06	0.00	20
0.50	47.65	316.44	12.76	1

Analyzing the results obtained in the preliminary experiments, the RVNS algorithm for the final experiment is configured with $\beta = 0.25$ and $k_{max} = 0.25$.

The final experiment is intended to compare the best variant of RVNS with the best previous method found in the state of the art [13]. Specifically, it consists of a random walk (RW) algorithm with Markov Chain Monte Carlo method. It is worth mentioning that we have not been able to contact to the authors of the previous work neither to obtain the set of instances nor an executable file of the algorithm. Therefore, we have reimplemented the previous algorithm following, in detail, all the steps described in the manuscript. Table 3 shows the results obtained by the proposed algorithm (RVNS) and the best previous method found (RW).

Table 3. Comparison of the best variant of RVNS with the best previous method found in the state of the art.

	Avg.	Time (s)	Dev (%)	#Best
RW	71.78	1070.35	25.98	5
RVNS	55.58	473.72	0.10	46

The results obtained clearly confirm the superiority of our proposal. In particular, RVNS is able to obtain 46 out of 50 best solutions, while RW only reaches the best solution in 5 out of 50 instances. Furthermore, the computing time for RVNS is half of the time required by RW. Finally, the average deviation of RVNS is close to zero, which indicates that, in those instances in which RVNS is not able to match the best solution, it stays close to it. However, the deviation of RW is higher, indicating that its results are not close to the best solution obtained by the RVNS.

We have finally conducted the nonparametric Wilcoxon Signed Test in order to confirm that there exists statistically significant differences between both algorithms. The p-value obtained is lower than 0.0001, which confirms the superiority of our proposal.

4   Conclusions

This work has proposed a Reduced VNS algorithm for detecting critical nodes in networks. The initial solution is generated by using a Greedy Randomized Adaptive Search Procedure whose greedy criterion is adapted from the social network field of research, which is named betweenness. The RVNS proposal is able to obtain ·better results than the best previous method found in the literature, which consists of a random walk algorithm in both quality and computing time. This results, supported by non-parametric statistical tests, confirms the superiority of the proposal. The adaptation of a social network metric to the problem under consideration in this work has led us to obtain high quality solutions, which reveals the relevance of the synergy among different fields of research.

References

1. Andersson, G., et al.: Causes of the 2003 major grid blackouts in North America and Europe, and recommended means to improve system dynamic performance. IEEE Trans. Power Syst. **20**(4), 1922–1928 (2005)
2. Brandes, U.: A faster algorithm for betweenness centrality. J. Math. Soc. **25**(2), 163–177 (2001)
3. Crucitti, P., Latora, V., Marchiori, M.: Model for cascading failures in complex networks. Phys. Rev. E **69**, 045104 (2004)
4. Duarte, A., Sánchez-Oro, J., Resende, M.G., Glover, F., Martí, R.: Greedy randomized adaptive search procedure with exterior path relinking for differential dispersion minimization. Inf. Sci. **296**, 46–60 (2015)
5. Erdős, P., Rényi, A.: On random graphs. Publ. Math. **6**, 290 (1959)
6. Feige, U., Mahdian, M.: Finding small balanced separators. In: Kleinberg, J.M. (ed.) STOC, pp. 375–384. ACM (2006)
7. Feo, T., Resende, M.: A probabilistic heuristic for a computationally difficult set covering problem. Oper. Res. Lett. **8**(2), 67–71 (1989)
8. Feo, T.A., Resende, M.G., Smith, S.H.: Greedy randomized adaptive search procedure for maximum independent set. Oper. Res. **42**(5), 860–878 (1994)
9. Garey, M., Johnson, D.: Computers and Intractability - A guide to the Theory of NP-Completeness. Freeman, San Fransisco (1979)
10. Hansen, P., Mladenovic, N.: Variable neighborhood search: principles and applications. Eur. J. Oper. Res. **130**(3), 449–467 (2001)
11. Hansen, P., Mladenović, N.: Variable neighborhood search. In: Burke, E., Kendall, G. (eds.) Search Methodologies, pp. 313–337. Springer, Boston (2014)
12. Hansen, P., Mladenović, N., Pérez, J.A.M.: Variable neighbourhood search: methods and applications. Ann. Oper. Res. **175**(1), 367–407 (2010)
13. Lee, J., Kwak, J., Lee, H.W., Shroff, N.B.: Finding minimum node separators: a Markov chain Monte Carlo method. In: 13th International Conference on Design of Reliable Communication Networks, DRCN 2017, pp. 1–8, March 2017
14. Mladenovic, N., Petrovic, J., Kovacevic-Vujcic, V., Cangalovic, M.: Solving spread spectrum radar polyphase code design problem by tabu search and variable neighbourhood search. Eur. J. Oper. Res. **151**(2), 389–399 (2003)
15. Mohamed-Sidi, M.: K-separator problem. (Problème de k-Sèparateur). Ph.D. thesis, Telecom & Management SudParis, Èvry, Essonne, France (2014)

16. Quintana, J.D., Sánchez-Oro, J., Duarte, A.: Efficient greedy randomized adaptive search procedure for the generalized regenerator location problem. Int. J. Comput. Intell. Syst. **9**(6), 1016–1027 (2016)

17. Wachs, M., Grothoff, C., Thurimella, R.: Partitioning the internet. In: Martinelli, F., Lanet, J.L., Fitzgerald, W.M., Foley, S.N. (eds.) CRiSIS, pp. 1–8. IEEE Computer Society (2012)

18. Whitaker, R.: A fast algorithm for the greedy interchange for large-scale clustering and median location problems. INFOR: Inf. Syst. Oper. Res. **21**(2), 95–108 (1983)

19. Yen, J.Y.: Finding the k shortest loopless paths in a network. Manag. Sci. **17**(11), 712–716 (1971)

A Variable Neighborhood Search with Integer Programming for the Zero-One Multiple-Choice Knapsack Problem with Setup

Yassine Adouani[1]([✉]), Bassem Jarboui[2], and Malek Masmoudi[3,4,5]

[1] Laboratory of Modeling and Optimization for Decisional,
Industrial and Logistic Systems (MODILS), Faculty of Economics
and Management Sciences of Sfax, University of Sfax, Sfax, Tunisia
adouaniyassine@gmail.com
[2] Emirates College of Technology, Abu Dhabi, United Arab Emirates
[3] University of Lyon, Saint Etienne, France
[4] University of Saint Etienne, Jean Monnet, Saint-Etienne, France
[5] LASPI, IUT de Roanne, Roanne, France

Abstract. This study proposes a new cooperative approach to the Multiple-Choice Knapsack problem with Setup (MCKS) that effectively combines variable neighborhood search (VNS) with an integer programing (IP). Our approach, based on a local search technique with an adaptive perturbation mechanism to assign the classes to knapsack, and then if the assignment is identified to be promising by comparing its result to the upper bound, we applied the IP to select the items in knapsack. For the numerical experiment, we generated different instances for MCKS. In the experimental setting, we compared our cooperative approach to the Mixed Integer Programming provided in literature. Experimental results clearly showed the efficiency and effectiveness of our cooperative approach with -0.11% as gap of the objective function and 13 s vs. 2868 s as computation time.

Keywords: Knapsack problem · Setup · Cooperative approach

1 Introduction

The 0-1 Multiple-choice Knapsack Problem with Setup (MCKS) is described as a knapsack problem with additional setup variables discounted both in the objective function and the constraint. Practical applications of the MCKS may be seen in production scheduling problems involving setups and machine preferences. A case study of knapsack problem with setup (KPS) is provided in Della et al. [12]. To extend the KPS to MCKS, we consider that items from the same family (or class) could be processed in multiple periods.

The MCKS is NP-hard problem, since it is a generalization of the standard knapsack problem (KP) [23]. MCKS reduces to a KP when considering one class, and no setup variables. The KPS is a particular case of MCKS, when the number of period is equal to one (T = 1) [3, 7, 19], etc. To the best of our knowledge, Yang [34] is the

A. Sifaleras et al. (Eds.): ICVNS 2018, LNCS 11328, pp. 152–166, 2019.
https://doi.org/10.1007/978-3-030-15843-9_13

unique author who dealt with MCKS. He provided an exact method based on a branch and bound for the MCKS, but it has no availability of benchmark instances in the literature.

To deal with the different variants of KP, exact techniques are introduced in the literature such as branch and bound algorithm [10, 20], lagrangian decomposition [5], and dynamic programming [26]. Chebil and Khemakhem [7] provided an improved dynamic programming algorithm for KPS. Akinc [2] studied approximated and exact algorithms to solve fixed charge knapsack problem. Michel et al. [24] developed an exact method based on a branch and bound algorithm to solve KPS. Della et al. [12] provided an exact approach for the 0-1 knapsack problem with setups. Al-Maliky et al. [3] studied a sensitivity analysis of the setup knapsack problem to perturbation of arbitrary profits or weights. Dudzinski and Walukiewicz [10] studied exact methods such as branch-and-bound and dynamic programming for KP and its generalizations. Martello and Toth [22] discussed an upper bound using Lagrangian relaxation for multiple knapsack problem (MKP). Pisinger [27] presented an exact algorithm using a surrogate relaxation to get an upper bound, and dynamic programming to get the optimal solution. Sinha and Zoltners [31] used two dominance rules for the linear multiple-choice KP to provide an upper bound for the multiple-choice knapsack problem.

Approximated algorithms have been also developed such as reactive local search techniques [17], tabu search [16], particle swarm optimization [4], genetic algorithm [8], iterated local search [25], etc. Khemakhem and Chebil [19] provided a tree search based combination heuristic for KPS. Freville and Plateau [14] provided a greedy algorithm and reduction methods for multiple constraints 0-1 linear programming problem. Tlili et al. [32] proposed an iterated variable neighborhood descent hyper heuristic for the quadratic multiple knapsack problems.

The hybridization technique between exact and metaheuristics approaches have been performed by many researchers during the last few decades. This technique provides interesting results as it takes advantages of both types of approaches [18]. A classifications of algorithms combining local search techniques and exact methods are provided in [11, 29]. The focus in these papers is particularly on the so called cooperative algorithms using exact methods to strengthen local search techniques. Fernandes and Lourenco [13] applied cooperative approach to solve different combinatorial optimization problems. Vasquez and Hao [33] proposed a new cooperative approach combining linear programming and tabu search to solve the MKP problem. They considered a two-phased algorithm that first uses Simplex to solve exactly a relaxation of the problem and then explore efficiently the solution neighborhood by applying a tabu search approach. Several works of literature have considered a combination of cooperative approach combining variable neighborhood search with exact technique. Prandtstetter and Raidl [28] applied a cooperative approach that combines an integer linear programming with variable neighborhood search for the car sequencing problem. Burke et al. [6] studied a cooperative approach of Integer Programming and Variable Neighborhood Search for Highly-Constrained Nurse Rostering Problems. Lamghari et al. [21] proposed a cooperative method based on linear programming and variable neighborhood descent for scheduling production in open-pit mines. To the best of the our knowledge, the combination of VNS with exact technique has never been considered for KPS problem.

The remainder of this paper is organized as following: Sect. 2 contains the mathematical formulations of MCKS. In Sect. 3, we propose a cooperative approach combining Variable Neighborhood Search and integer programming for MCKS. The experimental results and their interpretations are reported in Sect. 4. In Sect. 5, we conclude the paper and give possible and future research ideas.

2   Problem Description

The Multiple Choice Knapsack Problem is defined by knapsack capacity $b \in N$ with a set of T divisions (periods), where each division $t \in \{1, \ldots, T\}$, and a set of N classes of items. Each class $i \in \{1, \ldots, N\}$ consists of n_i items. Let f_{it}, negative number, denote the setup cost of class i in division t, and let d_i, a positive number, denote the setup capacity consumption of class i. Each item $j \in \{1, \ldots, n_i\}$ of a class i has a profit $c_{ijt} \in N$ and a capacity consumption $a_{ij} \in N$. For classes and items assignment to divisions of knapsack, we consider two sets of binary decision variables y_{it} and x_{ijt}, respectively. The variable y_{it} is equal to 1 if division t includes items belonging to class i and 0 otherwise. The variable x_{ijt} is equal to 1 if item j of class i is included in division t and 0 otherwise. We propose the following mathematical formulation for the MCKS:

$$\text{Max } z = \sum_{t=1}^{T} \sum_{i=1}^{N} \left(f_{it} y_{it} + \sum_{j=1}^{n_i} c_{ijt} x_{ijt} \right) \tag{1}$$

$$\sum_{t=1}^{T} \sum_{i=1}^{N} \left(d_i y_{it} + \sum_{j=1}^{n_i} a_{ij} x_{ijt} \right) \leq b \tag{2}$$

$$x_{ijt} \leq y_{it}; \forall i \in \{1, \ldots, N\}, \forall j \in \{1, \ldots, n_i\}, \forall t \in \{1, \ldots, T\} \tag{3}$$

$$\sum_{t=1}^{T} x_{ijt} \leq 1; \forall i \in \{1, \ldots, N\}, \forall j \in \{1, \ldots, n_i\} \tag{4}$$

$$x_{ijt}, y_{it} \in \{0, 1\}; \forall i \in \{1, \ldots, N\}, \forall j \in \{1, \ldots, n_i\}, \forall t \in \{1, \ldots, T\} \tag{5}$$

Equation (1) represents the objective function that is to maximize the profit of selected items minus the fixed setup costs of selected classes. Constraint (2) guarantees that the sum of the total weight of selected items and the class setup capacity consumption does not exceed the knapsack capacity b. Constraint (3) requires that each item is selected only if it belongs to a class that has been setup. Constraint (4) guarantees that each item is selected and assigned to one division at most. Constraint (5) ensures that the decision variables are binary.

Using CPLEX 12.7 to solve MCKS shows its limitation due to the complexity of the problems. We show later in the experimental results (Sect. 4) that by using CPLEX, only 27 instances of MCKS among 120 are solved to the optimality in less than 1 h CPU time. For the rest, the computation terminates with an out of memory or is stopped at 1 h. Thus we decided to invest in the development of a cooperative approach combining variable neighborhood search and integer programming. We explain our new approach in the next section.

3   Cooperative Approach for MCKS

Local search techniques have proven their efficiency in several combinatorial problems and have been used within cooperative approaches for several problems [11, 13, 28]. Particularly, the Variable Neighborhood Search (VNS) is a method based on a systematic change of the neighborhood structures. It is introduced by Maldenovic and Hansen [36] and has proven its efficiency on different scheduling problems: location routing [37], car sequencing problem [28], etc.

This paper contains a new cooperative approach combining VNS with IP. The main idea of our cooperative approach is to decompose the original problem in to two subproblems. The first problem is to assign classes to the divisions of knapsack (determine the setup variables y_{it}) using a VNS approach allowing the transformation of MCKS into classical KP. Two movements have been considered within the VNS approach: local search procedure (LS) and a perturbation mechanism. The second problem is to solve the classical KP by considering the IP that determines the values of x_{ijt} with a very short computation time. For efficiency issue, we apply the IP only if the search space is identified to be promising by comparing its result to an upper bound that we provided later. Note the found values of y_{it} and x_{ijt} yield a feasible solution to MCKS.

The approach starts with a construction heuristic called reduction-based heuristic (RBH). Then, the obtain solution is improved by using a Local search technique with integer programming (LS&IP) procedure. At each iteration, Perturb&IP and LS&IP are successively applied to the current solution. The current solution is update if the resulting solution is better than the current one. The algorithm works until a termination condition is satisfied. Algorithm 1 shows the whole framework of our approach.

VNS&IP $(Data, t\_max, k_{max})$

$S_0 \leftarrow$ **RBH**(Data);
$S_1 \leftarrow$ **LS&IP**(S_0); /*Local search*/
$S \leftarrow S_0$
While $(t < t\_max)$
  $k \leftarrow 1$;
  Do
    $S_2 \leftarrow$ **PERTURB&IP** (S_1) /* Random neighbor*/
    $S_3 \leftarrow$ **LS&IP**(S_2)
    If $(S_3 > S)$ then $S \leftarrow S_3$; $k \leftarrow 1$;
    else $k \leftarrow k+1$;
  While $(k=k_{max})$
  $t \leftarrow CPU\_time()$;
  return S ;

In the sequel, we detail the construction heuristic RBH, the calculation of the upper bound for IP that condition the application of IP after each local search move.

3.1   Initial Feasible Solution

To generate the initial solution, we proposed a construction heuristic that we call RBH. For illustration, we considered the MCKS problem and explain below the three successive phases of our RBH:

- First phase: We reduced the MCKS so that every class contains a single object ($n_i = 1, i \in \{1,\ldots,N\}$). This object is characterized by a weight a'_i and a profit c'_{it} with $a'_i = \sum_{j=1}^{n_i} a_{ij}$ and $c'_{it} = \sum_{j=1}^{n_i} c_{ijt} i \in \{1,\ldots,N\}; t \in \{1,\ldots,T\}$. Consequently, the reduced MCKS ($MCKS_{red}$) can be expressed mathematically as follows:

$$\text{Max } z' = \sum_{t=1}^{T} \sum_{i=1}^{N} \left(c'_{it}x_{it} + f_{it}y_{it}\right) \tag{6}$$

s.c.

$$\sum_{t=1}^{T} \sum_{i=1}^{N} (a'_i x_{it} + d_i y_{it}) \leq b; i \in \{1,\ldots,N\}; t \in \{1,\ldots,T\} \tag{7}$$

$$0 \leq x_{it} \leq y_{it}; \forall i \in \{1,\ldots,N\}, \forall j \in \{1,\ldots,n_i\}, \forall t \in \{1,\ldots,T\} \tag{8}$$

$$\sum_{t=1}^{T} x_{it} \leq 1; i \in \{1,\ldots,N\}; t \in \{1,\ldots,T\} \tag{9}$$

$$y_{it} \in \{0,1\}; i \in \{1,\ldots,N\}; t \in \{1,\ldots,T\} \tag{10}$$

- Second phase: we relaxed constraint (8) so that $0 \leq x_{it} \leq 1$ and $y_{it} \in \{0,1\}$. The relaxed model of $MCKS_{red}$ is solved using IP, which gives the values of y_{it}. We constructed the set of classes $Y_t = Y_t^1 \cup Y_t^0$, with $Y_t^1 = \{i \in \{1,\ldots,N\}/y_{it} = 1\}$ and $Y_t^0 = \{i \in \{1,\ldots,N\}/y_{it} = 0\}$.

- Third phase: We considered the following IP for the MCKS[Y] as a KP problem:

$$\text{IP} : \text{Max } Z_t = \sum_{t=1}^{T} \sum_{i \in Y_t^1} \sum_{j=1}^{n_i} c_{ijt} x_{ijt} + \theta_t \tag{11}$$

s.c.

$$\sum_{t=1}^{T} \sum_{i \in Y_t^1} \sum_{j=1}^{ni} a_{ij} x_{ijt} \le b - \gamma \tag{12}$$

$$x_{ijt} \in \{0, 1\}; \forall i \in Y_t^1; j \in \{1, \dots, n_i\} \tag{13}$$

Where $\theta_t = \sum_{i \in Y_t^1} f_{it}$, and $\gamma = \sum_{i \in Y_t^1} d_i \forall t \in \{1, \dots, T\}$.

We solved the $MCKS[Y]$ problems and noted also IP using CPLEX solver. The MCKS solution is represented by set of variables $Y = \{y_{it}, i = 1, \dots, N; t = 1, \dots, T\}$, and a set of variables $X = \{x_{ijt}, i = 1, \dots, N; j = 1, \dots, n_i; t = 1, \dots, T\}$.

In addition to RBH, we considered two other heuristics: Linear Programming based Heuristic (LPH) [15, 35] and Greedy Heuristic (GH) [1, 30]. In our problem the LPH heuristic is composed of two main phases: In the first phase, the relaxation of the MCKS (binary y_{it} and continues variables x_{ijt}) is solved to determine the variables y_{it}. In the second phase, the reduced MCKS is solved by using CPLEX solver to determine the variables x_{ijt}. The GH heuristic is to build iteratively a feasible solution. In our problem this heuristic is composed of two main phases. In the first phase, the variables y_{it} are fixed randomly. In the second phase, the partial feasible solution obtained in the previous phase is completed by inserting the items one by one until saturation of the knapsack from the set of items that are listed in the decreasing order of their ratio $r_{ijt} = c_{ijt}/a_{ij}$.

3.2 Upper Bound for IP

Dantzig [9] provided an upper bound for KP. We adapted this upper bound to our problem and provided a new upper bound for each division t of MCKS. This upper bound was used to decide whether to apply IP or not after the local search in order to explore only fruit full search spaces. We applied the following successive steps to obtain this upper bound:

- **Step1**: Let I denote the set of items of classes $i \in Y_t^1$ sorted in descending order of their efficiency ratio $r_{ijt} = \frac{c_{ijt}}{a_{ij}} \forall i \in Y_t^1; \forall j \in \{1, \dots, n_i\}$.
- **Step2**: Assign items from I one by one until saturation of the knapsack, i.e., Stop at item $i'j'$ that cannot be inserted due to capacity saturation of $MCKS[Y]$.
- **Step3**: The upper bound of division t is:

$$UB_t = \sum_{i \in Y_t^1} f_{it} + \sum_{i,j \in I'} c_{ijt} + \frac{b-C}{a_{i'j'}} c_{i'j't}, \text{ where } C = \sum_{t=1}^{T} \left(\sum_{i \in Y_t^1} d_i + \sum_{i,j \in I'} a_{ij} \right) \text{ with } I' \text{ the set}$$

of assigned items, and (b − C) the residual capacity for division t.

3.3 Local Search with IP

In the local search phase, two neighborhood structures, SWAP&IP and INSERT&IP operators are employed with in the LS&IP framework.

LS&IP (data, S_{best})

    Input: Instance data & best solution found
    Output: A feasible solution S''
    /* S_{best}: best solution found by **RBH** (first iteration) or by **perturb&IP** */
    Do
            $improve \leftarrow 0$;
            $S1 \leftarrow SWAP\&IP\ (S_{best})$
            $S2 \leftarrow INSERT\&IP\ (S1)$
            If $(f(S2) > f(S_{best}))$
                $S_{best} \leftarrow S2$; $improve \leftarrow 1$;
            EndIf
    While (*improve* == **1**)
    Return S_{best}

SWAP&IP. A Swap-based local search requires the definition of a neighborhood structure using simple moves so as to produce a set of neighbor solutions which permits to explore more search spaces and thus provide high quality solutions. The considered swap process consists of permuting two variables $y_{it} \in Y_t^1$ and $y_{jk} \in Y_k^1 (i, j \in \{1, \ldots, N\}; t \in \{1, \ldots, T\}, k \in \{t+1, \ldots, T+1\}$. Where $T + 1$ is a fictive knapsack that contains all the nonselected classes. We changed the value of setup variables from 1 to 0 and vice versa. A new $MCKS[Y]$ was obtained. In order to save computational effort, before applying IP, we calculated the sum of upper bounds of the new divisions t and k ($UB_{t,k} = UB_t + UB_k$) and compared it with the total profit of the two divisions before Swap move ($Z_{t,k} = Z_t + Z_k$). In case $UB_{t,k} > Z_{t,k}$. We applied IP to optimally solve the new classical knapsack $MCKS[Y]$ and the best solution was taken as a new initial solution for a next swap process. In case $UB_{t,k} \leq Z_{t,k}$, the search space was not promising as no better solution could be obtained, thus IP(t, k) was not applied and we proceeded to the next step. The procedure is terminated once no improvement is obtained. Algorithm 3 details the *SWAP&IP* procedure.

SWAP&IP (data, S)

Input: Instance data & initial solution
Output: A feasible solution
   do
      Improve ← 0;
        For t ← 1 *to* T **do**
            n ← $card(Y_t^1)$; /* Number of classes in division t */
            For x ← 1 *to* n **do**
               i ← $Y_t^1[x]$;
               For k ← $t + 1$ *to* $T + 1$ **do**
                  m ← $card(Y_k^1)$; /* Number of classes in division k */
                  For j ← 1 *to* m **do** /* Swap class i by each class j */
                     j ← $Y_k^1[x]$;
                     y_{it} ← 0; y_{jk} ← 0; y_{jt} ← 1; y_{ik} ← 1;
                     If $\left(UB_{t,k} > Z_{t,k}\right)$ **then**
                        $Z'_{t,k}$ ← $IP(t,k)$;
                        If $\left(Z'_{t,k} > Z_{t,k}\right)$ **then**
                           Store the best *S';Improve* ← 1;
                        EndIf
                     EndIf
                   EndFor
                 EndFor
               S ← S'; /* New starting solution */
            EndFor
        EndFor
      While (improve==1)
      Return S

INSERT&IP. The Insert-based local search is based on a neighborhood search which generates a new solution by removing the class i from knapsack t (change the value of the setup variable $y_{it} \in Y_t^1$ from 1 to 0) and then inserting it into another knapsack k, $k \in \{1, \ldots, T+1\}$, The IP(t, k) is applied if $(UB_{t,k} > Z_{t,k})$ by the same way than in the *SWAP&IP* procedure. The best solution is taken as a new initial solution for the next insert-based local search. The procedure is terminated once no improvement is obtained. Algorithm 4 details the *INSERT&IP* procedure.

INSERT&IP(data, S)

Input: Instance data & initial solution
Output: A feasible solution
  do
      $Improve \leftarrow 0$;
      For $i \leftarrow 1$ *to* N **do**
         $kp \leftarrow \{t \in 1, ..., T \ / \ y_{it} = 1\}$;
         $n \leftarrow card(kp)$; /* Number of divisons that contain class i */
         For $x \leftarrow 1$ *to* n **do**
            $t \leftarrow kp[x]$;
            For $k \leftarrow 1$ *to* $T + 1$ **do**
               If ($y_{ik} = 0$) **then**
                  $y_{it} \leftarrow 0$; $y_{ik} \leftarrow 1$; /* Insert i in k and delete it from t */
                  If $\left(UB_{t,k} > Z_{t,k}\right)$ **then**
                     $Z'_{t,k} \leftarrow IP(t) + IP(k)$;
                     If $(Z'_{t,k} > Z_{t,k})$ **then**
                        Store the best S'; $Improve \leftarrow 1$;
                     EndIf
                  EndIf
               EndIf
            EndFor
            $S \leftarrow S'$; /* New starting solution */
         EndFor
      EndFor
    While ($improve == 1$)
    Return S

PERTURB&IP. The design of the perturbation mechanism is crucial for the performance of the algorithm. If the mechanism provides too small perturbation, local search may return to the previously visited local optimum points and no further improvement can be obtained. The mechanism consists of strongly perturbing a part of the current solution to jump the local optima and obtain a new starting solution. Two phases were applied iteratively in order to simulate this jumping principle: The first is a select of k randomly chosen items (setup variables y_{it}) and the second is the IP which is applied to solve the classical knapsacks $MCKS[Y]$. The resulting solution is accepted according to the following condition if $(f(s') > \eta f(s))$, where that is constant value between 0 and 1. The perturbation method was terminated when the total number of applied moves (perturbation length) equals to the p_max. Algorithm 5 provides a description of the new local search method.

Perturb&IP *(data, S)*
Input: Instance data & best solution found S
Output: A randomly feasible solution
$p \leftarrow 1;$
k←Number of selected classes in best solution S
Do
Select a random set of k classes , from N
Randomly assigned the y_{it}variables
Apply IP to fix the x_{ijt} variables
If $(f(s') > \eta \, f(s))$ then
Store the best S' /* best solution found */
$p \leftarrow 1;$
Else
p←p+1;
End if
while $p \leq$ p_max
return S'

4   Computational Results

For computation, our approach was implemented and run using C language and CPLEX 12.7 solver on a 2.4 GHZ intel B960 computer with 4 GB of memory. Due to the unavailability of benchmark instances in the literature, we tested our cooperative approach *VNS&IP* on a set of randomly generated instances of MCKS with a total number of periods T in $\{5, 10, 15, 20\}$, total number of classes N in $\{10, 20, 30\}$, and total number of items n_i for each class i in $[90, 110]$ (Available at https://goo.gl/4fz6fg). We generated *120* instances in total: *10* instances for each combination (T, N). We designed a random generation scheme, as presented in [2], where:

- a_{ij} is selected with a uniform distribution in $[10, 10000]$.
- $c_{ijt} = a_{ij} + e_1$, e_1 is selected with a uniform distribution in $[0, 10]$.
- $b = 0, 5 * \sum_{i=1}^{N} \sum_{j=1}^{n_i} a_{ij}$.
- $d_i = \sum_{j=1}^{n_i} a_{ij} * e$.
- $f_{it} = - \sum_{j=1}^{n_i} c_{ijt} * e$, e is selected with a uniform distribution in $[0.15, 0.25]$.

The Gap report the standard deviation between IP and VNS&IP that is calculated as follows: $\mathrm{Gap}(\%) = 100 * \left(\frac{\mathrm{IP_{sol}} - \mathrm{VNS} \, \& \, IP_{sol}}{\mathrm{IP_{sol}}} \right)$.

4.1   Parameter Setting

Generally, when using approximate algorithms to solve optimization problems, it is well known that different parameter settings for the approach lead to different quality results. The parameters for VNS&IP are as follows: time_max, the maximal time measured in seconds and its fixed to T, where T is the number of periods (divisions). k_{max}, the maximum number of consecutive failed iterations is fixed to N, where N is the number of classes. The perturbation length p_max is fixed to T. that is constant value between 0 and 1 to relax the acceptance condition is fixed to 0.8. It is worth pointing out that a different adjustment of method's parameters would give important findings. But this better adjustment would sometimes lead to heavier execution time requirements. The set of values chosen in our experiment represents a satisfactory trade-off between quality solution and running time.

4.2   Computational Results

Before the experimentation, the effect on performance of the main components of our algorithm is assessed, mainly the construction Heuristic RBH and the combination of the two local search techniques *LS&IP* and *PERTURB&IP*.

In order to evaluate the performance of RBH, we compared it to HG and LPH heuristics explained in Sect. 3.1. The RBH, HG and LPH heuristics are tested on all the instances of MCKS. Table 1 shows the numerical results on average. The first column contains the name of the heuristic. The second column contains the average of computational time. We noted that LPH is stopped at a limit of computation time equal to 500 s. The third column contains the gap between the heuristic solution and the IP solution: $Gap(\%) = 100 * \left(\frac{CPLEX_{sol} - Heuristic_{sol}}{CPLEX_{sol}} \right)$.

Table 1. Comparison between RBH, HG and LPH: average of MCKS instances.

Heuristic	CPU (s)	Gap (%)
RBH	**0.63**	**1.46%**
LPH	304	5.34%
GH	0.51	8.13%

Table 1 shows that RBH outperforms the other construction heuristics in terms of computation time and quality solution.

It is important to give information about the impact of the LS&IP and PERTURB&IP on the performance of VNS&IP. We consider the application of our cooperative approach with RBH, RBH + LS&IP and RBH + LS&IP + PERTURB&IP (VNS&IP). Table 2 shows a comparison between these three combinations in terms of average Gap (%) with the IP for the four set instances regarding the number of periods (divisions). Each line presents the average of 10 instances. The first two columns present the number of divisions (or periods) T and the number of classes N. The next three columns show the corresponding average gap between **RBH** and IP, the average gap between **RBH + LS&IP** and IP, and the average gap between **RBH + LS&IP + PERTURB&IP** (VNS&IP) and IP.

$$\text{Gap}(\%) = 100 * \left(\frac{\text{IP}_{\text{sol}} - \text{Heuristic}_{\text{sol}}}{\text{IP}_{\text{sol}}} \right).$$

Table 2. Effect of VNS&IP components

Instances		RBH	RBH + LS&IP	RBH + LS&IP + PERTURB&IP
T	N			
5	30	0.95	0.23	−0.054
10		1.23	0.39	−0.012
15		1.82	0.45	−0.125
20		1.99	0.42	−0.155

Table 2 shows that by adding *LS&IP*, we observe an important advantage, for all the set of instances, compared to using only *RBH*. However, by adding *PERTURB&IP*, we observe a higher improvement with a gap that increases when the number of knapsacks increases. For the experimentations below, we considered the best combination with RBH as construction heuristic, *LS&IP* as local search techniques and *PERTURB&IP* as perturbation mechanism.

Table 3 summarizes the results obtained by *VNS&IP* and IP when solving the MCKS. Each line presents the average of *10* instances. The first two columns present the number of divisions (or periods) T and the number of classes N. The next three columns show the corresponding average of results provided by CPLEX, the average of results provided by the cooperative approach *VNS&IP* and the average of the best upper bounds, of all the remaining open nodes in the branch-and-cut tree, provided by CPLEX 12.7 ($CPLEX_{UB}$). The notations sol and CPU report the solution found and the computational time, respectively. We note that CPLEX is stopped at a limit of computation time equal to 1 h. Finally, the columns Gap_{IP} and Gap_{UB} report the gap between CPLEX and V*NS&IP*, calculated as follows: $Gap_{IP}(\%) = 100 * \left(\frac{\text{CPLEX}_{\text{sol}} - \text{VNS}_{\&IP\text{sol}}}{\text{CPLEX}_{\text{sol}}} \right)$, and the gap between CPLEX$_{UB}$ and *VNS&IP*, calculated as follows: $Gap_{UB}(\%) = 100 * \left(\frac{\text{CPLEX}_{\text{UB}} - \text{VNS}_{\&IP\text{sol}}}{\text{CPLEX}_{\text{UB}}} \right)$, respectively.

Table 3 shows that *VNS&IP* outperforms IP with a gap on average equal to −0.11%. In detail, the gap on average is about −0.06% for $T = 5$, −0.005% for $T = 10$, −0.18% for $T = 15$, and −0.18% for $T = 20$. The CPU on average for *VNS&IP* is about 13 s, which is very low in comparison to the average of CPU for CPLEX that is equal to 2868 s. For more detailed results, we note that *VNS&IP* provides a solution equal to the one provided by CPLEX for *51* instances and provides better solutions than CPLEX for 65 instances (available at https://goo.gl/w44aUs). Table 2 shows that the gap between *VNS&IP* and $CPLEX_{UB}$ is 0.001% on average.

Table 3. Numerical results for MCKS instances.

T	N	CPLEX		VNS&IP			UB	
		CPLEX$_{obj}$	CPU	VNS&IP$_{sol}$	CPU	Gap$_{IP}$ (%)	CPLEX$_{UB}$	Gap$_{UB}$ (%)
5	10	1772249	1735	1773409	6	−0.066	1773431	0.001
	20	3571514	2863	3573719	6	−0.063	3573771	0.001
	30	5398429	2267	5401333	6	−0.054	5401369	0.001
10	10	1795187	2587	1795188	11	0.000	1795221	0.002
	20	3602956	3439	3603067	10	−0.003	3603090	0.001
	30	5445060	2937	5445715	11	−0.012	5445752	0.001
15	10	1793209	2819	1795262	15	−0.118	1795311	0.003
	20	3605797	3333	3617045	15	−0.315	3617079	0.001
	30	5471310	3255	5478013	15	−0.125	5478052	0.001
20	10	1793091	2745	1796768	20	−0.208	1796796	0.002
	20	3609105	3481	3615497	20	−0.180	3615547	0.001
	30	5454676	2961	5463066	20	−0.155	5463115	0.001

Among the 120 instances of MCKS, CPLEX finds the optimal solutions for 27 instances, slightly outperforms the *VNS&IP* for 4 instances, and for the remaining it terminates with error: exceeds the capacity of RAM memory or exceeds the CPU time limit. the majority of instances solved at optimality are with $T = 5$ (*12* with $T = 5$, *8* with $T = 10$, *2* with $T = 15$ and *5* with $T = 20$). In addition, we can see that MCKS becomes more difficult when increasing the number of divisions T. In fact, the number of times that CPLEX terminates with exceeding the capacity of RAM or exceeding the time limit increases from *18* with $T = 5$ to *25* with $T = 20$.

5 Conclusion

In this paper, we consider the multiple choice knapsack problem with setup (MCKS). This problem can be used to model a wide range of concrete industrial problems, including order acceptance and production scheduling. We proposed a new cooperative approach that combines VNS and IP for the MCKS. Our cooperative approach denoted VNS&IP is tested on a wide set of instances that are generated for MCKS. The results showed that CPLEX was able to optimally solve only 22.5% of these problems; the rest had unknown optimal values. The experimental results showed that VNS&IP produced good quality (optimal and near-optimal solutions) solutions in a short amount of time and allowed for the enhancement of the solution provided by CPLEX in 65 instances. Considering the promising performance of the VNS&IP method presented in this work, further studies, some of which are currently underway in our laboratory, are needed to further extend the use of the space reduction technique to other general and critical problems.

References

1. Akcay, Y., Li, H., Xu, S.H.: Greedy algorithm for the general multidimensional knapsack problem. Ann. Oper. Res. **150**, 7–29 (2007)
2. Akinc, U.: Approximate and exact algorithms for the fixed-charge knapsack problem. Eur. J. Oper. Res. **170**, 363–375 (2006)
3. AlMaliky, F., Hifi, M., Mhalla, H.: Sensitivity analysis of the setup knapsack problem to perturbation of arbitrary profits or weights. Int. Trans. Oper. Res. **25**, 637–666 (2018)
4. Bansal, J.C., Deep, K.: A modified binary particle swarm optimization for knapsack problems. Appl. Math. Comput. **218**, 11042–11061 (2012)
5. Billionnet, A., Soutif, E.: An exact method based on Lagrangian decomposition for the 0-1 quadratic knapsack problem. Eur. J. Oper. Res. **157**, 565–575 (2004)
6. Burke, E.K., Li, J., Qu, R.: A hybrid model of integer programming and variable neighborhood search for highly-constrained nurse rostering problems. Eur. J. Oper. Res. **2003**, 484–493 (2010)
7. Chebil, K., Khemakhem, M.: A dynamic programming algorithm for the knapsack problem with setup. Comput. Oper. Res. **64**, 40–50 (2015)
8. Chu, P., Beasley, J.: A genetic algorithm for the multidimensional knapsack problem. J. Heuristics **4**, 63–86 (1998)
9. Dantzig, G.B.: Discrete variable extremum problems. Oper. Res. **5**, 266–277 (1957)
10. Dudziński, K., Walukiewicz, S.: Exact methods for the knapsack problem and its generalizations. Eur. J. Oper. Res. **28**, 3–21 (1987)
11. Dumitrescu, I., Stutzle, T.: Combinations of local search and exact algorithms. Appl. Evol. Comput. **2611**, 211–223 (2003)
12. Della, C.F., Salassa, F., Scatamacchia, R.: An exact approach for the 0-1 knapsack problem with setups. Comput. Oper. Res. **80**, 61–67 (2017)
13. Fernandes, S., Lourenco, H.: Hybrid combining local search heuristics with exact algorithms. In: Algoritmos Evolutivos y Bioinspirados, Spain, pp. 269–274 (2007)
14. Freville, A., Plateau, G.: Heuristics and reduction methods for multiple constraints 0-1 linear programming problems. Eur. J. Oper. Res. **24**, 206–215 (1986)
15. Haddar, B., Khemakhem, M., Rhimi, H., Chabchoub, H.: A quantum particle swarm optimization for the 0-1 generalized knapsack sharing problem. Nat. Comput. **15**, 153–164 (2016)
16. Hanafi, S., Fréville, A.: An efficient tabu search approach for the 0-1 multidimensional knapsack problem. Eur. J. Oper. Res. **106**, 659–675 (1998)
17. Hifi, M., Michrafy, M., Sbihi, A.: A reactive local search-based algorithm for the multiple-choice multidimensional knapasck problem. Comput. Optim. Appl. **33**, 271–285 (2006)
18. Jourdan, L., Basseur, M., Talbi, E.G.: Hybridizing exact methods and metaheuristics. Eur. J. Oper. Res. **199**, 620–629 (2009)
19. Khemakhem, M., Chebil, K.: A tree search based combination heuristic for the knapsack problem with setup. Comput. Ind. Eng. **99**, 280–286 (2016)
20. Kolesar, P.J.: A branch and bound algorithm for the knapsack problem. Manag. Sci. **13**, 723–735 (1967)
21. Lamghari, A., Dimitrakopoulos, R., Ferland, J.A.: A hybrid method based on linear programming and variable neighborhood descent for scheduling production in open-pit mines. J. Global Optim. **63**, 555–582 (2015)
22. Martello, S., Toth, P.: Solution of the zero-one multiple knapsack problem. Eur. J. Oper. Res. **4**, 276–283 (1980)

23. Martello, S., Toth, P.: Knapsack Problems: Algorithms and Computer Implementations. Wiley, New York (1990)
24. Michel, S., Perrot, N., Vanderbeck, F.: Knapsack problems with setups. Eur. J. Oper. **196**, 909–918 (2009)
25. Penna, P.H., Subramanian, A., Ochi, L.S.: An iterated local search heuristic for the heterogeneous fleet vehicle routing problem. J. Heuristics **19**, 201–232 (2013)
26. Pferschy, U., Rosario, S.: Improved dynamic programming and approximation results for the knapsack problem with setups. Int. Trans. Oper. Res. **25**, 667–682 (2018)
27. Pisinger, D.: An exact algorithm for large multiple knapsack problems. Eur. J. Oper. Res. **114**, 528–541 (1999)
28. Prandtstetter, M., Raidl, G.R.: An integer linear programming approach and a hybrid variable neighborhood search for the car sequencing problem. Eur. J. Oper. Res. **191**, 1004–1022 (2008)
29. Puchinger, J., Raidl, G.R.: Combining metaheuristics and exact algorithms in combinatorial optimization: a survey and classification. In: Mira, J., Álvarez, J.R. (eds.) IWINAC 2005. LNCS, vol. 3562, pp. 41–53. Springer, Heidelberg (2005). https://doi.org/10.1007/11499305_5
30. Richard, L., Eleftherios, M.: New greedy-like heuristics for the multidimensional 0-1 knapsack problem. Oper. Res. **27**, 1101–1114 (1979)
31. Sinha, A., Zoltners, A.A.: The multiple-choice knapsack problem. Oper. Res. **27**, 503–515 (1979)
32. Tlili, T., Yahyaoui, H., Krichen, S.: An iterated variable neighborhood descent hyper-heuristic for the quadratic multiple knapsack problem. Softw. Eng. Artif. Intell. **612**, 245–251 (2016)
33. Vasquez, M., Hao, J.K.: A hybrid approach for the 0-1 multidimensional knapsack problem. In: Proceedings of the International Joint Conference on Artificial Intelligence, Washington, pp. 328–333 (2001)
34. Yang, Y.: Knapsack problems with setup. Dissertation, Auburn University (2006)
35. Zhang, C.W., Ong, H.L.: Solving the biobjective zero-one knapsack problem by an efficient LP-based heuristic. Eur. J. Oper. Res. **159**, 545–557 (2004)
36. Maldenovic, N., Hansen, P.: Variable neighborhood search. Comput. Oper. Res. **24**, 1097–1100 (1997)
37. Jarboui, B., Derbel, H., Hanafi, S., Maldenovic, N.: Variable neighborhood search for location routing. Comput. Oper. Res. **40**, 47–57 (2013)

A VNS-Based Algorithm with Adaptive Local Search for Solving the Multi-Depot Vehicle Routing Problem

Sinaide Nunes Bezerra[1], Marcone Jamilson Freitas Souza[2],
Sérgio Ricardo de Souza[1(✉)], and Vitor Nazário Coelho[3]

[1] Federal Center of Technological Education of Minas Gerais (CEFET-MG),
Av. Amazonas 7675, Nova Gameleira, Belo Horizonte, MG 30510-000, Brazil
sinaide@hotmail.com, sergio@dppg.cefetmg.br
[2] Federal University of Ouro Preto (UFOP),
Campus Universitário, Morro do Cruzeiro, Ouro Preto, MG 35400-000, Brazil
marcone@ufop.edu.br
[3] Fluminense Federal University (UFF), Institute of Computing, Niterói, RJ, Brazil
vncoelho@gmail.com

Abstract. The Multi-Depot Vehicle Routing Problem (MDVRP) is a variant of the Vehicle Routing Problem (VRP) that consists in designing a set of vehicle routes to serve all customers, such that the maximum number of vehicle per depot, the vehicle capacity and the maximum time for each route are respected. The objective is to minimize the total cost of transportation. This paper presents an algorithm, named VNSALS, based on the Variable Neighborhood Search (VNS) with Adaptive Local Search (ALS) for solving it. The main procedures of VNSALS are perturbation, ALS and cluster refinement. The perturbation procedure of VNS is important to diversify the solutions and avoid getting stuck in local optima. The ALS procedure consists in memorizing the results found after applying a local search and in using this memory to select the most promising neighborhood for the next local search application. The choice of the neighborhood is very important to improve the solution in heuristic methods because the complexity of the local search is high and expensive. On the other hand, customer's reallocation keeps the clusters more balanced. VNSALS is tested in classical instances of MDVRP for evaluating its efficiency and the results are presented and discussed.

Keywords: Multi-Depot Vehicle Routing Problem ·
Adaptive Local Search · Variable Neighborhood Search

1 Introduction

Vehicle routing problem (VRP) is a classical optimization problem with several variants. Researchers are dedicated to studying it all over the world, applying the most diverse techniques for its solution, as the literature points out. In a

A. Sifaleras et al. (Eds.): ICVNS 2018, LNCS 11328, pp. 167–181, 2019.
https://doi.org/10.1007/978-3-030-15843-9_14

solution of VRP, each vehicle leaves the depot and executes a route over a certain
number of customers and returns to the depot, insuring that the total demand
on the route does not exceed the vehicle capacity. In some cases, a maximum
route duration or distance constraint is enforced and the problem may involve a
homogeneous or heterogeneous fleet [6,11,16,17].

This work has its focus on the Multi-Depot Vehicle Routing Problem
(MDVRP) with homogeneous fleet. MDVRP is a variant of the classical VRP in
which there is more than one depot. MDVRP is solved in [6] with Tabu Search. In
[12], the authors apply an algorithm based on the Adaptive Large Neighborhood
Search (ALNS). Genetic Algorithm (GA) is used in [17] and a hybrid algorithm
based on Iterated Local Search (ILS) is applied in [16]. An exact method is
proposed in [5]. Other hybrid metaheuristic algorithms combining Greedy Randomized Adaptive Search Procedure (GRASP), ILS and Simulated Annealing
(SA) are proposed in [1]. In [11], the authors present a parallel coevolutionary
algorithm based in evolution strategy and in [4] an algorithm based on the General VNS [8] is proposed. A recent survey of exact and heuristic methods for
solving MDVRP can be found in [10].

In this current paper, we proposes a new algorithm, inspired on the Adaptive Guided Variable Neighborhood Search algorithm from [2] and similar to
the General VNS algorithm of [4], for solving MDVRP. The new algorithm uses
an adaptive local search method rather than a local search based on the Randomized Variable Neighborhood Descent (RVND) method, as used in [4]. The
idea behind VNS is that switching the neighborhood structure after the current
neighborhood structure trapped in local optima may help VNS to escape from
the local optima. Thus, applying different neighborhood structures can generate
different search trajectories, which help in escaping from the current point as well
as dealing with problems related to landscape changes that usually occur during the solving process. However, the sequence of the neighborhood structures
in VNS has a critical impact on the algorithm performance, which is usually
dependent on the problem and/or its instances. This implies that not only different instances require different sequences of the neighborhood structures but
also different stages of the solving process [2]. In this scenario, we create a simply
ranking to classify the neighborhoods that will be chosen for refining the current
solution.

The remaining of this paper is organized as follows. Section 2 describes
MDVRP. Section 3 presents the neighborhoods used for exploring the solution
space of MDVRP. Section 4 introduces the proposed algorithm, named Variable
Neighborhood Search with Adaptive Local Search (VNSALS). The calibration of
the VNSALS parameters are described in Sect. 5, followed by the experimental
results that are shown and discussed in Sect. 6. Finally, Sect. 7 concludes this
work.

2   Multi-Depot Vehicle Routing Problem

The Multi-Depot Vehicle Routing Problem (MDVRP) consists in determining a
set of vehicle routes such that [10]:

(i) each vehicle route starts and ends at the same depot;
(ii) each customer is serviced exactly once by a vehicle;
(iii) the total demand of each route does not exceed the vehicle capacity; and
(iv) the total cost of the distribution is minimized.

Figure 1 represents a MDVRP problem with thirteen customers, in the form $\mathcal{V}^{CST} = \{1, 2, \ldots, 13\}$ and two depots, named $\mathcal{V}^{DEP} = \{14, 15\}$. In this representation, depot 14 has two routes that serves customers 2, 3, 4, 5, 9 and 12; depot 15 has three routes, serving customers 1, 6, 7, 8, 10, 11 and 13.

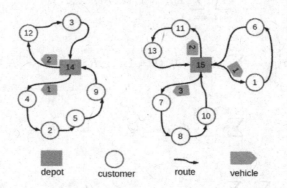

Fig. 1. Example of MDVRP with thirteen customers and two depots.

MDVRP is defined as follows [10]. Let $\mathcal{G} = (\mathcal{V}, \mathcal{A})$ be a complete graph, where $\mathcal{V}$ is a set of nodes and $\mathcal{A}$ is a set of arcs. The set of nodes are partitioned into two subsets: the set of customers to be served, given by $\mathcal{V}^{CST} = \{1, 2, \ldots, N\}$, and the set of depots $\mathcal{V}^{DEP} = \{N + 1, N + 2, \ldots, N + M\}$, with $\mathcal{V} = \mathcal{V}^{DEP} \cup \mathcal{V}^{CST}$ and $\mathcal{V}^{DEP} \cap \mathcal{V}^{CST} = \varnothing$. There is a non-negative cost c_{ij} associated with each arc $(i, j) \in \mathcal{A}$. For each customer, there is a non-negative demand d_i and there is no demand at the depot nodes. In each depot, there are a fleet of K identical vehicles, each with capacity Q. The service time at each customer i is t_i, while the maximum route duration time is set to T. A conversion factor w_{ij} might be needed to transform the cost c_{ij} into time units. In this work, however, the cost is the same as the time and distance units, so $w_{ij} = 1$.

MDVRP consists in designing a set of vehicle routes serving all customers, such that the maximum number of vehicle per depot, vehicle-capacity and maximum duration time in the route are respected, and the total cost of transportation is minimized. The MDVRP mathematical formulation requires the definition of the binary decision variable x_{ijk}, which is equal to 1 when vehicle k visits node j immediately after node i, and 0 otherwise. Auxiliary binary variables y_i are also used in the subtour elimination constraints [10,11]. The mathematical model is given by:

$$\min \sum_{i=1}^{N+M} \sum_{j=1}^{N+M} \sum_{k=1}^{K} c_{ij} x_{ijk} \tag{1}$$

Subject to

$$\sum_{i=1}^{N+M} \sum_{k=1}^{K} x_{ijk} = 1 \ (j = 1, \ldots, N) \tag{2}$$

$$\sum_{j=1}^{N+M} \sum_{k=1}^{K} x_{ijk} = 1 \ (i = 1, \ldots, N) \tag{3}$$

$$\sum_{i=1}^{N+M} x_{ihk} - \sum_{j=1}^{N+M} x_{hjk} = 0 \ (k = 1, \ldots, K; \ h = 1, \ldots, N+M) \tag{4}$$

$$\sum_{i=1}^{N+M} \sum_{j=1}^{N+M} d_i x_{ijk} \leq Q \ (k = 1, \ldots, K) \tag{5}$$

$$\sum_{i=1}^{N+M} \sum_{j=1}^{N+M} (c_{ij} w_{ij} + t_i) x_{ijk} \leq T \ (k = 1, \ldots, K) \tag{6}$$

$$\sum_{i=N+1}^{N+M} \sum_{j=1}^{N} x_{ijk} \leq 1 \ (k = 1, \ldots, K) \tag{7}$$

$$\sum_{j=N+1}^{N+M} \sum_{i=1}^{N} x_{ijk} \leq 1 \ (k = 1, \ldots, K) \tag{8}$$

$$y_i - y_j + (N+M) x_{ijk} \leq N + M - 1$$
$$\text{for } 1 \leq i \neq j \leq N \text{ and } 1 \leq k \leq K \tag{9}$$

$$x_{ijk} \in \{0, 1\} \ \forall \ i, j, k \tag{10}$$
$$y_i \in \{0, 1\} \quad \forall \ i \tag{11}$$

In this formulation, the objective, shown in Expression 1, is to minimize the total cost. Constraints (2) and (3) guarantee that each customer is served by exactly one vehicle. Flow conservation is guaranteed through constraints (4). Constraints (5) and (6) refer to the vehicle capacity and the total route cost, respectively. Vehicle availability is verified by constraints (7) and (8). Subtour elimination is provided by constraints (9). Finally, constraints (10) and (11) define x and y as binary variables.

3 Neighborhoods for MDVRP

We used seven neighborhoods widely applied in the literature to explore the solution space of this problem: Swap(1, 1), Swap(2, 1), Shift(1, 0), Shift(2, 0), 2-Opt, 2-Opt* and Reverse [15,17,19]. These neighborhoods are described in the sequel.

Swap(1, 1) - permutation between a customer v_j from a route r_k and a customer v_t from a route r_l. In Fig. 2(a), the clients 1 and 13 were swapped in the same depot. In Fig. 2(b), the client 13 from depot 14 and client 13 from depot 15 are swapped.

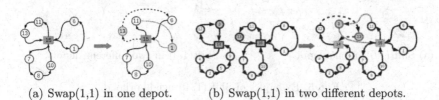

(a) Swap(1,1) in one depot.   (b) Swap(1,1) in two different depots.

Fig. 2. Examples of neighborhood Swap(1, 1).

Swap(2, 1) - permutation of two adjacent customers v_j and v_{j+1} from a route r_k by a customer v_t from a route r_l. In Fig. 3(a), the adjacent clients 4 and 2 were exchanged with client 12. In Fig. 3(b), clients 12 and 3 from depot 15 were exchanged with client 13 belongs to the depot 14.

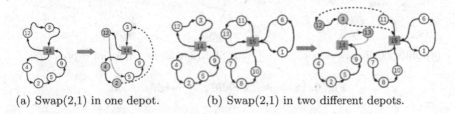

(a) Swap(2,1) in one depot.   (b) Swap(2,1) in two different depots.

Fig. 3. Examples of Swap(2, 1) neighborhood.

Swap(2, 2) - permutation between two adjacent customers v_j and v_{j+1} from a route r_k by another two adjacent customers v_t and v_{t+1}, $\forall v_t, v_{t+1} \in \mathcal{V}^{CST}$, belonging to a route r_l. In Fig. 4(a), the adjacent clients 5 and 9 were exchanged with client 12 and 3. In Fig. 4(b), clients 5 and 9 from depot 14 were exchanged with client 8 and 10 belonging to the depot 15.

Shift(1, 0) - transference of a customer v_j from a route r_k to a route r_l. In Fig. 5(a), the client 4 was moved from one route to the other one. The same situation occurs in Fig. 5(b) where client 9 was moved to the depot 15.

172   S. Nunes Bezerra et al.

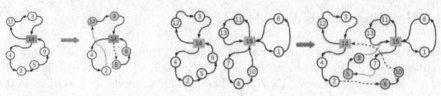

(a) Swap(2,2) in one depot.   (b) Swap(2,2) in two different depots.

Fig. 4. Examples of Swap(2, 2) neighborhood.

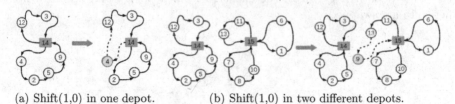

(a) Shift(1,0) in one depot.   (b) Shift(1,0) in two different depots.

Fig. 5. Examples of Shift(1, 0) neighborhood.

Shift(2, 0) - transference of two adjacent customers v_j and v_{j+1} from a route r_k to a route r_l. In Fig. 6(a), the adjacent clients 5 and 9 were moved from one route to the other one. The same occurs in Fig. 6(b), where clients 5 and 9 from depot 14 were moved to the depot 15.

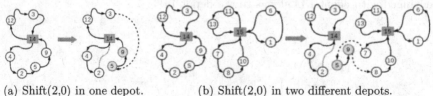

(a) Shift(2,0) in one depot.   (b) Shift(2,0) in two different depots.

Fig. 6. Examples of Shift(2, 0) neighborhood.

2-Opt - Two non-adjacent arcs are deleted and another two ones are added so that a new route is generated. In Fig. 7, the arcs $(4, 2)$ and $(9, 12)$ are deleted, while the arcs $(4, 9)$ and $(2, 12)$ are inserted, changing the sub-route to $(14, 4, 9, 5, 2, 12, 3, 14)$.

The 2-Opt* neighborhood is based on the deletion and reinsertion of two arc pairs from two different routes. This neighborhood is sometimes called crossover neighborhood [18]. For 2-Opt* in two distinct routes r_1 and r_2, let u and v be arcs from r_1 and v and y be arcs from r_2. There are two alternatives, as follows:

- *Alternative 1*: replace (u, x) and (v, y) by (u, v) and (x, y);
- *Alternative 2*: replace (u, x) and (v, y) by (u, y) and (x, v).

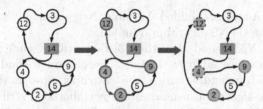

Fig. 7. Example of neighborhood 2-Opt.

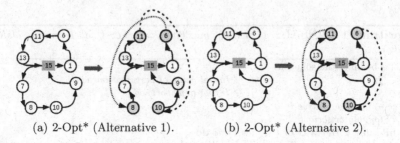

(a) 2-Opt* (Alternative 1).   (b) 2-Opt* (Alternative 2).

Fig. 8. Examples of 2-Opt* neighborhoods.

Figure 8 represents 2-Opt* cases.

Reverse - This move reverses the route direction. It is represented in Fig. 9.

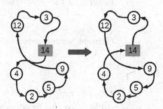

Fig. 9. Example of Reverse neighborhood.

The Swap(1, 1), Swap(2, 1), Shift(1, 0) and Shift(2, 0) neighborhoods may be in route (intra-route), between different routes (inter-route) and occur in the same depot (intra-depot) or between different depots (inter-depots); 2-Opt and Reverse may occur in intra-depot/intra-route; and 2-Opt* may occur in intra-depot/inter-route.

4 Description of the Proposed Algorithm

In this work we proposed an algorithm, named Variable Neighborhood Search with Adaptive Local Search (VNSALS), for solving MDVRP. VNSALS is inspired in the Adaptive Guided Variable Neighborhood Search algorithm from [2] and in the GVNS algorithm from [4].

VNSALS is a VNS-based algorithm [8] that utilizes some problem-specific knowledge and uses an adaptive learning mechanism to find the most suitable neighborhood structure during the searching process. In the adaptive learning part, the algorithm memorizes the neighborhood structure that made an improvement in the solution quality [3]. The pseudo-code of VNSALS is described in Algorithm 1.

Algorithm 1. VNSALS (*iterMax, maxTime, maxLevel, alsTraining, alsReset, Nshake*)

1: Let s be an initial solution;
2: $p \leftarrow 1$;   // Initial value of perturbation level
3: $k \leftarrow 1$;   // Initial neighborhood
4: $iter \leftarrow 0$; $itAdaptive \leftarrow 0$; $s' \leftarrow s$
5: // Stopping criterion
6: **while** $iter < IterMax$ or $t < maxTime$ **do**
7: **if** $itAdaptive < (alsTraining * IterMax)$ **then**
8:   $s'' \leftarrow RandomNeighborhood(s', success, \mathcal{N})$;
9: **else**
10:   $s'' \leftarrow ALS(s', success, \mathcal{N})$;
11: **end if**
12: **if** $f(s'') < f(s)$ **then**
13:   $s \leftarrow s''$;
14:   $k \leftarrow 1$;   // Return to the first neighborhood
15:   $p \leftarrow 1$;   // Return to the first level
16:   $iter \leftarrow 0$;
17: **else**
18:   $p \leftarrow p + 1$;   // Change perturbation level
19:   $iter \leftarrow iter + 1$;
20: **end if**
21: **if** $p > maxLevel$ **then**
22:   $k \leftarrow k + 1$;   // Change neighborhood
23:   $p \leftarrow 1$;   // Return to the first level
24: **end if**
25: **if** $k > Nshake$ **then**
26:   $k \leftarrow 1$; $p \leftarrow 1$;   // Reset neighborhood and level
27: **end if**
28: $itAdaptive \leftarrow itAdaptive + 1$
29: **if** $itAdaptive >= (alsReset * IterMax)$ **then**
30:   $resetRankNeighborhood()$;
31:   $itAdaptive \leftarrow 0$;
32: **end if**
33: $s' \leftarrow Perturbation(s, k, p, \mathcal{N})$;
34: $s' \leftarrow Split(s')$;
35: $s' \leftarrow ClusterRefinement(s')$;
36: $s' \leftarrow Split(s')$;
37: **end while**
38: **return** s;

A solution s of MDRVP is represented by a list vector, inspired and adapted from [14]. Each position of this vector indicates a depot and each list indicates the visit routes to be performed by vehicles from that depot.

As an example, let $\mathcal{V}^{DEP} = \{14, 15\}$ and $\mathcal{V}^{CST} = \{1, 2, \ldots, 13\}$ be a set of two depots and thirteen customers, respectively, and $\mathcal{R}_{DEP}$ be the set of routes per depot. The solution represented in Fig. 1 is $\mathcal{R}_{14} = \{r_1, r_2\}$, where $r_1 = [14\ 4\ 2\ 5\ 9\ 14\ 12\ 3\ 14]$ and $r_2 = [15\ 2\ 6\ 15\ 11\ 13\ 15\ 7\ 8\ 10\ 15]$.

In this solution, there are two routes in the depot 14. At the first one, the vehicle leaves the depot and visits the customers 4, 2, 5 and 9, in this order, and then returns to the depot. In the second route, a vehicle leaves the depot, visits the customers 12 and 3 and returns to the depot. In the depot 15 there are 3 routes, which are represented in the vector r_2.

The strategy used for generating an initial solution (line 1 of Algorithm 1) is "cluster first and then route". In this way, the customers are assigned to the depots in a balanced way using the Gillet and Johnson algorithm [7]. Initially, the heuristic determines the distance between each consumer $j \in \mathcal{V}^{CST}$ not yet assigned to the two nearest depots a_1 and $a_2 \in \mathcal{V}^{DEP}$, with rate $vr_j = a_1/a_2$, $\forall j = N + 1, N + 2, \cdots, N + M$. Then, this rate is sorted ascending according to the values vr_j, assigning the consumer to the nearest feasible depot. This process repeats until all customers are allocated in only one depot. In order to generate a feasible solution, the number of vehicles can be greater than K in this phase. After this construction phase, the VNSALS algorithm tries to refine this constructed solution.

The main procedures of the VNSALS algorithm are *Perturbation*, *Cluster-Refinement* and the Adaptive Local Search (ALS), introduced at lines 33, 35, 8 and 10 of Algorithm 1, respectively. These procedures are necessary because only search in neighborhoods are insufficient to lead to an optimal solution [19].

Perturbation: In the perturbation phase (showed in Algorithm 2), the solution s undergoes a shake move in a given neighborhood $\mathcal{N}_k(s)$, generating a new solution s'. The neighborhood moves are applied in different depots (inter-depot), where the routes, for example, r_1 belong to d_1 and r_2 to d_2 and the number of moves is represented by *Nshake*. The moves are applied in this order: Shift(1, 0), Swap(2, 1), Shift(2, 0), Swap(2, 2), Swap(1, 1).

In order to generate a shake move in a solution s belonging to the neighborhood $\mathcal{N}_k$, each move in the *Perturbation* phase is applied p times, at most, where the value of p is a random integer between 1 and *maxLevel* (see line 33 of Algorithm 1).

Algorithm 2. Perturbation $(s, k, p, \mathcal{N})$

1: **for** $(i = 1; i \le p; i + +)$ **do**
2:     Generate the neighbor $s' \in \mathcal{N}_k(s)$;
3: **end for**
4: **return** s';

ClusterRefinement: The aim is to reduce the total distance between customers and depot assigned. After criteria are established, the idea is select the more expensive customer in depot $\mathcal{A}$ and move it to another generic depot $\mathcal{F}$. We create three criteria to select the customer, where $dist_{\mathcal{A},i}$ is the distance from depot $\mathcal{A}$ to customer i; d_i and t_i are the demand and service time of the customer i, respectively; $\mathcal{S}_1$ the set of customers assigned to depot $\mathcal{A}$; and $\mathcal{S}_{\mathcal{F}}$ the set of customers assigned to depot $\mathcal{F}$:

(i) Let θ_1 the cost of customer i given by $\theta_1 = \max\{\alpha \times dist_{\mathcal{A},i} + \beta \times d_i \mid i \in S_1\}$ and π_1 given by $\pi_1 = \min\{\alpha \times dist_{\mathcal{F},i} + \beta \times d_i \mid i \in S_{\mathcal{F}}, \mathcal{F} \in \mathcal{V}^{DEP}\}$. Then select the depot $\mathcal{F}$ given by π_1 and move the customer i to depot $\mathcal{F}$.

(ii) Let θ_2 be the cost given by $\theta_2 = (\sum_{i=1}^{|\mathcal{S}_1|}(\alpha \times dist_{\mathcal{A},i} + \beta \times d_i))/|\mathcal{S}_1|$. Select all customers $\in \mathcal{S}_1$ with cost $\theta_1 > \theta_2$ and save them in $\mathcal{S}_3$. For any customer in $\mathcal{S}_3$, find a depot $\mathcal{F} \in \mathcal{V}^{DEP}$, with cost π_1 and move the selected customer to depot $\mathcal{F}$;

(iii) Find the most expensive customer $i \in \mathcal{S}_1$ given by $\theta_3 = \max\{\alpha \times dist_{\mathcal{A},i} + \beta \times d_i\}/c_{\mathcal{A}}$, where $c_{\mathcal{A}}$ is the cost of depot $\mathcal{A}$, according to Eq. (1). Let $c_{\mathcal{F}}$ be the cost of depot $\mathcal{F}$ and $\pi_2 = (\pi_1/c_{\mathcal{F}})$. Then select the depot $\mathcal{F}$ given by π_2 and move the customer i to depot $\mathcal{F}$.

For any execution (line 35 in Algorithm 1) only one of these criteria is randomly selected and applied in solution. In this work, we considered $\alpha = 1$ and $\beta = 1$.

Local Search: The local search is used to refine the solution by neighborhoods described in Sect. 3. Only one neighborhood is applied in s'. During the training, each neighborhood is selected randomly (Algorithm 3) and always that its local search improves, the value associated with the i-th neighborhood is updated (accumulated). In vector *success* is accumulated the improvement of any neighborhood. For example, let $\mathcal{N}_1$, $\mathcal{N}_2$ and $\mathcal{N}_3$ be neighborhood structures. Consider that during the training $\mathcal{N}_1$ improved the solution three times, $\mathcal{N}_2$ five times and $\mathcal{N}_3$ two times. After the training (Algorithm 1, line 10), the probabilities of neighborhoods are $\{3/10, 5/10, 2/10\} = \{0.3, 0.5, 0.2\}$, respectively. In this phase, roulette wheel selection is used to select the neighborhood which will be applied in the current solution (Algorithm 4). After any times (*alsReset*), the vector *success* is reset at line 30 of Algorithm 1 (*resetRankNeighborhood*()) and the training process is restarted.

Algorithm 3. RandomNeighborhood(s, *success*, $\mathcal{N}$)

1: $i \leftarrow rand(|\mathcal{N}|)$; {Select a neighborhood $\mathcal{N}_i \in \mathcal{N}$}
2: $s' \leftarrow LocalSearch(\mathcal{N}_i, s)$; {Apply local search to neighborhood $\mathcal{N}_i$}
3: **if** $f(s') < f(s)$ **then**
4:     $s \leftarrow s'$;
5:     $success(i) \leftarrow success(i) + 1$; {Increment success vector associated with $\mathcal{N}_i$}
6: **end if**
7: **return** $s, success$;

Algorithm 4. ALS(s, *success*, $\mathcal{N}$)

```
1: i ← rouletteWheel(success); {Select a index of neighborhood}
2: s' ← LocalSearch(Nᵢ, s); {Apply local search to the neighborhood selected}
3: if f(s') < f(s) then
4:     s ← s';
5: end if
6: return s;
```

Split Algorithm: After perturbation, the solution is represented by a giant tour from each depot, without route delimiters. It is basically a single sequence made of all customers assigned to a depot. For example, the sequence for depot 14 in Fig. 1 is $\{4, 2, 5, 9, 12, 3\}$. Individual routes are created from this giant tour with the Split algorithm [13], which can optimally extract feasible routes from a single sequence. After Split algorithm, we apply Reverse neighborhood, because it can improve the solution [19].

If the solution s'' returned by the local search is better than the current solution s, then s is updated (line 13 of Algorithm 1) and the perturbation level returns to its minimum value (line15 of Algorithm 1). Otherwise, the perturbation level is increased (line 18 of Algorithm 1). If the perturbation level exceeds the level value ($maxLevel$), then the search moves to the next neighborhood (line 22 of Algorithm 1) and the perturbation level returns to the minimum level (line 23 of Algorithm 1).

5   Parameter Tuning

The VNSALS algorithm has six parameters to be tuned: (i) maximum number of iterations ($iterMax$); (ii) maximum time of processing ($maxTime$); (iii) number of neighborhoods for shaking ($Nshake$); (iv) maximum level of perturbations ($maxLevel$); (v) number for training ($alsTraining$) and (vi) reset adaptive ($alsReset$). To make a fare calibration of the parameters we used an automated algorithm, called IRACE (Iterated Racing for Automatic Algorithm Configuration) [9]. This algorithm was designed to provide the most appropriate parameters for an optimization algorithm and a set of instances. IRACE runs as a package of the R software, that is a free environment for statistical computing and graphics.

Table 1. Grouping of instances.

Group	\multicolumn{13}{c}{1}												\multicolumn{9}{c}{2}								
Instance	p01	p02	p03	p04	p05	p06	p07	p12	p13	p14	p15	p16	p17	p08	p09	p10	p11	p18	p19	p20	p21	p22	p23
n	50	50	75	100	100	100	100	80	80	80	160	160	160	249	249	249	249	240	240	240	360	360	360
m	4	2	3	8	5	6	4	5	5	5	5	5	5	14	12	8	6	5	5	5	5	5	5
d	4	4	5	2	2	3	4	2	2	2	4	4	4	2	3	4	5	6	6	6	9	9	9
β	800	400	1125	1600	1000	1800	1600	800	800	800	3200	3200	3200	6972	8964	7968	7470	7200	7200	7200	16200	16200	16200

In order to calibrate the parameters of the proposed algorithm, the instances were grouped in sets according to theirs sizes, determined by the value

$\gamma = n \times m \times d$. Table 1 shows two groups of instances with $\gamma \leq 3200$ and $\gamma > 3200$. A set of instances of each group was chosen for testing by IRACE software. The following instances were chosen: Group 1 (p02, p04, p06, p14, p17); Group 2 (p09, p19 and p21). We set the following values for the parameters: $iterMax = \{400, 450, 500\}$; $Nshake = \{3, 4, 5\}$; $maxLevel = \{3, 4, 5\}$; $alsTraining = \{0.1, 0.15\}$; $alsReset = \{0.2, 0.3, 0.4\}$ and $maxTime = \{30, 60\}$. The best values returned by IRACE were: $iterMax = 500$, $Nshake = 5$, $maxLevel = 3$, $alsTraining = 0.15$, $alsReset = 0.3$ and $maxTime = 60$ min.

6   Computational Experiments

The VNSALS algorithm was coded in C++ and tested in a computer with Intel Core i5-2310M, 2.90 GHz, 4 GB RAM, under operational system Ubuntu 16.04 64 bits and compiler G++ version 5.4.

The Courdeau's instances of MDVRP from [6] were used to verify the performance of the VNSALS algorithm. These instances have $n = 50$ to $n = 360$ customers; $d = 2$ to $d = 9$ depots; $m = 2$ to $m = 14$ vehicles; and load $q = 60$ to $q = 500$.

Six algorithms from literature were used for comparing the results of the VNSALS algorithm. These algorithms are: ALNS [12], HGSADC [17], ILS-RVND-SA [16], HGSADC+ [19], CoES [11], and GVNS [4]. The computer configurations used to test these algorithms, as well as the number of runs of each algorithm in each instance, are shown in Table 2.

Table 2. Computer configurations used to test the algorithms.

Algorithm	Runs	Computer
ALNS	10	Pentium IV 3.0 GHz
HGSADC	10	Opteron 2.4 GHz scaled for a Pentium IV 3.0 GHz
ILS+RVND+SA	10	Intel CoreTM i7 with 2.93 GHZ and 8 GB of RAM
GRASPxILS	5	Intel CoreTM 2 Quad CPU Q8400 @ 2.66 GHz
HGSADC+	10	Opteron 2.4 GHz scaled for a Pentium IV 3.0 GHz
GVNS	30	Intel Core i3-2370M, 2.40 GHz, 4GB RAM
VNSALS (our algorithm)	30	Intel Core i5-2310M, 2.90 GHz, 4GB RAM

Table 3 presents the best results found by the algorithms ALNS, HGSADC, ILS-RVND-SA, HGSADC+, CoES, GVNS, shown in their respective original articles, and the results concerning the application of the proposed VNSALS algorithm. For VNSALS, the average values, the best solutions, as well as the execution times, are presented. Each instance was run 30 times, using the parameter values chosen by IRACE (which are described in Sect. 5) and the parameter $maxTime = 60$ min. The average results and the best results of VNSALS are 1.88% and 0.99%, respectively, higher in relation to the values of the Best Known Solutions (BKS). Compared to the HGSADC+ algorithm, which has the best results for MDVRP, VNSALS presents a total cost 1.85% and 0.96% higher in relation to the average and best values, respectively. These results show that VNSALS is a competitive algorithm against the best algorithms for the MDVRP solution in the literature.

Table 3. Results of the algorithms.

Inst	n	m	d	q	T	BKS	ALNS	HGSADC	ILS-RVND-SA	HGSADC+	CoES	GVNS	VNSALS Average	VNSALS Best	T(min) Average
p01	50	4	4	80	∞	**576.87**	576.87	576.87	576.87	576.87	576.87	591.09	579.12	576.87	0.37
p02	50	2	4	160	∞	**473.53**	473.53	473.53	473.53	473.53	475.06	476.66	476.72	473.53	0.40
p03	75	3	5	140	∞	**640.65**	641.19	641.19	641.19	640.65	643.57	641.19	642.81	641.19	1.01
p04	100	8	2	100	∞	**999.21**	1006.09	1001.04	1001.04	1000.66	1011.42	1025.44	1017.53	1003.72	10.69
p05	100	5	2	200	∞	**750.03**	752.34	750.03	750.21	750.03	752.39	757.46	756.55	751.15	6.85
p06	100	6	3	100	∞	**876.50**	883.01	876.5	876.5	876.5	877.86	889.79	888.23	880.42	5.17
p07	100	4	4	100	∞	**881.97**	889.36	884.43	881.97	881.97	893.36	898.31	895.08	884.04	4.88
p08	249	14	2	500	310	**4372.78**	4421.03	4397.42	4393.7	4383.63	4474.23		4577.46	4503.76	60.29
p09	249	12	3	500	310	**3858.66**	3892.5	3868.59	3864.22	3860.77	3904.92		3987.11	3928.40	60.13
p10	249	8	4	500	310	**3631.11**	3666.85	3636.09	3634.72	3631.71	3680.02		3774.47	3702.20	59.17
p11	249	6	5	500	310	**3546.06**	3573.23	3548.25	3546.15	3547.37	3593.37		3627.77	3567.71	55.67
p12	80	5	2	60	∞	**1318.95**	1319.13	1318.95	1318.95	1318.95	1318.95	1326.85	1321.46	1318.95	1.96
p13	80	5	2	60	200	**1318.95**	1318.95	1318.95	1318.95	1318.95	1318.95		1323.61	1318.95	0.75
p14	80	5	2	60	180	**1360.12**	1360.12	1360.12	1360.12	1360.12	1360.12		1363.09	1360.12	0.58
p15	160	5	4	60	∞	**2505.42**	2519.64	2505.42	2505.42	2505.42	2549.65	2567.62	2551.49	2505.42	29.31
p16	160	5	4	60	200	**2572.23**	2573.95	2572.23	2572.23	2572.23	2572.23		2590.57	2572.23	4.12
p17	160	5	4	60	180	**2709.09**	2709.09	2709.09	2710.21	2709.09	2733.8		2720.61	2709.09	2.28
p18	240	5	6	60	∞	**3702.85**	3736.53	3702.85	3702.85	3702.85	3781.66	3796.04	3797.59	3762.64	60.16
p19	240	5	6	60	200	**3827.06**	3838.76	3827.06	3827.55	3827.06	3827.06		3851.20	3839.36	14.62
p20	240	5	6	60	180	**4058.07**	4064.76	4058.07	4058.07	4058.07	4094.86		4080.18	4069.21	6.81
p21	360	5	9	60	∞	**5474.84**	5501.58	5476.41	5474.84	5474.84	5668.97		5711.17	5669.61	60.61
p22	360	5	9	60	200	**5702.16**	5722.19	5702.16	5705.84	5702.16	5708.78	-	5736.91	5714.45	46.53
p23	360	5	9	60	180	**6078.75**	6092.66	6078.75	6078.75	6080.43	6159.9		6116.5	6089.9	20.6
Total cost						**61235.86**	61533.36	61284.00	61273.88	61253.86	61978.00		62387.20	61842.91	

Table 4 compares the average gaps of the ALNS, HGSADC, ILS-RVND-SA, HGSADC+, CoES, GVNS and VNSALS algorithms. A blank line means that the respective algorithm was not tested in the respective instance. In this table, the average gap is calculated by Eq. (12):

$$gap_i^{\text{avg}} = \frac{\overline{f}_i^{\text{VNSALS}} - f_i^\star}{f_i^\star} \tag{12}$$

where $f_i^\star$ represents the value of the best known solution (BKS) from literature relative to instance i and $\overline{f}_i^{\text{VNSALS}}$ represents the average value produced by the VNSALS algorithm in this instance. From Table 4, we may observed that the best results achieved by the VNSALS algorithm are close to the best known values. In terms of variability of the final solutions, the gap varies from 0.19% to 4.68%, with average gap of 1.49%. The VNSALS results are better than those of the GVNS algorithm, which is a VNS-based algorithm without the ALS procedure and the cluster refinement.

Table 4. Comparison of the algorithms with respect to the average gaps

Inst	ALNS	HGSADC	ILS-RVND-SA	HGSADC+	CoES	GVNS	VNSALS
p01	0.00	0.00	0.00	0.00	0.00	2.47	0.39
p02	0.00	0.00	0.00	0.00	0.32	0.66	0.67
p03	0.08	0.08	0.08	0.00	0.46	0.08	0.34
p04	0.69	0.18	0.18	0.15	1.22	2.63	1.83
p05	0.31	0.00	0.02	0.00	0.31	0.99	0.87
p06	0.74	0.00	0.00	0.00	0.16	1.52	1.34
p07	0.84	0.28	0.00	0.00	1.29	1.85	1.49
p08	1.10	0.56	0.48	0.25	2.32		4.68
p09	0.88	0.26	0.14	0.05	1.20		3.33
p10	0.98	0.14	0.10	0.02	1.35		3.95
p11	0.77	0.06	0.00	0.04	1.33		2.30
p12	0.01	0.00	0.00	0.00	0.00	0.60	0.19
p13	0.00	0.00	0.00	0.00	0.00		0.35
p14	0.00	0.00	0.00	0.00	0.00		0.22
p15	0.57	0.00	0.00	0.00	1.77	2.48	1.84
p16	0.07	0.00	0.00	0.00	0.00		0.71
p17	0.00	0.00	0.04	0.00	0.91		0.43
p18	0.91	0.00	0.00	0.00	2.13	2.52	2.56
p19	0.31	0.00	0.01	0.00	0.00		0.63
p20	0.16	0.00	0.00	0.00	0.91		0.54
p21	0.49	0.03	0.00	0.00	3.55		4.32
p22	0.35	0.00	0.06	0.00	0.12		0.61
p23	0.23	0.00	0.00	0.03	1.33		0.62
Average	0.41	0.07	0.05	0.02	0.90		1.49

7   Conclusions

In this paper, we presented a VNS-based algorithm, named VNSALS (Variable Neighborhood Search with Adaptive Local Search), for solving the Multi-Depot Vehicle Routing Problem (MDVRP). The main characteristic of VNSALS is the existence of an Adaptive Local Search (ALS) algorithm that classifies the neighborhoods and performs cluster refinement, in order to minimize the cost associated with depots. In the ALS procedure, the neighborhood order is not a parameter, because it is chosen adaptively by roulette wheel.

VNSALS was tested in classical instances of MDVRP and its results were compared with those of the best known values and other six algorithms of the literature.

The computational experiments showed that the VNSALS algorithm is better than the other VNS-based algorithm of [4]. However, the proposed algorithm was not better than the other algorithms with which it was compared. New improvements need to be made to improve its performance.

Acknowledgements. The authors would like to thank the CAPES Foundation, the Brazilian Council of Technological and Scientific Development (CNPq), the Minas Gerais State Research Foundation (FAPEMIG), the Federal Center of Technological Education of Minas Gerais (CEFET-MG), and the Federal University of Ouro Preto (UFOP) for supporting this research.

References

1. Allahyari, S., Salari, M., Vigo, D.: A hybrid metaheuristic algorithm for the multi-depot covering tour vehicle routing problem. Eur. J. Oper. Res. **242**(3), 756–768 (2015)
2. Aziz, R.A., Ayob, M., Othman, Z., Ahmad, Z., Sabar, N.R.: An adaptive guided variable neighborhood search based on honey-bee mating optimization algorithm for the course timetabling problem. Soft Comput. **21**(22), 6755–6765 (2017)
3. Aziz, R., Ayob, M., Othman, Z., Sarim, H.: Adaptive guided variable neighborhood search. J. Appl. Sci. **13**(6), 883–888 (2013)
4. Bezerra, S.N., de Souza, S.R., Souza, M.J.F.: A GVNS algorithm for solving the multi-depot vehicle routing problem. Electron. Notes Discrete Math. **66**, 167–174 (2018). 5th International Conference on Variable Neighborhood Search
5. Contardo, C., Martinelli, R.: A new exact algorithm for the multi-depot vehicle routing problem under capacity and route length constraints. Discrete Optim. **12**, 129–146 (2014)
6. Cordeau, J.F., Gendreau, M., Laporte, G.: A tabu search heuristic for periodic and multi-depot vehicle routing problems. Networks **30**(2), 105–119 (1997)
7. Gillett, B.E., Johnson, J.G.: Multi-terminal vehicle-dispatch algorithm. Omega **4**(6), 711–718 (1976)
8. Hansen, P., Mladenović, N., Moreno Pérez, J.A.: Variable neighbourhood search: methods and applications. 4OR **6**(4), 319–360 (2008)
9. López-Ibáñez, M., Dubois-Lacoste, J., Cáceres, L.P., Birattari, M., Stützle, T.: The IRACE package: iterated racing for automatic algorithm configuration. Oper. Res. Perspect. **3**, 43–58 (2016)
10. Montoya-Torres, J.R., Franco, J.L., Isaza, S.N., Jiménez, H.F., Herazo-Padilla, N.: A literature review on the vehicle routing problem with multiple depots. Comput. Ind. Eng. **79**, 115–129 (2015)
11. de Oliveira, F.B., Enayatifar, R., Sadaei, H.J., Guimarães, F.G., Potvin, J.Y.: A cooperative coevolutionary algorithm for the multi-depot vehicle routing problem. Expert Syst. Appl. **43**, 117–130 (2016)
12. Pisinger, D., Ropke, S.: A general heuristic for vehicle routing problems. Comput. Oper. Res. **34**(8), 2403–2435 (2007)
13. Prins, C.: A simple and effective evolutionary algorithm for the vehicle routing problem. Comput. Oper. Res. **31**(12), 1985–2002 (2004)
14. Salhi, S., Imran, A., Wassan, N.A.: The multi-depot vehicle routing problem with heterogeneous vehicle fleet: formulation and a variable neighborhood search implementation. Comput. Oper. Res. **52**(PB), 315–325 (2014)
15. Subramanian, A., Drummond, L., Bentes, C., Ochi, L., Farias, R.: A parallel heuristic for the vehicle routing problem with simultaneous pickup and delivery. Comput. Oper. Res. **37**(11), 1899–1911 (2010)
16. Subramanian, A., Uchoa, E., Ochi, L.S.: A hybrid algorithm for a class of vehicle routing problems. Comput. Oper. Res. **40**(10), 2519–2531 (2013)
17. Vidal, T., Crainic, T.G., Gendreau, M., Lahrichi, N., Rei, W.: A hybrid genetic algorithm for multidepot and periodic vehicle routing problems. Oper. Res. **60**(3), 611–624 (2012)
18. Vidal, T., Crainic, T.G., Gendreau, M., Prins, C.: Heuristics for multi-attribute vehicle routing problems: a survey and synthesis. Eur. J. Oper. Res. **231**(1), 1–21 (2013)
19. Vidal, T., Crainic, T.G., Gendreau, M., Prins, C.: Implicit depot assignments and rotations in vehicle routing heuristics. Eur. J. Oper. Res. **237**(1), 15–28 (2014)

Skewed Variable Neighborhood Search Method for the Weighted Generalized Regenerator Location Problem

Lazar Mrkela[1] and Zorica Stanimirović[2]([✉])

[1] Belgrade Metropolitan University, Tadeuša Košćuška 63, 11000 Belgrade, Serbia
lazar.mrkela@metropolitan.ac.rs
[2] Faculty of Mathematics, University of Belgrade,
Studentski trg 16/IV, 11000 Belgrade, Serbia
zoricast@matf.bg.ac.rs

Abstract. This paper deals with the Weighted Generalized Regenerator Location Problem (WGRLP) that arises in the design of optical telecommunication networks. During the transmission of optical signal, its quality deteriorates with the distance from the source, and therefore, it has to be regenerated by installing regenerators at some of the nodes in the network. The WGRLP involves weights assigned to potential regenerator locations, reflecting the costs of regenerator deployment. The objective of WGRLP is to minimize the sum of weights assigned to locations with installed regenerators, while ensuring a good quality communication among terminal nodes. As telecommunication networks usually involve large number of nodes, an efficient optimization method is required to deal with real-life problem dimensions. In this paper, a Skewed Variable Neighborhood Search method (SVNS) is proposed as solution approach for the WGRLP. The designed SVNS uses adequate data structures for solution representation and efficient procedures for objective function update, feasibility check, and solution repair. Computational results on the WGRLP data set from the literature show that the proposed SVNS reaches all known optimal solutions on small and medium size instances in short running times and outperforms existing heuristic approaches for the WGRLP. In addition, SVNS is tested on large scale WGRLP instances not considered in the literature so far. The presented computational results indicate the potential of SVNS as solution method for WGRLP and related network design problems.

Keywords: Weighted Generalized Regenerator Location Problem ·
Skewed Variable Neighborhood Search · Optical networks ·
Telecommunication

This research was partially supported by Serbian Ministry of Education, Science and Technological Development under the grant no. 174010.

A. Sifaleras et al. (Eds.): ICVNS 2018, LNCS 11328, pp. 182–201, 2019.
https://doi.org/10.1007/978-3-030-15843-9_15

1   Introduction

When designing an optical telecommunication network, the most important requirement is the reliability of signal transmission from an origin to destination node. The optical signal is transmitted through the set of links connecting two terminal nodes in the network and its quality deteriorates as the distance from the origin increases. In optical networks, the signal can only travel a maximum distance before it starts loosing its quality, mostly due to attenuation and transmission impairments. In order to regenerate the signal, special devices - *regenerators* must be installed at some of the nodes in the network. Regenerators transform the deteriorated optical signal to electronic one, which is then regenerated, and finally, the recovered electronic signal is converted back to optical form. As regenerator deployment is highly expensive, it is necessary to install the lowest number of regenerators in order to reduce the total costs in optical network.

The described optimization problem has been identified in the literature as the Regenerator Location Problem (RLP). More precisely, the objective of RLP is to minimize the number of regenerators to be deployed at some of the nodes in the given network, such that communication between all terminal nodes is ensured with good signal quality. Regenerator Location Problem was introduced by Chen et al. [2], who proposed three heuristics and exact Branch-and-Cut approach as solution methods for the RLP. In addition, a correspondence between the RLP and the Maximum Leaf Spanning Tree Problem (MLSTP) was established in [2]. Due to this correspondence, a strategic oscillation procedure for the MLSTP proposed in [17] can also be used as solution approach to RLP. Duarte et al. [4] designed a Greedy Randomized Adaptive Search Procedure (GRASP) and a Biased Random Key Genetic Algorithm for the RLP.

However, the RLP may not address accurately all constraints required by a service provider. In practice, it is rarely required that all nodes in the network must communicate with each other and the set of potential locations for installing regenerators is usually restricted. In order to overcome this situation, a Generalized Regenerator Location Problem (GRLP) is introduced [1]. In GRLP, candidate locations for installing regenerators belong to the subset of nodes in the observed telecommunication network, and it is not necessary that all nodes in the network must communicate with each other. Chen et al. [1] proposed a Branch-and-Cut algorithm (BnC), as well as two constructive heuristics (GH1 and GH2) and a local search procedure for solving the GRLP. Quintana et al. [15] designed an efficient GRASP metaheuristic for the same problem.

The weighted variant of the GRLP problem (WGRLP) was proposed by Chen et al. [1]. WGRLP takes into account the fact that the costs of locating regenerators at different nodes of a network may vary, which is mostly due to real estate costs. For example, the deployment of regenerators in urban areas may be much more expensive than in rural areas. Chen et al. [1] assigned weights to each candidate regenerator location, reflecting the costs of regenerator deployment, while other assumptions of WGRLP are the same as in GRLP. Differently from GRLP that minimizes the number of deployed regenerators, the objective of WGRLP is to minimize the sum of weights (costs) of installing regenerators at chosen locations. In order to solve WGRLP, Chen et al. [1] adapted Branch-and-Cut algorithm and two constructive heuristics previously designed for GRLP to the weighted variant of the problem. The adapted BnC, GH1, and GH2 approaches were tested on instances from Set 2 with up to 150 nodes, obtained by modifying instances from Set1 for GRLP.

As optical networks usually include large number of nodes, it is necessary to design a metaheuristic method that will efficiently provide solutions for large scale problem WGRLP instances. In this paper, we propose a variant of Variable Neighborhood Search metaheurstic for the WGRLP, known as Skewed Variable Neighborhood Search (SVNS), which allows the exploration of the valleys far from the incumbent solution [6,7]. The proposed SVNS is tested on WGRLP instances from the Set 2 introduced in [1], and the results are compared with the results of BnC, GH1, and GH2 presented in [1]. Following the strategy described in [1], we modify large scale GRLP instances from the Set 3 used in [1]. The newly generated sets of large scale WGRLP instances are denoted as Set 4 and Set 5, and the proposed SVNS is also benchmarked on these challenging WGRLP test examples. The obtained computational results clearly indicate the potential of SVNS method when solving the WGRLP.

The rest of the paper is organized as follows. Section 2 contains the description of the considered WGRLP. The proposed SVNS heuristic is explained in details in Sect. 3. Computational results are presented and analyzed in Sect. 4. Finally, in Sect. 5, some conclusions and directions for future work are given.

2   Problem Description

Let $G = (V, E)$ be a given network, where V represents the set of nodes and E denotes the set of edges. The set V consists of two subsets: the set of potential locations for installing regenerators $S \subseteq V$ and the set of terminal nodes $T \subseteq V$ that must communicate with each other. Note that T and S are disjoint sets, i.e., $T \cap S = \emptyset$ and $T \cup S = V$. For each edge $(i, j) \in E$, its length $l(i, j) \geq 0$ is known. A weight $w_r > 0$ is assigned to each candidate location $r \in S$, corresponding to the costs of regenerator deployment at this location. Parameter $d_{max} > 0$ represents maximal distance that a signal can traverse before its quality deteriorates.

A path P from origin node $i \in V$ to destination node $j \in V$ is defined as the array of nodes $i, v_1, \ldots, v_m, j$, where $v_i \in V$, $i = 1, \ldots, m$ are the nodes traversed from origin i to reach destination j, assuming that there is an edge connecting each pair of subsequent nodes in P. The length of path P is calculated as $l(P) = l(i, v_1) + l(v_1, v_2) + \ldots + l(v_m, j)$. If $l(P) \leq d_{max}$ holds, a signal can traverse directly from i to j without being regenerated. Otherwise, regenerator devices must be installed at one or more internal nodes of the path P to ensure the quality of signal transmission.

Let $L \subseteq S$ denote the set of nodes with installed regenerators in the network. Let us consider a path P that contains regenerators $\{r_1, \ldots, r_k\} \subseteq L$ installed at some of the internal nodes of P. The distance between two terminal nodes i and j along the path P is calculated as $d(P) = \max\{l(P_{i,r_1}), l(P_{r_1,r_2}), ..., l(P_{r_k,j})\}$, where $P_{k,l}$ denotes a subpath of P connecting k and l, $k, l \in \{i, r_1, ..., r_k, j\}$. The objective of WGRLP is to minimize the sum of weights of chosen regenerator locations $\sum_{r \in L} w_r$, such that for each pair of terminal nodes $i, j \in T$, there exists a path P for which $d(P) \leq d_{max}$ holds.

If the weights of candidate nodes are neglected, i.e., if $w_r = 1$ for all $r \in S$, the objective function is equal to the number of installed regenerators, and therefore, the WGRLP reduces to the Generalized Regenerator Location Problem (GRLP). In addition, if $S = T = V$ holds, we obtain Regenerator Location Problem (RLP) as a special case of the GRLP. As RLP is proved to be NP-hard in [2,5], it is obvious that both GRLP and WGRLP are also NP-hard optimization problems.

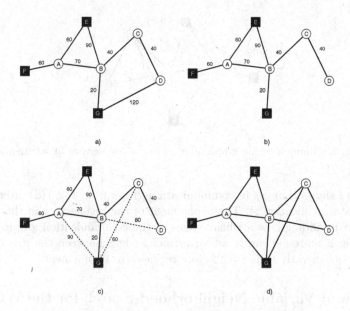

Fig. 1. The construction of communication graph B

In order to construct an efficient algorithm for the WGRLP, the considered graph $G = (V, E)$ is transformed into a simpler graph B, denoted as *communication graph* [1]. Initially, graph B has the same set of nodes V and the same set of edges E as the graph G. Then, all edges of length greater than d_{max} are removed, as it is not possible to transmit good quality signal along them. The next step consists of adding an artificial edge for each non-adjacent pairs of nodes. The length of an artificial edge connecting nodes $i, j \in V$ is equal to the length of the corresponding shortest path between them. Then, the artificial edges of length greater than d_{max} are also removed. Finally, a communication graph $B = (V, E')$ is obtained, in which information on length of the edges are discarded. If there is an edge in the graph B connecting two nodes from the set

T, this pair of nodes is considered *connected*, meaning that the direct communication between them is possible. Otherwise, this pair of nodes is marked as *not directly connected (NDC)*.

Figure 1 illustrates the described steps in constructing communication graph B on a small example. Terminal nodes are represented as black filled squares, while empty circles denote potential locations for installing regenerators. Continuous lines in Fig. 1 correspond to the edges of the initial graph G, while dashed lines represent artificial edges added when constructing the communication graph B. From Fig. 1(d) it can be seen that the solution represented by communication graph B is not feasible. In order to construct a feasible solution, all NDC pairs must be connected with new artificial edges in the communication graph.

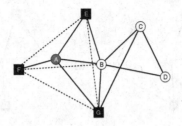

Fig. 2. Communication graph after adding a regenerator at location A

Figure 2 shows changes in communication graph from Fig. 1(d) after adding a regenerator at location A. The deployment of the regenerator in the network may result in adding new artificial edges to the communication graph. Dashed lines in Fig. 2 represent newly added artificial edges between the pair of nodes that were not directly connected before regenerator deployment.

3   Skewed Variable Neighborhood Search for the WGRLP

Variable Neighborhood Search (VNS) is a simple and effective metaheuristics proposed by Mladenović and Hansen [14]. Basic variant of VNS method consists of two phases, which alternate together with neighborhood change step, until a stopping criterion is satisfied. The first phase, denoted as *Shaking*, is used to drive away the search from the incumbent solution in a stochastic manner. The solution obtained in Shaking phase is subject to *Local search*, which tries to find an improvement in a deterministic manner. Starting from the basic VNS algorithm, many variants of VNS have proposed: Reduced VNS (RVNS), Variable Neighborhood Descent (VND), Variable Neighborhood Decomposition Search (VNDS), General VNS (GVNS), Skewed VNS (SVNS), Primal-dual VNS, Parallel VNS, etc. [6,7]. In the literature, VNS has been proposed as solution approach to many combinatorial and global optimization problems, including the problems related to the design of telecommunication networks, such as: topological design of a yotta-bit-per-second multidimensional optical network [3], optical routing in networks using latin routers [13], real-life network problem of France Telecom R&D [9], the design of SDH/WDM networks [8,11,12], single path multicommodity flow problem in telecommunication networks [10], etc.

Having in mind numerous examples of successful VNS applications to the network design problems, we have developed Skewed VNS, as a variant of VNS metaheuristic to solve the considered WGRLP. The most important characteristic of Skewed VNS is that the local optimum obtained in the Local Search phase is accepted even if it is worse than the current solution. The level of acceptable degradation is regulated by a parameter. In the following subsections, all aspects of the proposed SVNS approach are explained in detail.

3.1 Solution Representation

The solution of WGRLP is represented as array of integers. Each element of the array corresponds to a location with installed regenerator. Objective function is simply calculated by summing up weights of all locations with deployed regenerators. The solution is feasible if all NDC pairs in the corresponding communication graph are connected.

In our SVNS implementation, communication graph is represented by the adjacency list. Vertices of the graph are stored in a list, while adjacent nodes of each vertex are stored in a Java HashSet. The use of hashing helps us to examine if the considered edge exists in the graph and to search trough adjacent vertices of the given vertex in an efficient manner. Similarly, NDC pairs are stored in a Java HashMap that enables us to efficiently check if the considered pair of nodes is a NDC pair and if it is connected. Each NDC pair is stored with the assigned boolean value: *true* if NDC pair is connected, and *false* otherwise. Therefore, a solution to WGRLP is feasible if all values in the hash map are set to *true*.

The array of indices of installed regenerators is stored in a stack data structure. The index of a regenerator is pushed on the stack after it is being added to the solution [15]. The use of stack data structure increases the efficiency of procedures for adding and removing regenerator.

3.2 Generating Initial Solution

An initial solution for SVNS is constructed by a greedy procedure presented in Algorithm 1. The procedure starts from an empty solution and adds one by one regenerator until a feasible solution is constructed. For each candidate regenerator location i, the procedure calculates the total number of NDC pairs that would be connected if a regenerator is deployed at i. The calculated value is denoted as the *gain* of the candidate regenerator location i. A regenerator is then installed at the location which has maximal gain to cost ratio, and the index of the location with installed regenerator is deleted from the list of candidate locations. The described steps are repeated until a solution becomes feasible.

Algorithm 1. Greedy solution construction

$x \leftarrow$ empty solution;
$candidates \leftarrow$ the list of all potential regenerator locations;
while solution x is infeasible **do**
    for i in $candidates$ **do**
        $gain_i \leftarrow$ the count of NDC pairs that would be connected
                if a regenerator is installed at location i;
    $max \leftarrow$ location with maximal gain to cost ratio;
    add regenerator at location max;
    remove location max from $candidates$;

3.3   The Structure of the Proposed SVNS for WGRLP

The structure of the proposed SVNS is outlined in Algorithm 2. Input parameters for SVNS are: maximal number of iterations without improvement I_{max}, parameter k_{max} used to control neighborhood change, and α_{max} that defines the acceptable degradation level of the objective function value.

The proposed SVNS uses neighborhood structures N_1 and N_2. Neighborhood N_1 is obtained by removing one randomly chosen regenerator from the solution. Neighborhood N_2 is based on a swap operation that removes regenerator from one location and adds regenerator to another location with smaller installation costs. Neighborhood N_1 is used in Shaking phase, while neighborhood N_2 is explored in Local search part.

The SVNS algorithm starts with the greedy procedure that generates initial solution x. Initially, the current best solution x_{best} is set to x, while the counter of non-improving iterations I_{count} is set to 0. The main SVNS loop consists of three steps: *Shaking*, *Local search*, and *Move or not*. These steps are repeated within the neighborhood change loop while $k \leq k_{max}$. Initially, k is set to 1.

In Shaking step, one randomly chosen regenerator is removed from x, resulting in a neighbor solution x'. If $k > k_{max}/2$ holds, solution x' is created by deleting two randomly chosen regenerators from x. In this way, we aim to provide stronger diversification of solutions for larger values of k and to help the algorithm in escaping from a local optimum trap. The obtained neighbor solution x' is corrected to be feasible, if necessary. Procedure REPAIR is performed by adding regenerators to a solution until it becomes feasible. In order to choose regenerators to be added, we have used the idea from the construction phase of the well-known GRASP heuristic [16]. First, the list of regenerator locations that can be added to a solution is created and the corresponding values of gain to cost ratio are calculated. Then, a restricted list of candidate locations is obtained as follows. Location i is included in the restricted candidate list if $r_i \geq r_{min} + \gamma(r_{max} - r_{min})$ holds. Here, r_i denotes gain to cost ratio of location i, while r_{max} and r_{min} are the maximum and minimum values of the gain to cost ratio among all locations in the candidate list, respectively. In addition, the procedure uses parameter γ that depends on k and takes real values between 0 and 1. A regenerator to be added is chosen randomly from the created restricted candidate list. Once a regenerator has been added to a solution, it is deleted from the candidate list. The described steps are repeated until a feasible solution is obtained. Algorithm 3 shows the steps of our REPAIR procedure.

Algorithm 2. The proposed Skewed VNS for the WGRLP

Input: I_{max}, k_{max}, α_{max}
$x \leftarrow$ solution constructed by the greedy procedure;
$x_{best} \leftarrow x$;
$I_{count} \leftarrow 0$;
while $I_{count} \leq I_{max}$ **do**
    $k \leftarrow 1$;
    $improvement \leftarrow$ **false**;
    $I_{count} \leftarrow I_{count} + 1$;
    $\alpha = I_{count} * \alpha_{max}/I_{max}$;
    while $k \leq k_{max}$ **do**
        $x' \leftarrow$ randomly choose solution from $N_1(x)$;   //Shaking step
        if $k > k_{max}/2$ **then**
            $x' \leftarrow$ randomly choose solution from $N_1(x')$;
        if x' is not feasible **then**
            $x' \leftarrow$ REPAIR (x', k, k_{max});
        $x' \leftarrow$ CLEAN (x');   //Local search step
        $x' \leftarrow$ LOCALSEARCH in $N_2(x')$;
        if $f(x') - \alpha d(x', x) < f(x)$ **then**   //Move or not step
            $x \leftarrow x'$;
            $k \leftarrow 1$;
        else
            $k \leftarrow k + 1$;
        if $f(x') < f(x_{best})$ **then**
            $x_{best} \leftarrow x'$;
            $improvement \leftarrow$ **true**;
            $I_{count} \leftarrow 0$;

Local search phase consists of procedures CLEAN and LOCALSEARCH. The procedure CLEAN goes trough the list of all regenerators in the solution and tries to remove useless ones. A regenerator is considered *useless* if it can be removed without affecting solution's feasibility. The resulting solution x' is further subject to a simple LOCALSEARCH procedure that tries to find an improvement in the neighborhood N_2 of x'. The procedure LOCALSEARCH tries to remove an installed regenerator in x' and to deploy new regenerator at some other, cheaper location. The swap move is performed only if it preserves solution's feasibility. The first improvement strategy is used, meaning that LOCALSEARCH finishes after an improvement of solution x' is found, and x' is replaced with the new, improved solution.

Algorithm 3. Repair procedure

---

   procedure REPAIR(x, k, k_{max})
      if $k > k_{max}/2$ **then** $k \leftarrow k - k_{max}/2$
      $\gamma \leftarrow 1/k$;
      while x is not feasible **do**
         candidate_list $\leftarrow$ the list of regenerator locations that can be added;
         calculate gain to cost ratio for each location from *candidate_list*;
         sort *candidate_list* in decreasing order according to gain to cost ratio;
         $r_{máx} \leftarrow$ the maximum gain to cost ratio among *candidate_list*;
         $r_{min} \leftarrow$ the minimum gain to cost ratio among *candidate_list*;
         restricted_list $\leftarrow$ empty list;
         for each i in *candidate_list* **do**
            if $r_i \geq r_{min} + \gamma(r_{max} - r_{min})$ **then**
               add location i to *restricted_list*;
         randomly choose regenerator j from *restricted_list*;
         add regenerator j to the solution;
         remove j from *candidate_list*;

In Move or not step, the algorithm will move to the solution x' produced by the Local search phase if inequality $f(x') - \alpha d(x', x) < f(x)$ is satisfied. Here, $f(x)$ stands for the objective function value of solution x, while $d(x', x)$ denotes the distance between solutions x' and x, defined as the number of regenerators included ether in x' or in x, but not in both of them. More precisely, the distance $d(x', x)$ represents the cardinality of the symmetric difference of the sets of installed regenerators in solutions x' and x, respectively. An adequate value of parameter α must be chosen in order to accept the exploration of search space regions distant from x in cases when x' is worse than x, but not too far away. Otherwise, the search will always leave x and the SVNS would turn to a multi-start heuristic. In our SVNS implementation, the value of parameter α depends on the number of iterations without improvement as follows: $\alpha = I_{count} \cdot \alpha_{max}/I_{max}$, meaning that α takes positive values up to α_{max}. If $f(x') - \alpha d(x', x) < f(x)$ holds, x is replaced with x', k is set to 1, and the algorithm continues the search in the neighborhood N_1; otherwise k is set to $k + 1$. In addition, in case of $f(x') < f(x_{best})$, the current best solution x_{best} is updated with x' and the counter of non-improving iterations I_{count} is reset to 0.

The described steps are repeated until the maximal number of iterations without improvement I_{max} is reached. The value of termination criteria parameter I_{max} usually depends on the problem size and it is determined through the set of parameter tuning experiments.

3.4 Efficient Objective Function Update

Local search phase in the proposed SVNS is based on two main operations: adding regenerator to a solution and removing regenerator from a solution. The most time consuming part in this phase is checking the feasibility of the newly obtained solution after each of the two operations. Therefore, an efficient implementation of feasibility check may reduce the overall computational time significantly.

Algorithm 4. Adding a regenerator

Input: location i where a new regenerator is deployed
for each pair (u, v) of nodes adjacent to i **do**
    if u and v are both end nodes **then**
        if pair (u, v) is not connected **then**
            add pair (u, v) to the list of pairs of location i;
            mark pair (u, v) as connected in the NDC pairs map;
            increase count of connected NDC pairs by one;
    else if edge (u, v) does not exists in the graph **then**
        add edge (u, v) to the graph;
        add edge (u, v) to the list of artificial edges of location i;
    increase objective function by the cost of location i;
    push the i to the solution stack;

 Adding a new regenerator in a solution may result in inserting additional artificial edges in the communication graph. If a newly added artificial edge forms a NDC pair, the total number of connected NDC pairs is increased by one, and the considered NDC pair is marked as *connected*. At the same time, objective function value is increased by the cost of the newly added regenerator. Removing a regenerator from a solution may cause the deletion of some artificial edges from the communication graph. If the removed artificial edge had formed a NDC pair, the total number of connected NDC pairs is decreased by one and the observed NDC pair is marked as *not connected*. At the same time, objective function value is decreased by the installation cost (weight) of the removed regenerator. The main problem when implementing an efficient feasibility check is to decide which artificial edges should be inserted after adding a new regenerator and which ones should be deleted when a regenerator is removed.

When adding a regenerator to the solution, all pairs of vertices adjacent to the newly added regenerator are examined. If there is no edge between some pair of adjacent vertices, an artificial edge connecting them is added to the graph. Note that the feasibility of the resulting solution depends on the order of regenerator deployment. The procedure of adding a regenerator is given by Algorithm 4.

The procedure of removing a regenerator from the top of the stack (the last added regenerator) is outlined in Algorithm 5. The procedure creates the list of pairs of nodes connected by the considered regenerator and the list of artificial edges that were added when the considered regenerator was deployed. Each pair of nodes from the first list is marked as *not connected* and the number of connected NDC pairs is decreased by one. Each artificial edge from the second list is being removed from the graph. Finally, the objective function value is decreased by the installation costs of regenerator that has been removed.

Algorithm 5. Removing the last added regenerator

$i \leftarrow$ pop location from the solution stack;
pairs $\leftarrow$ list of pairs connected by location i;
edges $\leftarrow$ list of artificial edges added by location i;
for each pair (u, v) in *pairs* **do**
    mark pair (u, v) as not connected in the NDC pairs map;
    decrease the number of connected NDC pairs by one;
clear list *pairs*;
for each edge (u, v) in *edges* **do**
    remove edge (u, v) from the graph;
clear list *edges*;
decrease objective function by the cost of regenerator at location i;

Algorithm 6. Remove arbitrary regenerator from the stack

Input: location i from which regenerator is being removed
removed $\leftarrow$ empty stack;
while true **do**
    tmp $\leftarrow$ location from the top of the solution stack;
    remove regenerator from the location *tmp*;
    if *tmp* = i **then**
        break;
    push *tmp* to the *removed* stack;
while *removed* is not empty **do**
    tmp $\leftarrow$ location from the top of the *removed* stack;
    add regenerator to location *tmp*;

Algorithm 6 shows the procedure of removing arbitrary regenerator from a solution. This procedure consists of three steps. First, all regenerators added after the considered one are removed from the solution and saved in a list, preserving the order of their deployment. Next, the considered regenerator is removed in the same way as described in Algorithm 5. Finally, regenerators saved in the list are added back to the solution. Preliminary computational experiments showed that the described strategy is more efficient than destroying whole solution and then adding all regenerators back.

4 Experimental Analysis

In our computational study, we use three data sets:

Set 2 - the set of small and medium size WGRLP instances proposed in [1], obtained from the Set 1 of unicost GRLP instances from the same paper. These instances are characterized by two parameters: the number of nodes $n = |V|$ in the network and the percentage $p(\%)$ of terminal nodes among them. For each combination of $n \in \{50, 75, 100, 125, 150\}$ and $p \in \{25, 50, 75\}$, a group of ten instances is generated with different weights assigned to potential regenerator nodes. Weights are randomly chosen from the set $\{2, 3, 4\}$.

Set 4 - the newly generated set of large scale WGRLP instances, obtained by modifying large unicost GRLP instances from the Set 3 used in [1]. We have followed the same strategy as the one used by the authors of [1] to generate the Set 2 from the Set 1. The newly obtained Set 4 contains 10 instances for each $n \in \{175, 200, 300, 400, 500\}$ and $p \in \{25, 50, 75\}$, while the weights of potential regenerator are also randomly chosen from the set $\{2, 3, 4\}$.

Set 5 - the new set of large scale WGRLP instances obtained in the same way as instances from the Set 4, but with the wider range of weights of potential regenerators. In this data set, weights are random values from $\{1, 2, 3, 4, 5, 6, 7, 8\}$.

All computational experiments were performed on a machine with Intel Core i3-4170 CPU on 3.7 GHz with 8 GB of RAM. The proposed SVNS method is implemented in Java programming language. As in the case of other metaheuristic methods, the performance of the proposed SVNS is sensitive to the parameter values. Therefore, before evaluating its performance, a set of parameter tuning experiments was performed in order to find adequate values of k_{max}, α_{max}, and I_{max}. The results of parameter tuning tests on the chosen subset of instances are given in Subsect. 4.1, while Subsect. 4.2 contains the results of SVNS with the obtained parameter values on all three data sets.

4.1   Parameter Tuning

The set of preliminary tests were performed to tune parameters k_{max}, α_{max}, and I_{max} in SVNS. Parameter tuning tests were performed on the subset of instances from each data set. For each problem dimension n, one randomly chosen instance with medium value of $p = 50\%$ is included in this subset.

First, we have considered five different values of parameter k_{max}: 2, 4, 6, 8, and 10. In these experiments, α_{max} and I_{max} were fixed to 0.3 and 10, respectively. For each considered value of k_{max} and each instance, SVNS was run five times. The results are presented in Table 1 as follows. Each part of Table 1, related to one tested formula for k_{max}, consists of three columns containing: the best SVNS solution obtained in five runs - *best*, the average percentage gap from the best solution - *gap(%)*, and the average CPU time that SVNS needed to reach its best solution - *t(s)*. Three horizontal sections of Table 1 contain the results obtained on the subset of instances from the Set 2, Set 4, and Set 5, respectively. The last row in each horizontal section, denoted by **Average**, shows average values calculated over the data presented in each column of the section, and the best average solution values are bolded.

The results presented in Table 1 indicate that the overall differences in average best solution value, average time and average gap are rather small when using different formulae for k_{max}. As it was expected, the value $k = 10$ leaded to the best average solution values for all three sets, but the corresponding running

Table 1. Parameter tuning experiments for k_{max}

Instance	$k_{max} = 2$			$k_{max} = 4$			$k_{max} = 6$			$k_{max} = 8$			$k_{max} = 10$		
n	best	gap(%)	t(s)	best	gap(%)	t(s)	best	gap(%)	t(s)	best	gap(%)	t(s)	best	gap(%)	t(s)
Set 2															
50	28.00	0.00	2.77	28.00	0.00	2.63	28.00	0.00	2.04	28.00	0.00	2.26	28.00	0.00	1.19
75	24.00	0.00	7.07	24.00	0.00	2.48	24.00	0.00	2.08	24.00	0.00	1.68	24.00	0.00	2.64
100	26.00	6.92	3.45	26.00	3.85	14.47	26.00	2.31	18.24	26.00	3.85	30.72	26.00	2.31	24.33
125	45.00	2.67	4.02	45.00	3.56	4.80	45.00	1.33	8.96	45.00	0.89	9.51	45.00	0.89	10.22
150	14.00	7.14	3.52	14.00	7.14	13.10	14.00	1.43	10.65	14.00	2.86	10.53	14.00	2.86	25.45
Average	**27.40**	3.35	4.17	**27.40**	2.91	7.50	**27.40**	1.01	8.39	**27.40**	1.52	10.94	**27.40**	1.21	12.77
Set 4															
175	29.00	7.59	14.76	31.00	3.23	23.04	29.00	5.52	27.99	29.00	4.14	45.73	29.00	4.14	44.21
200	28.00	3.57	3.04	29.00	0.69	0.81	29.00	0.00	5.37	29.00	0.69	1.06	28.00	2.86	5.83
300	58.00	4.14	94.42	57.00	1.40	163.69	55.00	2.91	324.89	56.00	3.21	333.24	56.00	0.71	273.78
400	58.00	3.10	52.20	58.00	2.07	110.45	57.00	3.51	228.83	56.00	1.07	295.37	54.00	3.70	457.07
500	18.00	7.78	17.26	20.00	0.00	1.88	18.00	6.67	42.45	18.00	0.00	60.74	20.00	0.00	2.12
Average	38.20	5.24	36.34	39.00	1.48	59.98	37.60	3.72	125.91	37.60	1.82	147.23	**37.40**	2.28	156.60
Set 5															
175	65.00	4.31	29.92	66.00	1.82	43.63	65.00	2.15	45.16	65.00	1.54	78.64	65.00	3.08	77.24
200	10.00	8.00	0.77	10.00	6.00	2.41	10.00	0.00	15.71	10.00	0.00	21.16	10.00	0.00	25.81
300	46.00	2.61	27.73	46.00	2.17	68.07	46.00	1.30	37.12	47.00	0.85	66.89	46.00	1.30	285.24
400	60.00	4.33	82.71	59.00	2.71	453.83	59.00	2.71	330.51	58.00	4.14	504.75	59.00	3.39	637.26
500	23.00	0.87	53.34	22.00	5.45	55.74	22.00	5.45	149.17	22.00	1.82	85.43	22.00	2.73	123.17
Average	40.80	4.02	38.89	40.60	3.63	124.74	**40.40**	2.32	115.53	**40.40**	1.67	151.37	**40.40**	2.10	229.75

Table 2. Parameter tuning experiments for α_{max}

Instance	$\alpha_{max} = 0.2$			$\alpha_{max} = 0.4$			$\alpha_{max} = 0.5$			$\alpha_{max} = 0.6$			$\alpha_{max} = 0.7$		
n	best	gap(%)	t(s)	best	gap(%)	t(s)	best	gap(%)	t(s)	best	gap(%)	t(s)	best	gap(%)	t(s)
Set 2															
50	28.00	0.00	0.86	28.00	0.00	1.04	28.00	0.00	0.45	28.00	0.00	1.07	28.00	0.00	0.93
75	24.00	0.00	1.19	24.00	0.00	0.62	24.00	0.00	1.44	24.00	0.00	0.61	24.00	0.00	1.41
100	26.00	3.85	9.02	26.00	4.62	9.42	26.00	6.92	4.46	26.00	4.62	12.60	26.00	3.85	10.94
125	45.00	0.89	18.53	45.00	0.89	6.67	45.00	0.44	8.26	45.00	0.89	11.16	45.00	1.33	9.89
150	14.00	2.86	7.73	14.00	1.43	8.31	14.00	5.71	4.56	14.00	4.29	7.41	14.00	2.86	8.40
Average	**27.40**	1.52	7.47	**27.40**	1.39	5.21	**27.40**	2.62	3.83	**27.40**	1.96	6.57	**27.40**	1.61	6.31
Set 4															
175	29.00	4.83	51.28	30.00	1.33	36.61	30.00	2.67	35.18	29.00	6.21	65.82	29.00	7.59	39.41
200	29.00	0.69	6.34	29.00	0.69	1.44	28.00	2.14	28.32	29.00	0.00	5.16	29.00	0.00	3.85
300	55.00	2.91	342.03	57.00	0.70	504.14	56.00	4.64	186.73	55.00	5.82	240.49	57.00	2.11	201.46
400	54.00	2.96	395.66	56.00	5.36	133.58	56.00	1.79	275.71	54.00	8.15	313.02	57.00	2.46	259.35
500	18.00	3.33	138.75	18.00	5.56	50.55	18.00	6.67	65.20	18.00	4.44	98.22	19.00	4.21	21.47
Average	**37.00**	2.94	186.81	38.00	2.73	145.26	37.60	3.58	118.23	**37.00**	4.92	144.54	38.20	3.27	105.11
Set 5															
175	65.00	1.23	137.36	65.00	3.38	78.33	65.00	2.46	56.61	65.00	2.46	194.27	67.00	0.30	82.20
200	10.00	2.00	9.76	10.00	0.00	23.52	10.00	0.00	14.35	10.00	4.00	11.92	10.00	0.00	33.95
300	46.00	0.87	168.58	46.00	1.30	161.41	46.00	0.87	47.60	46.00	1.74	97.93	47.00	1.28	24.74
400	57.00	3.86	510.83	60.00	1.67	248.48	60.00	1.33	129.63	60.00	2.67	518.45	60.00	2.00	641.77
500	22.00	1.82	104.32	22.00	1.82	178.18	22.00	0.00	136.14	22.00	4.55	29.90	22.00	3.64	67.82
Average	**40.00**	1.96	186.17	40.60	1.63	137.98	40.60	0.93	76.87	40.60	3.08	170.49	41.20	1.44	170.10

times were significantly longer, especially in the case of Set 5. For this reason, we have further considered $k = 6$ and $k = 8$ that leaded to best average solution values on all instances from the Set 2 and Set 5, while average solution values for instances from the Set 4 were slightly worse than in the case of $k = 10$. Finally, we have chosen $k = 8$, as it provided two times smaller average gap for instances from the Set 4 and not significantly longer running times when compared to $k = 6$.

Table 3. Parameter tuning experiments for I_{max}

Instance n	$I_{max} = 5 + n/20$ best	gap(%)	t(s)	$I_{max} = 10 + n/20$ best	gap(%)	t(s)	$I_{max} = 15 + n/30$ best	gap(%)	t(s)	$I_{max} = 20 + n/30$ best	gap(%)	t(s)	$I_{max} = 25 + n/30$ best	gap(%)	t(s)
Set 2															
50	28.00	0.00	1.21	28.00	0.00	2.67	28.00	0.00	1.42	28.00	0.00	0.78	28.00	0.00	0.80
75	24.00	0.00	1.08	24.00	0.00	1.61	24.00	0.00	1.74	24.00	0.00	3.39	24.00	0.00	1.09
100	26.00	2.31	14.30	26.00	2.31	21.70	26.00	4.62	15.22	26.00	2.31	10.30	26.00	0.00	33.29
125	45.00	0.89	22.79	45.00	0.44	28.80	45.00	0.00	18.70	45.00	0.44	8.09	45.00	0.44	19.07
150	14.00	5.71	14.94	14.00	2.86	13.15	14.00	0.00	17.01	14.00	0.00	13.46	14.00	0.00	13.04
Average	**27.40**	1.78	10.87	**27.40**	1.12	13.59	**27.40**	0.92	10.82	**27.40**	0.55	7.20	**27.40**	0.09	13.46
Set 4															
175	29.00	4.14	50.15	29.00	1.38	88.99	29.00	3.45	60.59	29.00	2.07	119.54	29.00	1.38	70.81
200	28.00	2.86	5.28	29.00	0.00	2.52	29.00	0.00	3.40	28.00	2.86	31.61	28.00	2.86	10.50
300	55.00	2.18	549.99	55.00	2.18	594.47	55.00	1.82	429.08	55.00	2.55	474.03	55.00	2.55	839.47
400	56.00	0.71	569.14	54.00	2.22	1059.06	55.00	2.91	787.10	54.00	5.19	1087.68	55.00	2.91	607.48
500	18.00	5.56	85.27	18.00	0.00	649.10	18.00	0.00	303.75	18.00	1.11	139.99	18.00	1.11	406.73
Average	37.20	3.09	251.97	37.00	1.16	478.83	37.20	1.64	316.78	**36.80**	2.75	370.57	37.00	2.16	387.00
Set 5															
175	65.00	1.85	60.04	65.00	2.15	105.69	65.00	0.92	117.92	65.00	0.62	141.52	65.00	0.31	147.53
200	10.00	2.00	18.82	10.00	0.00	33.13	10.00	2.00	14.48	10.00	0.00	18.70	10.00	0.00	23.53
300	46.00	0.00	273.21	46.00	0.43	316.29	46.00	0.87	57.57	46.00	0.87	173.41	46.00	0.43	190.11
400	57.00	3.51	821.24	59.00	1.02	736.27	59.00	2.71	880.77	56.00	5.00	723.59	58.00	2.76	1500.11
500	22.00	0.00	348.58	22.00	0.00	268.97	22.00	2.73	425.49	22.00	0.00	183.02	22.00	0.00	274.78
Average	40.00	1.47	304.38	40.40	0.72	292.07	40.40	1.85	299.25	**39.80**	1.30	248.05	40.20	0.70	427.21

In Table 2, the results of computational experiments with parameter α are given. We have considered 5 different values of α, ranging from $\alpha = 0.2$ to $\alpha = 0.7$. In these tests, parameter values $k_{max} = 8$ and $I_{max} = 10$ were used. For each considered value of α and each instance, SVNS was run five times. The results in Table 2 are presented in the same way as in Table 1. The value $\alpha = 0.2$ showed the best performance in the sense of solution quality, as it produced the best solution function values for all three subsets of instances. Among other tested values for α, only $\alpha = 0.6$ provided best solutions on two out of three considered subsets. Although running times for $\alpha = 0.6$ were shorter compared to the ones required for $\alpha = 0.2$, the average gaps for $\alpha = 0.6$ were quite high, indicating possible problems with algorithm's stability for this value of α. Therefore, α was set to 0.2 in further experiments with SVNS.

Finally, we have considered five different formulae for I_{max} depending on the problem size n. In our tests with I_{max}, we set $k_{max} = 8$ and $\alpha_{max} = 0.2$, and executed SVNS five times on each considered instance. Table 3 contains the results of experiments with I_{max} has the same structure as Tables 1 and 2. Based on the results from Table 3, we have chosen formula $I_{max} = 20 + n/30$, as it leaded to the best average solution values for all three subsets. Other four considered formulae for I_{max} provided the best average solutions only for instances from the Set 2. In addition, in the case of $I_{max} = 20 + n/30$, our SVNS had the smallest average running time for the Set 2 and Set 5, and reasonably short average running time for the Set 4. The average gap values obtained with $I_{max} = 20 + n/30$ were also low for all three considered subsets of instances. For these reasons, $I_{max} = 20 + n/30$ is chosen for stopping criterion in our SVNS implementation.

4.2   Computational Results

The proposed SVNS was evaluated on all instances from the three data sets described above. Based on the results of parameter tuning tests, presented in Sect. 4.1, parameter k_{max} is set to 8, α_{max} takes the value of 0.2, while $I_{max} = 20 + n/30$. On each considered instance, SVNS was run five times.

In order to investigate the effects of accepting slightly worse solution in SVNS, we have also considered the variant of VNS obtained from the proposed SVNS by setting the value of parameter α to 0 in Move or Not step. The resulting VNS implementation is actually basic variant of VNS method, denoted as BVNS. It was also executed five times on each instance with the same values of parameters k_{max} and $I_{max} = 20 + n/30$ as in SVNS, while α_{max} is set to 0.

In Table 4, we report the results of our SVNS method on 15 groups of WGRLP instances from the Set 2, as well as the results obtained by BVNS, exact Branch-and-Cut algorithm (BnC), and two heuristics (GH1 and GH2) from [1] on the same data set. The first two columns contain parameters n and $p(\%)$ for each group of instances. As in paper [1], remaining columns in Table 4 contain average results obtained by corresponding method on all instances from the same group. Optimal solution *opt.sol* obtained by exact BnC method and its execution time $t(s)$ on an Intel Core 2 Duo with 3 GHz and 3.25 GB RAM are given in the third and fourth column, respectively. The next four columns contain the best results provided by GH1 and GH2 heuristics and the corresponding running times, obtained on the same platform as BnC. In the next three columns, we report the results related to the performance of our SVNS: the best SVNS solution obtained in five runs - *best*, the average percentage gap from the optimal solution - *gap(%)*, the average CPU time that SVNS needed to reach its best solution - $t(s)$. The last three columns contain the results of BVNS implementation, presented in the same way as the results of SVNS. The average optimal values on each group of instances are bolded.

As it can be seen from Table 4, the proposed SVNS method reaches all optimal solutions obtained by exact BnC method on instances from the Set 2. It can be noticed that the objective function values of optimal solutions decrease as the percentage of terminal nodes p increases, due to the fact that smaller number of regenerators is required to ensure good quality signal transmission between terminal nodes. On all groups of instances from the Set 2, the average gap values of SVNS solutions from the optimal ones are very small (from 0% up to 0.64%), which indicates the stability of the proposed SVNS approach. The average times in which SVNS reaches its best solutions range from 1.97 up to 80.92 s. Greedy heuristics GH1 and GH2 from [1] are much faster compared SVNS, but they both showed poor performance regarding solution's quality. GH1 reached all optimal solutions for 3 out of 15 groups of instances from the Set 2, while GH2 provided optimal solutions for only one group. When comparing the performance of BVNS and SVNS, it can be noticed that the BVNS failed to provide optimal

Table 4. Computational results of the proposed SVNS and comparisons with the results of BnC, GH1, GH2 and BVNS for instances from Set 2

Size		BnC		GH1		GH2		SVNS			BVNS		
n	p(%)	opt.sol	t(s)	best	t(s)	best	t(s)	best	gap(%)	t(s)	best	gap(%)	t(s)
50	25	**21.3**	0.33	21.4	0.00	21.4	0.00	**21.3**	0.00	2.29	**21.3**	0.00	2.52
50	50	**21.3**	0.52	**21.3**	0.00	21.4	0.00	**21.3**	0.00	3.81	**21.3**	0.00	2.93
50	75	**11.3**	0.21	11.5	0.00	11.5	0.00	**11.3**	0.00	1.97	**11.3**	0.00	1.61
75	25	**28.2**	2.00	28.4	0.00	28.4	0.00	**28.2**	0.00	2.57	**28.2**	0.00	3.99
75	50	**25.7**	4.52	25.8	0.00	26.1	0.00	**25.7**	0.07	8.82	**25.7**	0.00	9.92
75	75	**13.9**	2.28	14.2	0.00	14.5	0.00	**13.9**	0.00	4.71	**13.9**	0.00	5.73
100	25	**35.0**	9.73	**35.0**	0.00	35.1	0.00	**35.0**	0.04	15.42	**35.0**	0.28	15.04
100	50	**28.6**	26.81	28.9	0.20	28.8	0.10	**28.6**	0.05	29.03	**28.6**	0.69	19.06
100	75	**18.7**	12.54	18.9	0.30	19.1	0.10	**18.7**	0.19	16.27	**18.7**	0.11	12.52
125	25	**42.6**	41.11	**42.6**	0.10	**42.6**	0.40	**42.6**	0.00	27.03	42.7	0.12	19.91
125	50	**31.6**	181.41	32.3	0.60	32.5	0.90	**31.6**	0.44	80.92	**31.6**	0.52	61.33
125	75	**20.1**	58.80	20.9	0.20	20.8	0.10	**20.1**	0.47	42.64	**20.1**	0.75	33.86
150	25	**41.9**	159.69	42.1	0.30	42.0	0.30	**41.9**	0.00	16.73	42.2	0.04	15.88
150	50	**32.3**	728.19	32.8	1.80	33.3	1.40	**32.3**	0.59	21.25	**32.3**	0.75	23.28
150	75	**21.4**	199.15	21.9	0.50	22.1	0.80	**21.4**	0.64	37.67	**21.4**	0.87	50.10

solutions for two groups of instances ($n = 125$, $p = 25$ and $n = 150$, $p = 25$). The values in columns *gap(%)* show that, on average, SVNS showed better stability than BVNS in providing the best solutions. The average gap calculated over all instances from the Set 2 was 0.16% for SVNS, compared to 0.3% for BVNS. On average, the running times of BVNS and SVNS were similar for instances in the Set 2: BVNS was faster for instances with $n = 50, 100, 125$ and SVNS for $n = 75, 150$.

Computational results on the Sets 4 and 5, each containing 15 groups of large size WGRLP instances, are presented in Tables 5 and 6, respectively. These newly generated WGRLP instances are considered for the first time in the literature and no optimal solutions for these instances are known. Therefore, Tables 5 and 6 contain only results of the SVNS and BVNS methods, presented in the same way as in Table 4.

The results given in Table 5 show that BVNS method had better performance compared to SVNS regarding solution quality on the data Set 4. On average, BVNS obtained the best known solutions for 13 out of 15 groups of instances, while SVNS reached the best known solutions for 7 groups of instances only. On the other hand, SVNS showed the advantages over BVNS in the sense of running times. For all groups of instances from the Set 4, SVNS was up to 1.45 times faster than BVNS. The difference in running times is more obvious for larger problem dimensions. In addition, the average gap from the best known solution, calculated over all instances in this data set, was slightly lower for SVNS (1.66%) than for BVNS (1.71%).

198   L. Mrkela and Z. Stanimirović

Table 5. Computational results of the proposed SVNS and BVNS on large scale instances from Set 4

Size		SVNS			BVNS		
n	p(%)	best	gap(%)	t(s)	best	gap(%)	t(s)
175	25	**43.60**	0.55	28.44	**43.60**	0.37	33.55
175	50	34.40	0.92	67.02	**34.30**	0.64	77.74
175	75	21.60	0.68	25.50	**21.40**	1.13	22.65
200	25	**48.00**	0.49	58.62	**48.00**	0.40	64.49
200	50	32.80	0.88	52.09	**32.70**	1.06	58.89
200	75	**25.00**	1.81	53.50	**25.00**	0.96	54.07
300	25	50.90	0.61	223.31	**50.80**	0.78	346.04
300	50	**39.90**	1.70	314.77	**39.90**	1.58	448.48
300	75	**28.90**	1.85	202.92	**28.90**	1.95	282.75
400	25	50.90	1.82	614.73	**50.60**	1.26	674.45
400	50	40.60	1.98	606.37	**40.30**	2.65	626.66
400	75	**29.50**	4.20	390.84	29.60	3.26	479.98
500	25	**54.30**	1.57	757.32	54.50	2.20	786.44
500	50	44.80	2.89	690.11	**44.70**	4.01	927.09
500	75	27.30	3.83	424.06	**26.80**	4.95	699.36

Table 6. Computational results of the proposed SVNS and BVNS on large scale instances from Set 5

Size		SVNS			BVNS		
n	p(%)	best	gap(%)	t(s)	best	gap(%)	t(s)
175	25	**56.10**	0.00	17.06	**56.10**	0.00	14.30
175	50	**37.30**	0.44	40.62	**37.30**	0.34	43.34
175	75	**19.70**	0.07	17.19	**19.70**	0.07	27.31
200	25	**64.10**	0.00	42.48	**64.10**	0.10	27.35
200	50	**31.30**	0.19	40.17	**31.30**	0.00	35.42
200	75	**22.90**	0.05	29.99	**22.90**	0.05	20.83
300	25	**55.10**	0.33	240.05	**55.10**	0.05	231.45
300	50	**37.80**	0.50	170.46	**37.80**	0.29	244.32
300	75	**22.80**	1.15	173.30	22.90	1.01	148.42
400	25	**52.60**	0.59	384.67	**52.60**	0.26	290.12
400	50	34.90	0.78	433.95	**34.70**	0.25	458.64
400	75	**21.80**	0.62	227.74	**21.80**	0.47	263.24
500	25	**58.50**	0.92	630.76	58.70	0.66	555.41
500	50	**34.10**	1.15	542.47	34.30	0.91	563.93
500	75	**18.40**	0.34	408.37	**18.40**	0.26	345.23

According to results presented in Table 6, the proposed SVNS was superior over BVNS in the sense of solution quality on the data Set 5. On average, SVNS provided the best-known solutions for 14 out of 15 groups of instances, while BVNS reached the best-known solutions for 12 groups of instances from the Set 5. Average running times of BVNS and SVNS were similar on this data set. BVNS showed slightly better stability than SVNS, as the average gap values from the best known solution for all instances from the Set 5, were 0.30% and 0.46% in the case of BVNS and SVNS, respectively. From the data presented in Tables 4, 5 and 6, it can be also noticed that for instances with larger percentage of terminal nodes p, both SVNS and BVNS produce solutions with lower objective function values in shorter CPU times, as the obtained best solutions include smaller number of installed regenerators.

5   Conclusion

In this article, we considered a variant of Generalized Regenerator Location Problem (GRLP) dealing with optimization of optical telecommunication networks. As the quality of optical signal decreases with the distance from the origin node, it has to be regenerated by installing regenerator devices at some of the given locations. Having in mind high costs of regenerator deployment, the goal of GRLP is to install minimal number of regenerators, while ensuring a good quality signal transmission between terminal nodes. In practice, the costs of establishing regenerators at different locations vary, mostly due to real estate costs. This study deals with more realistic variant of GRLP, denoted as the Weighted Generalized Regenerator Location Problem (WGRLP). The considered WGRLP includes weights for each potential regenerator location, which correspond to the regenerator deployment costs. The objective of WGRLP is to minimize the sum of weights of chosen regenerator locations, while problem constraints are the same as in GRLP.

In order to solve WGRLP instances of real-life dimensions, we designed a variant of Variable Neighborhood Search (VNS) method, known as Skewed VNS (SVNS). The elements of the proposed SVNS were adapted to the characteristics of the considered WGRLP. Adequate data structures and efficient procedures for objective function update, feasibility check, and solution repair were implemented. A set of preliminary experiments was performed in order to tune SVNS parameters. The values obtained from parameter tuning tests were used in computational experiments with SVNS on the three data sets: the Set 2 from the literature containing small and medium size WGRLP instances and two newly generated data sets of large scale WGRLP instances, denoted as the Set 4 and Set 5. The proposed SVNS quickly reached all known optimal solutions on instances from the Set 2 and outperformed two existing constructive heuristics for WGRLP regarding solution's quality. On large scale WGRLP instances from the Sets 4 and 5, which were not preciously considered, SVNS was successful in providing its best solutions in short running times. In addition, basic variant of VNS is considered, obtained from SVNS implementation by using standard Move or Not step. By analyzing and comparing the results of SVNS and BVNS on all three data sets, it can be concluded that both VNS approaches showed good performance regarding solution quality, stability, and running times. In general, SVNS provided solutions of better quality compared to BVNS on instances from the Set 2 and Set 5, while BVNS was better than SVNS on instances from the Set 4. On instances from the Set 2 and Set 5, both VNS and SVNS had similar average running times, while SVNS was faster than BVNS in the case of Set 4. For all considered data sets, it was noticed that the objective function values of optimal/best known solutions decreased with the increase in the percentage of terminal nodes p, as smaller number of regenerators was required to ensure good quality signal transmission in the network.

Future work may be directed to combining SVNS or BVNS with an exact method in order to provide optimal solutions or to improve upper bounds for large scale WGRLP instances. We also plan to extend the considered WGRLP by involving capacities of edges and demands of terminal nodes and to develop adequate solution methods starting from SVNS implementation proposed in this study.

References

1. Chen, S., Ljubić, I., Raghavan, S.: The generalized regenerator location problem. INFORMS J. Comput. **27**(2), 204–220 (2015)
2. Chen, S., Ljubić, I., Raghavan, S.: The regenerator location problem. Networks **55**(3), 205–220 (2010)
3. Dégila, J.R., Sanso, B.: Topological design optimization of a Yottabit-per-second lattice network. IEEE J. Sel. Areas Commun. **22**(9), 1613–1625 (2004)
4. Duarte, A., Martí, R., Resende, M.G., Silva, R.M.: Improved heuristics for the regenerator location problem. Int. Trans. Oper. Res. **21**(4), 541–558 (2014)
5. Flammini, M., Marchetti-Spaccamela, A., Monaco, G., Moscardelli, L., Zaks, S.: On the complexity of the regenerator placement problem in optical networks. IEEE/ACM Trans. Netw. **19**(2), 498–511 (2011)
6. Hansen, P., Mladenović, N.: Variable neighborhood search. In: Burke, E.K., Graham, R.D. (eds.) Search Methodologies: Introductory Tutorials in Optimization and Decision Support Techniques, pp. 313–337. Springer, New York (2014). https://doi.org/10.1007/978-1-4614-6940-7_12
7. Hansen, P., Mladenović, N., Pérez, J.A.: Variable neighbourhood search: methods and applications. Ann. Oper. Res. **175**(1), 367–407 (2010)
8. Höller, H., Melián, B., Voß, S.: Applying the pilot method to improve VNS and GRASP metaheuristics for the design of SDH/WDM networks. Eur. J. Oper. Res. **191**(3), 691–704 (2008)
9. Loudni, S., Boizumault, P., David, P.: On-line resources allocation for ATM networks with rerouting. Comput. Oper. Res. **33**(10), 2891–2917 (2006)
10. Masri, H., Krichen, S., Guitouni, A.: A multi-start variable neighborhood search for solving the single path multicommodity flow problem. Appl. Math. Comput. **251**, 132–142 (2015)
11. Melián, B.: Using memory to improve the VNS metaheuristic for the design of SDH/WDM networks. In: Almeida, F., et al. (eds.) HM 2006. LNCS, vol. 4030, pp. 82–93. Springer, Heidelberg (2006). https://doi.org/10.1007/11890584_7

12. Melián-Bautista, B., Höller, H., Voß, S.: Designing WDM networks by a variable neighborhood search. J. Telecommun. Inf. Technol. **4**, 15–20 (2006)

13. Meric, L., Pesant, G., Pierre, S.: Variable neighbourhood search for optical routing in networks using latin routers. Annales des Télécommunications - Ann. Telecommun. **59**(3), 261–286 (2004)

14. Mladenović, N., Hansen, P.: Variable neighborhood search. Comput. Oper. Res. **24**(11), 1097–1100 (1997)

15. Quintana, J., Sánchez-Oro, J., Duarte, A.: Efficient greedy randomized adaptive search procedure for the generalized regenerator location problem. Int. J. Comput. Intell. Syst. **9**(6), 1016–1027 (2016)

16. Resende, M.G.C., Ribeiro, C.C.: GRASP: greedy randomized adaptive search procedures. In: Burke, E.K., Kendall, G. (eds.) Search Methodologies: Introductory Tutorials in Optimization and Decision Support Techniques, pp. 287–312. Springer, Boston (2014). https://doi.org/10.1007/978-1-4614-6940-7_11

17. Sánchez-Oro, J., Duarte, A.: Beyond unfeasibility: strategic oscillation for the maximum leaf spanning tree problem. In: Puerta, J.M., Gámez, J.A., Dorronsoro, B., Barrenechea, E., Troncoso, A., Baruque, B., Galar, M. (eds.) CAEPIA 2015. LNCS (LNAI), vol. 9422, pp. 322–331. Springer, Cham (2015). https://doi.org/10.1007/978-3-319-24598-0_29

Using a Variable Neighborhood Search to Solve the Single Processor Scheduling Problem with Time Restrictions

Rachid Benmansour[1,3]([:envelope:]) [iD], Oliver Braun[2], Saïd Hanafi[3][iD],
and Nenad Mladenovic[4][iD]

[1] SI2M Laboratory, Institut National de Statistique et d'Economie Appliquée,
Rabat, Morocco
r.benmansour@insea.ac.ma
[2] Trier University of Applied Sciences, Environmental Campus Birkenfeld,
55761 Birkenfeld, Germany
o.braun@umwelt-campus.de
[3] LAMIH - UMR CNRS 8201, Université Polytechnique des Hauts-de-France,
Valenciennes, France
said.hanafi@univ-valenciennes.fr
[4] Serbian Academy of Sciences and Arts, Belgrade, Serbia
nenad@mi.sanu.ac.rs

Abstract. We study the single-processor scheduling problem with time restrictions in order to minimize the makespan. In this problem, n independent jobs have to be processed on a single processor, subject only to the following constraint: During any time period of length $\alpha > 0$ the number of jobs being executed is less than or equal to a given integer value B. It has been shown that the problem is NP-hard even for $B = 2$. We propose the two metaheuristics variable neighborhood search and a fixed neighborhood search to solve the problem. We conduct computational experiments on randomly generated instances. The results indicate that our algorithms are effective and efficient regarding the quality of the solutions and the computational times required for finding them.

Keywords: Scheduling · Time restrictions · Single processor ·
NP-hard · Variable neighborhood search · Fixed neighborhood search

1 Introduction

Scheduling problems are studied for two main reasons. On the one hand, they are very common problems in practice and can model industrial problems such as planning, assignment, transportation, and so on. On the other hand, many scheduling problems are NP-hard and therefore interest researchers to find effective and efficient solutions. In this paper, we study the single-processor scheduling problem with time restrictions (STR) in order to minimize the makespan. In this problem, n independent jobs have to be processed on a single processor,

© Springer Nature Switzerland AG 2019
A. Sifaleras et al. (Eds.): ICVNS 2018, LNCS 11328, pp. 202–215, 2019.
https://doi.org/10.1007/978-3-030-15843-9_16

such that during any time period of length $\alpha > 0$ the number of jobs being executed is less than or equal to a given integer value B. We assume that the jobs are simultaneously available for processing at the beginning of the planning horizon, and their processing times are fixed and known in advance.

Formally, the problem can be stated as follows. Given is a set $N = \{1, 2, \ldots, n\}$ of n independent jobs with their processing times p_i, $i \in N$. Only one job can be executed at any point in time and the jobs cannot be preempted. A feasible schedule is a permutation π of the set N, say $\pi = (\pi_1, \pi_2, \ldots, \pi_n)$, such that the jobs are placed sequentially on the real line: The initial job π_1 begins at time 0 and finishes at time p_{π_1}. For $i \geq 2$, the job π_i begins as soon as the job π_{i-1} is completed, such that the following constraint is always satisfied:

$$\forall x \in \mathbb{R}_{\geq 0}, \text{ the interval } [x, x + \alpha) \text{ can intersect at most } B \text{ jobs.}$$

This constraint reflects the condition that each job needs one of B additional resources for being processed and that a resource has to be renewed (e.g. preventive maintenance) in α time-units after the processing of a job has been finished. The objective function is to minimize C_{max}, where the *makespan* C_{max} corresponds to the time at which the last job π_n is finished. In this paper we assume also that all the data (B, α, p_i for $i \in N$) are integers. The scheduling problem discussed in this paper has practical applications in manufacturing systems. As an illustration of the STR problem, let imagine that each job being processed requires the use of one of B identical external resources. Furthermore, each external resource that has been used needs a certain amount of time α to be reset before it can be used again. Hence, it is never possible to process more than B jobs during any interval $[x, x + \alpha)$, $\forall x \in \mathbb{R}_{\geq 0}$.

The STR problem was studied at first by Braun et al. [6] where the authors provide a detailed worst-case analysis. They show that for $B = 2$, any feasible solution can be processed within a factor of $\frac{4}{3}$ of the optimum (plus the additional constant 1), and that for $B \geq 3$, this factor is equal to $2 - \frac{1}{B-1}$ of the optimum (plus the additional constant 3). Both proposed factors are best possible. Later, they improve the additional constant to $\frac{B}{B-1}$ and provide an analysis of the LPT-algorithm, where the jobs are ordered non-increasingly according to the Longest-Processing-Time-first (LPT) algorithm and show that LPT-ordered jobs can be processed within a factor of $2 - \frac{2}{B}$ of the optimum (plus 1) and that this factor is best possible [7]. The authors also state that an easy improvement of the LPT algorithm where the schedule is started with the smallest job and then LPT is performed, leads to an additional constant of only $\frac{1}{2}$. Zhang et al. [20] show that for $B \geq 5$ there exists a permutation of the jobs which can be processed within a factor of $5/4$ of the optimum (plus an additional small constant). When $B = 3$ or $B = 4$ the corresponding factor equals $\frac{B}{B-1}$.

Benmansour et al. [2] proposed a Mixed Integer Linear Programming formulation (MIP) based on time index variables and later another MIP model based on assignment and positional date variables (see also [2–4]).

The STR problem can be seen as a special case of the parallel machine scheduling problem with a single server (PSS) (see [5]). The PSS problem can be stated as follows. There are m identical parallel machines which must process n jobs. Each job has a known integer processing time p_i' and before its processing, it must be loaded on a machine which will take s_i units of time (the setup operation). There is only one server for all machines. During the loading operation, both the machine and the server are occupied.

In the literature this problem is denoted as $Pm, S1 \mid s_i, p_i' \mid C_{max}$, where Pm and $S1$ mean that there exist respectively m machines and one server ([1,8,10, 14,16]). The PSS problem was proven to be NP-hard [8]. As far as we know, no attempt has been made to solve large instances of neither the STR nor the PSS problem. Therefore, our approach is to propose a variable neighborhood search metaheuristic to solve these problems.

The remainder of this paper is as follows: The next two sections are dedicated to present variable neighborhood search and fixed neighborhood search respectively. In Sect. 4 we first present the comparison between the performances of the two algorithms. Then we present the performance of the VNS algorithm with respect to an exact method previously published in the literature. Finally, we present in the conclusion the avenues of research that can be considered to complete this work afterwards.

2   Variable Neighborhood Search

Variable neighborhood search (VNS) is a metaheuristic proposed by Mladenović and Hansen in 1997 [18]. VNS systematically changes neighborhood structures during the search for an optimal (or near-optimal) solution based on the following observations: (i) A local optimum relative to one neighborhood structure is not necessarily a local optimum for another neighborhood structure; (ii) A global optimum is a local optimum with respect to all neighborhood structures; (iii) Empirical evidence shows that for many problems all or a large majority of the local optima are relatively close to each other. Several implementations of VNS have been proposed and applied successfully to solve NP-hard problems in different domains such as: scheduling, routing, maintenance problems, etc. [17, 19,21]. For an overview on VNS applications and VNS variants see [11], and [13]. Among the variants most used, one finds the General variable neighborhood search (GVNS) [12] which uses variable neighborhood descent (VND) as a local search to explore several neighborhood structures at once.

VNS uses a finite set of neighborhood structures denoted as $\mathcal{N}_k$, where $k \in \{1, 2, \ldots, k_{max}\}$. The kth neighborhood of solution π, $\mathcal{N}_k(\pi)$, is a subset of the search space, which is obtained from the solution π by small changes. The VNS (Algorithm 1) includes an improvement phase in which a local search is applied and one so-called shaking phase used to hopefully resolve local minima traps. The local search and the shaking procedure, together with the neighborhood change step, are executed alternately until fulfilling a predefined stopping criterion. As the stopping criterion, most often, is used maximum CPU time allowed to be consumed by VNS.

Algorithm 1. VNS

Data: An instance of STR, neighborhood structures $\mathcal{N}_k$ for $k = 1, 2, \ldots, k_{max}$,
h diversification parameter, CPU time: CPU_{MAX}
Result: Solution π
Generate an initial solution π;
repeat
$\quad$ $k \leftarrow 1$;
$\quad$ **while** $k \leq k_{max}$ **do**
$\quad\quad$ $\pi' \leftarrow$ **Shaking**(π, k, h);
$\quad\quad$ $\pi'' \leftarrow$ **Local_Search**(π', k);
$\quad\quad$ **if** $f(\pi'') < f(\pi)$ **then**
$\quad\quad\quad$ $\pi \leftarrow \pi''$;
$\quad\quad\quad$ $k \leftarrow 1$;
$\quad\quad$ **else**
$\quad\quad\quad$ $k \leftarrow k + 1$;
$\quad\quad$ **end**
$\quad$ **end**
until $CPU \geq CPU_{MAX}$;

The Algorithm 1 integrates two important procedures; it includes a *Shaking* phase to escape from the local minima traps and a *Local search* procedure (intensification phase) to exploit the accumulated search experience.

In order compute the cost of a given sequence of the jobs π, denoted as $f(\pi)$, the following proposition (see [4, 20]) is used.

Proposition 1. *Let C_j be the completion time of the job placed at position j, and $p_{[j]}$ its corresponding processing time. Then the completion time is computed for all jobs as follows ([6]):*

$$C_j = \begin{cases} 0 & \text{if} \quad j = 0, \\ p_{[j]} + C_{j-1} & \text{if} \quad 1 \leq j \leq B, \\ p_{[j]} + \max\left(C_{j-1}, C_{j-B} + \alpha\right) & \text{if} \quad B+1 \leq j \leq n. \end{cases}$$

Note that $f(\pi)$ corresponds to the makespan of the sequence π (i.e. $f(\pi) = C_{\pi_n}$).

2.1 Initial Solution and Neighborhood Structures

It is clear that any permutation of the jobs π is a feasible solution for the STR problem. The jobs should be scheduled according to the procedure in Proposition 1 that provides also the value of objective function $f(\pi) = C_{\pi_n}$. The initial solution π is generated by using the LPT rule.

In order to obtain an efficient VNS algorithm we have to decide about three things [12,15]: which neighborhoods to use, how to use them in the search process, and finally which search strategy to use. It is worth mentioning that the best number of neighborhoods is often 2 [9]. Suppose π is a permutation of jobs in N. It is clear that π is a feasible solution to the STR problem. We proposed these neighborhood structures and chose the most effective among them after some preliminary tests.

- Neighborhood $\mathcal{N}_1(\pi) = Transpose(\pi)$: It consists of all permutations that can be obtained by swapping two adjacent jobs in π. Let π' be a neighborhood solution of π obtained by swapping π_k and π_{k+1}. Note that according to Proposition 1 $f(\pi') = f(\pi)$ for $k < B$.
- Neighborhood $\mathcal{N}_2(\pi) = Swap(\pi)$: The neighborhood set consists of all solutions obtained from the solution π swapping two random jobs of π.
- Neighborhood $\mathcal{N}_5(\pi) = Left\_Pivot(\pi)$: Given a job π_j we reverse the order of jobs before π_j.
- Neighborhood $\mathcal{N}_6(\pi) = Right\_Pivot(\pi)$: Given a job π_j we reverse the order of jobs after π_j.
- Neighborhood $\mathcal{N}_7(\pi) = Reverse(\pi)$: Given two jobs π_j and π_k we reverse the order of jobs being between those two jobs.

After performing several computational tests, the following neighborhood structures were chosen in the proposed VNS algorithm: $Transpose(\pi)$, $Swap(\pi)$ and $Reverse$ ($k_{max} = 3$).

2.2   Local Search and Shaking

The Local Search (Algorithm 2) starts with an initial solution π^0 and tries continually to construct a new improved solution from the current solution π by exploring its neighborhood $\mathcal{N}_k(\pi)$. The process continues to generate neighboring solutions until no further improvement can be made. A basic version of a local search is the descent method, also called iterative improvement, in which the current solution is replaced by its neighboring solution with lower cost. In our implementation, we use the first improvement search strategy (i.e. as soon as an improving solution π' in a neighborhood structure $\mathcal{N}_k(\pi)$ is detected it is set to be the new incumbent solution $(\pi \leftarrow \pi')$).

Algorithm 2. Local Search

Data: Solution π^0 and neighborhood structure $\mathcal{N}_k$
Result: Solution π
$\pi \leftarrow \pi^0$;
$Stop \leftarrow False$;
while $Stop = False$ **do**
    Select $\pi' \in \mathcal{N}_k(\pi)$ such that $f(\pi') < f(\pi)$;
    if π' *exists* **then**
       | $\pi \leftarrow \pi'$;
    else
       | $Stop = True$;
    end
end
return π

The aim of a shaking procedure used within a VNS algorithm is to hopefully escape from local minima traps. The simple shaking procedure consists of selecting a random solution from the current neighborhood of the current solution $\mathcal{N}_k(\pi)$.

Algorithm 3. Shaking

Data: Solution π and neighborhood structure $\mathcal{N}_k$,
h diversification parameter
Result: Solution π
for $j = 1$ *to* h **do**
    Select randomly $\pi' \in \mathcal{N}_k(\pi)$;
    $\pi \leftarrow \pi'$;
end
return π

To test the performances of the proposed algorithms, in this case VNS and FNS algorithms, we tested them on the same instances. The results demonstrate the effectiveness of both algorithms with a slight advantage for VNS algorithm, which is capable of producing high-quality solutions. On the other hand we compared the VNS algorithm with an exact model for the small instances of the problem. This model is not reported in this article but the interested reader can find it in [2]. The results show that VNS is able to produce very good results in a reasonable time. In the next section we present the Fixed neighborhood search algorithm (FNS).

3   Fixed Neighborhood Search

Fixed neighborhood search (FNS) is a step in between classical local search and variable metric on the one hand and VNS on the other [13]. Instead of generating initial solutions completely at random, the next starting point for local search in FNS is a randomly generated solution taken from the vicinity of the best one found so far (incumbent solution). Fixed neighborhood search is described in Algorithm 4.

Algorithm 4. Fixed neighborhood search algorithm

Data: Neighborhood structures $\mathcal{N}$ and $\mathcal{N}'$, π, CPU_{MAX}
Result: Solution π
$f_{best} \leftarrow \infty$
$t \leftarrow 0$
repeat
    Generate point $\pi' \in \mathcal{N}(\pi)$ at random;
    $x \leftarrow \textbf{Local\_Search}(\pi', \mathcal{N}')$;
    if $f(\pi') < f_{best}$ **then**
        $\pi \leftarrow \pi'$;
        $f_{best} = f(\pi)$;
    end
    $t \leftarrow CPU.Time$;
until $t \geq CPU_{MAX}$;
return π

In Algorithm 4, Neighborhood structures $\mathcal{N}$ and $\mathcal{N}'$ represent respectively $Reverse(\pi)$ and $Swap(\pi)$ structures. The effectiveness of this algorithm compared to the VNS algorithm did not push us to try to test other neighborhood structures.

4   Computational Results

All experiments were conducted on an Intel i7 2.8 GHz computer with 16GB of RAM memory under Windows 7 operating system and we used CPLEX 12.6 as the integer programming solver. The problem instances are generated as follows. The processing times p_i were generated from integer uniform distributions in $[1, 10]$ and $[1, 100]$, and the number of jobs $n \in \{10, 100, 500\}$. The number of resources B is chosen as $B \in \{2, 5, 10\}$. The length of the unit interval, i.e. the time that a resource needs to recover until it is available again, was chosen as $10, 100, 1000$, depending on the maximum possible processing time (e.g. when the processing times are out of the interval $[1, 10]$ we chose the length of the unit interval as 10). Furthermore, in order to make the problem instances even harder to solve, we chose the length α of the unit-interval 10 times larger than the maximum possible processing time. As an example, when the processing times are out of $[1, 10]$, we chose the length of the unit interval as $\alpha = 100$. The notation $NxUy$ means that the number of jobs is equal to x and that the processing times of the instance were generated from the integer uniform distribution $[1, y]$.

The algorithms solved to optimality all small and medium size instance in less than 10 s. For each value of $B \in \{2, 5, 10\}$ and of $\alpha \in \{10, 100\}$ (or $\alpha \in \{100, 1000\}$) we generate randomly 10 instances.

4.1   Comparison Between VNS and FNS Algorithms

Both algorithms were tested on the instances described above. The time limit was set to 10 s for each algorithm. The computational results are presented in the appendix. For each table the we report the values returned by the VNS algorithm and the FNS algorithm; respectively (f_{VNS}) and (f_{FNS}). The parameter l represents the instances.

Table 1. Comparison of the efficiency of VNS and FNS.

	N10U10	N10U100	N50U10	N50U100	N100U10	N100U100	N500U10	N500U100
=	97	93	75	65	75	17	47	33
+	100	50	27	71	33	100	100	90

In Table 1 the first row represents, in %, the number of times algorithms VNS and FNS find the same value (= symbol). For the remaining instances, in the second row we report, in %, the number of times VNS algorithm was better than FNS algorithm (+ symbol). For example, in column one, in 97% of all instances of type $N10U10$ VNS and FNS found the same value; for the remaining cases, VNS found a better solution than FNS.

In sum, for $n = 10$ both algorithms have the same performance with a minimal advantage for VNS algorithm. For $n = 50$, on average the two algorithms find the same result in 70% of cases. For the case N50U10 FNS algorithm finds better results than VNS in 63% of the time. In contrast, for the case N50U100 VNS algorithm finds better results than FNS in 71% of the time. For $n \geq 100$ VNS algorithm gives better results except for N100U10 instances type where FNS algorithms find in five instances only a better results.

In view of these results we decided to compare the results of the VNS algorithm with exact results obtained by the mathematical model.

4.2   Comparison Between VNS and the MIP Model

In the following tables, we present the results obtained by the MIP model and the VNS algorithm. The MIP model used for computational results was published in [3]. The columns of each table correspond respectively to the number of jobs n, the number of resources B, the length of the unit interval α, the *minimum* (min), the *average* (avg) and the *standard-deviation* (std) of the CPU time (in seconds) that was needed to compute optimal values using the MIP formulation and CPLEX 12.6: (CPU_MIP), the *minimum* (min), the *average* (avg) and the *standard-deviation* (std) of the GAP: (MIP_GAP), the relative deviation of the VNS algorithm (RD), and in the last column, the relative deviation of the VNS algorithm from the linear relaxation value of MIP (RDL). The last two metrics are calculated as follows:

$$RD = \frac{f_{VNS} - f_{MIP}}{f_{MIP}} \times 100$$

$$RDL = \frac{f_{VNS} - f_{LP}}{f_{LP}} \times 100$$

f_{VNS}, f_{MIP} and f_{LR} are the solution values obtained by VNS algorithm, the MIP model and the linear relaxation of the MIP model. The time limit for CPLEX was set to 3600 s for all instances. The time limit for VNS = 1 s. The results are given for each 10 instances: i.e. Each cell, in a table, represents the synthesis of the results of 10 instances. Note that ∗∗ means that CPLEX was not able to solve the problem.

In sum, the VNS algorithm found optimal solutions for all instances with $n = 10$. The algorithm can also be used to solve any type of instance because the MIP model is only suitable for small instances. The reason is that the quality of the solutions found by the VNS algorithm is very good and the time required to find these solutions is negligible. This brings us back to the question of choice of instances. In the future we will generate other instances according to other probability density function (normal, Beta, etc.) to see if our algorithm always remains efficient or not (Tables 2, 3, 4, 5, 6 and 7).

Table 2. $n = 10$ and $p_i \in [1, 10]$

n	B	α	CPU_MIP			MIP_GAP			RD			RDL		
			min	avg	std	min	avg	std	min	avg	std	min	avg	std
10	10	100	0.17	0.21	0.02	0	0	0	0	0	0	0	0	0
10	10	10	0.08	0.13	0.02	0	0	0	0	0	0	0	0	0
10	5	100	0.08	0.23	0.24	0	0	0	0	0	0	0	0.116	0.187
10	5	10	0.05	0.05	0.01	0	0	0	0	0	0	0	0	0
10	2	100	0.07	0.19	0.11	0	0	0	0	0	0	0	0.035	0.056
10	2	10	0.07	0.14	0.11	0	0	0	0	0	0	0	0.216	0.350

Table 3. $n = 10$ and $p_i \in [1, 100]$

n	B	α	CPU_MIP			MIP_GAP			RD			RDL		
			min	avg	std	min	avg	std	min	avg	std	min	avg	std
10	10	1000	0.13	0.21	0.05	0	0	0	0	0	0	0	0	0
10	10	100	0.11	0.13	0.01	0	0	0	0	0	0	0	0	0
10	5	1000	0.12	0.60	0.32	0	0	0	0	0	0	0	0.074	0.084
10	5	100	0.07	0.09	0.01	0	0	0	0	0	0	0	0	0
10	2	1000	0.08	0.55	0.40	0	0	0	0	0	0	0	0.022	0.025
10	2	100	0.07	0.44	0.31	0	0	0	0	0	0	0	0.156	0.162

Table 4. $n = 100$ and $p_i \in [1, 10]$

n	B	α	CPU_MIP			MIP_GAP			RD			RDL		
			min	avg	std	min	avg	std	min	avg	std	min	avg	std
100	10	100	1.80	911.38	1437.61	0	0.0203	0.042	1.440	1.810	0.211	1.491	1.866	0.229
100	10	10	0.32	0.56	0.30	0	0	0	0	0	0	0	0	0
100	5	100	1.34	2.80	1.38	0	0	0	0.199	0.401	0.124	0.244	0.433	0.116
100	5	10	0.45	1.28	0.68	0	0	0	0	0	0	0	0	0
100	2	100	0.65	0.95	0.19	0	0	0	0	0	0	0	0.007	0.005
100	2	10	0.93	1.12	0.11	0	0	0	0	0.0127	0.040	0	0.059	0.036

Table 5. $n = 100$ and $p_i \in [1, 100]$

n	B	α	CPU_MIP			MIP_GAP			RD			RDL		
			min	avg	std	min	avg	std	min	avg	std	min	avg	std
100	10	1000	9.14	49.24	48.96	0.098	0.102	0.001	**	**	**	1.189	1.466	0.286
100	10	100	0.45	0.51	0.05	0	0	0	0	0	0	0	0	0
100	5	1000	5.15	7.59	1.71	0.019	0.055	0.028	0.069	0.185	0.076	0.156	0.245	0.060
100	5	100	1.68	2.70	1.12	0	0.009	0.029	−0.093	−0.009	0.029	0	0	0
100	2	1000	1.12	1.43	0.22	0	0.003	0.002	−0.007	−0.001	0.002	0	0.001	0.001
100	2	100	1.27	2.11	0.99	0	0.011	0.010	−0.027	0.013	0.034	0	0.027	0.030

Table 6. $n = 500$ and $p_i \in [1, 10]$

n	B	α	CPU_MIP			MIP_GAP			RD			RDL		
			min	avg	std	min	avg	std	min	avg	std	min	avg	std
500	10	100	148.43	347.49	154.33	0	0.036	0.044	**	**	**	1.485	1.772	0.202
500	10	10	16.90	18.77	01.98	0	0	0	0	0	0	0	0	0
500	5	100	65.79	95.65	29.71	0	0.049	0.040	0.353	0.496	0.103	0.398	0.543	0.105
500	5	10	17.48	107.69	65.97	0	0.036	0.067	−0.180	−0.025	0.076	0	0.011	0.017
500	2	100	19.65	38.40	21.15	0	0.082	0.078	−0.186	−0.059	0.095	0.006	0.023	0.033
500	2	10	50.04	75.62	37.27	0	0.009	0.019	2.944	3.628	0.610	2.873	3.518	0.561

Table 7. $n = 500$ and $p_i \in [1, 100]$

n	B	α	CPU_MIP			MIP_GAP			RD			RDL		
			min	avg	std	min	avg	std	min	avg	std	min	avg	std
500	10	1000	275.70	275.70	**	0.016	0.016	**	**	**	**	1.578	1.779	0.150
500	10	100	27.83	27.83	**	0	0	**	**	**	**	0	0	0
500	5	1000	705.49	705.49	**	0.004	0.004	**	**	**	**	0.313	0.474	0.126
500	5	100	127.01	127.01	**	0.042	0.042	**	**	**	**	0	0.088	0.123
500	2	1000	29.25	29.25	**	0.213	0.213	**	**	**	**	0.004	0.016	0.015
500	2	100	214.92	214.91	**	0	0	**	**	**	**	2.733	3.416	0.600

5   Conclusion

This paper proposes a variable neighborhood search algorithm and a fixed neighborhood search algorithm to solve the single processor scheduling problem with time restrictions. Based on the generated instances, it turns out that the proposed VNS algorithm is very efficient both in terms of solution time and quality of the solutions. For example, large instances are solved in seconds while the exact model fails to solve instances of average size. Prospects are multiple for this new type of problem. A first track is to generalize the problem to non-homogeneous resources, i.e. α will depend on the (usage of the) resources and to consider other objective functions. Furthermore, we will propose a new set of benchmark instances to test the efficiency of the algorithm.

Appendix

See Tables 8, 9, 10, 11, 12, 13, 14 and 15.

Table 8. Instances of type $N10U10$

l	α	$B=10$		$B=5$		$B=2$	
		f_{VNS}	f_{FNS}	f_{VNS}	f_{FNS}	f_{VNS}	f_{FNS}
1	100	61	61	135	135	435	435
2		46	46	124	124	424	424
3		50	50	126	126	426	426
4		50	50	126	126	426	426
5		56	56	131	131	431	431
6		47	47	125	126	425	426
7		53	53	128	128	428	428
8		43	43	123	123	423	423
9		56	56	131	131	431	431
10		44	44	124	124	424	424
1	10	61	61	61	61	75	75
2		46	46	46	46	64	64
3		50	50	50	50	66	66
4		50	50	50	50	66	66
5		56	56	56	56	71	71
6		47	47	47	47	65	65
7		53	53	53	53	68	68
8		43	43	43	43	63	63
9		56	56	56	56	71	71
10		44	44	44	44	64	64

Table 9. Instances of type $N10U100$

l	α	$B=10$		$B=5$		$B=2$	
		f_{VNS}	f_{FNS}	f_{VNS}	f_{FNS}	f_{VNS}	f_{FNS}
1	1000	564	564	1321	1321	4322	4321
2		414	414	1220	1220	4220	4220
3		457	457	1234	1234	4234	4234
4		394	394	1206	1206	4206	4206
5		650	650	1339	1339	4339	4339
6		457	457	1241	1241	4241	4241
7		519	519	1261	1261	4261	4261
8		571	571	1306	1308	4308	4306
9		319	319	1171	1171	4171	4171
10		439	439	1225	1225	4225	4225
1	100	564	564	564	564	721	721
2		414	414	414	414	620	620
3		457	457	457	457	639	639
4		394	394	394	394	606	606
5		650	650	650	650	740	740
6		457	457	457	457	641	641
7		519	519	519	519	661	661
8		571	571	571	571	706	708
9		319	319	319	319	571	571
10		439	439	439	439	625	625

Table 10. Instances of type $N50U10$

l	α	$B=10$		$B=5$		$B=2$	
		f_{VNS}	f_{FNS}	f_{VNS}	f_{FNS}	f_{VNS}	f_{FNS}
1	100	469	468	972	972	2545	2545
2		463	461	963	960	2529	2529
3		466	467	964	964	2539	2539
4		463	463	963	962	2532	2532
5		467	465	967	966	2540	2540
6		463	463	963	967	2535	2535
,7		464	463	965	964	2534	2534
8		464	463	965	964	2536	2536
9		468	468	966	968	2539	2539
10		456	457	957	955	2519	2519
1	10	287	287	287	287	385	385
2		256	256	256	256	369	369
3		276	276	276	276	379	379
4		262	262	262	262	372	372
5		277	277	277	277	380	380
6		267	267	267	267	375	375
7		266	266	266	266	374	374
8		269	269	269	269	376	376
9		276	276	276	276	379	379
10		235	235	235	235	359	359

Table 11. Instances of type $N50U100$

l	α	$B=10$		$B=5$		$B=2$	
		f_{VNS}	f_{FNS}	f_{VNS}	f_{FNS}	f_{VNS}	f_{FNS}
1	1000	4589	4603	9599	9606	25327	25327
2		4596	4588	9604	9616	25300	25300
3		4567	4571	9552	9569	25235	25235
4		4670	4671	9671	9678	25426	25426
5		4601	4587	9585	9599	25311	25311
6		4536	4541	9534	9531	25195	25195
7		4595	4599	9596	9598	25280	25280
8		4543	4546	9522	9531	25205	25205
9		4527	4525	9513	9514	25161	25161
10		4591	4584	9591	9586	25288	25288
1	100	2646	2646	2646	2646	3727	3727
2		2590	2590	2590	2590	3700	3700
3		2460	2460	2460	2460	3635	3635
4		2833	2833	2833	2833	3826	3827
5		2612	2612	2612	2612	3711	3711
6		2387	2387	2387	2387	3595	3595
7		2553	2553	2553	2553	3680	3680
8		2408	2408	2408	2408	3606	3606
9		2316	2316	2316	2316	3561	3561
10		2569	2569	2569	2569	3688	3688

Table 12. Instances of type $N100U10$

l	α	$B = 10$		$B = 5$		$B = 2$	
		f_{VNS}	f_{FNS}	f_{VNS}	f_{FNS}	f_{VNS}	f_{FNS}
1	100	996	996	2031	2028	5193	5193
2		981	981	2007	2007	5146	5146
3		997	1001	2027	2028	5186	5186
4		995	995	2025	2022	5184	5184
5		984	989	2015	2012	5156	5156
6		991	990	2018	2019	5172	5172
7		999	997	2026	2025	5181	5181
8		991	986	2019	2018	5165	5165
9		995	995	2029	2026	5185	5185
10		996	991	2020	2021	5174	5174
1	10	584	584	584	584	783	783
2		489	489	489	489	736	736
3		569	569	569	569	776	776
4		565	565	565	565	774	774
5		510	510	510	510	746	746
6		541	541	541	541	762	762
7		560	560	560	560	771	771
8		527	527	527	527	755	755
9		567	567	567	567	775	775
10		545	545	545	545	764	764

Table 13. Instances of type $N100U100$

l	α	$B = 10$		$B = 5$		$B = 2$	
		f_{VNS}	f_{FNS}	f_{VNS}	f_{FNS}	f_{VNS}	f_{FNS}
1	1000	9816	9827	20126	20113	51661	51661
2		9780	9788	20061	20055	51514	51514
3		9780	9809	20064	20060	51526	51526
4		9734	9733	20034	20038	51444	51444
5		9838	9841	20125	20140	51685	51685
6		9793	9817	20082	20068	51563	51563
7		9805	9786	20128	20106	51628	51628
8		9791	9819	20123	20128	51630	51630
9		9797	9786	20081	20083	51573	51573
10		9752	9796	20043	20062	51455	51455
1	100	5317	5317	5317	5317	7562	7562
2		5025	5025	5025	5025	7416	7414
3		5047	5047	5047	5047	7426	7426
4		4886	4886	4886	4886	7344	7344
5		5365	5365	5365	5365	7585	7585
6		5120	5120	5120	5120	7463	7463
7		5250	5250	5250	5250	7529	7528
8		5255	5255	5255	5255	7530	7530
9		5138	5138	5138	5138	7473	7473
10		4906	4906	4906	4906	7355	7355

Table 14. Instances of type $N500U10$

l	α	$B = 10$		$B = 5$		$B = 2$	
		f_{VNS}	f_{FNS}	f_{VNS}	f_{FNS}	f_{VNS}	f_{FNS}
1	100	5248	5263	10475	10496	26281	26281
2		5239	5261	10472	10480	26267	26267
3		5251	5260	10484	10499	26291	26291
4		5249	5272	10475	10499	26271	26272
5		5255	5266	10487	10505	26299	26299
6		5245	5265	10458	10482	26232	26232
7		5258	5263	10477	10503	26282	26283
8		5252	5265	10476	10486	26282	26282
9		5246	5272	10467	10505	26270	26270
10		5249	5271	10478	10512	26296	26296
1	10	2759	2759	2759	2759	3909	3947
2		2731	2731	2731	2731	3901	3952
3		2779	2779	2779	2779	3915	3978
4		2740	2740	2740	2740	3897	3936
5		2795	2795	2795	2795	3925	3977
6		2661	2661	2661	2661	3852	3932
7		2761	2761	2761	2761	3903	3950
8		2761	2761	2761	2761	3906	3950
9		2737	2737	2738	2738	3909	3974
10		2789	2789	2789	2789	3915	3956

Table 15. Instances of type $N500U100$

l	α	$B = 10$		$B = 5$		$B = 2$	
		f_{VNS}	f_{FNS}	f_{VNS}	f_{FNS}	f_{VNS}	f_{FNS}
1	1000	52254	52469	104420	104634	262002	261997
2		52116	52342	104191	104566	261657	261657
3		52160	52396	104193	104415	261662	261643
4		51850	52145	103880	104115	260889	260891
5		51898	52246	103920	104155	261052	261052
6		52110	52467	104184	104578	261625	261632
7		52243	52457	104307	104482	261830	261826
8		52096	52247	104088	104400	261266	261265
9		52080	52271	104299	104591	261655	261658
10		51956	52211	103902	104091	260834	260845
1	100	25985	25985	25985	25985	38150	38606
2		25312	25312	25312	25312	37827	38427
3		25283	25283	25283	25283	37812	38250
4		23772	23772	23772	23772	37016	37493
5		24101	24101	24101	24101	37262	37642
6		25243	25243	25243	25243	37921	38339
7		25649	25649	25649	25649	37962	38540
8		24527	24527	24527	24527	37481	38025
9		25304	25304	25304	25307	37854	38289
10		23665	23665	23665	23672	36912	37555

References

1. Abdekhodaee, A.H., Wirth, A.: Scheduling parallel machines with a single server: some solvable cases and heuristics. Comput. Oper. Res. **29**(3), 295–315 (2002)
2. Benmansour, R., Braun, O., Artiba, A.: On the single-processor scheduling problem with time restrictions. In: 2014 International Conference on Control, Decision and Information Technologies (CoDIT), pp. 242–245. IEEE (2014). https://doi.org/10.1109/CoDIT.2014.6996900
3. Benmansour, R., Braun, O., Artiba, A.: Mixed integer programming formulations for the single processor scheduling problem with time restrictions. In: 45th International Conference on Computers & Industrial Engineering (2015)
4. Benmansour, R., Braun, O., Hamid, A.: Modeling the single-processor scheduling problem with time restrictions as a parallel machine scheduling problem. In: Proceedings MISTA, pp. 325–330 (2015)
5. Benmansour, R., Braun, O., Hanafi, S.: The single processor scheduling problem with time restrictions: complexity and related problems. J. Sched.https://doi.org/10.1007/s10951-018-0579-8
6. Braun, O., Chung, F., Graham, R.: Single-processor scheduling with time restrictions. J. Sched. **17**(4), 399–403 (2014)
7. Braun, O., Chung, F., Graham, R.: Worst-case analysis of the LPT algorithm for single processor scheduling with time restrictions. OR Spectrum **38**(2), 531–540 (2016)
8. Brucker, P., Dhaenens-Flipo, C., Knust, S., Kravchenko, S.A., Werner, F.: Complexity results for parallel machine problems with a single server. J. Sched. **5**(6), 429–457 (2002)
9. Glover, F.W., Kochenberger, G.A.: Handbook of Metaheuristics, vol. 57. Springer, Heidelberg (2006). https://doi.org/10.1007/978-1-4419-1665-5
10. Hall, N.G., Potts, C.N., Sriskandarajah, C.: Parallel machine scheduling with a common server. Discrete Appl. Math. **102**(3), 223–243 (2000)
11. Hansen, P., Mladenović, N.: Variable neighborhood search: principles and applications. Eur. J. Oper. Res. **130**(3), 449–467 (2001)
12. Hansen, P., Mladenović, N.: Variable neighborhood search. In: Burke, E., Kendall, G. (eds.) Search Methodologies, pp. 313–337. Springer, Boston (2014). https://doi.org/10.1007/978-1-4614-6940-7_12
13. Hansen, P., Mladenović, N., Todosijević, R., Hanafi, S.: Variable neighborhood search: basics and variants. EURO J. Comput. Optim. **5**(3), 423–454 (2017)
14. Kim, M.Y., Lee, Y.H.: Mip models and hybrid algorithm for minimizing the makespan of parallel machines scheduling problem with a single server. Comput. Oper. Res. **39**(11), 2457–2468 (2012)
15. Kirlik, G., Oguz, C.: A variable neighborhood search for minimizing total weighted tardiness with sequence dependent setup times on a single machine. Comput. Oper. Res. **39**(7), 1506–1520 (2012)
16. Kravchenko, S.A., Werner, F.: Parallel machine scheduling problems with a single server. Math. Comput. Model. **26**(12), 1–11 (1997)
17. Lei, D.: Variable neighborhood search for two-agent flow shop scheduling problem. Comput. Ind. Eng. **80**, 125–131 (2015)
18. Mladenović, N., Hansen, P.: Variable neighborhood search. Comput. Oper. Res. **24**(11), 1097–1100 (1997)
19. Todosijević, R., Benmansour, R., Hanafi, S., Mladenović, N., Artiba, A.: Nested general variable neighborhood search for the periodic maintenance problem. Eur. J. Oper. Res. **252**(2), 385–396 (2016)

20. Zhang, A., Ye, F., Chen, Y., Chen, G.: Better permutations for the single-processor scheduling with time restrictions. Optim. Lett. **11**(4), 715–724 (2017)
21. Zhao, L., Xiao, H., Kerbache, L., Hu, Z., Ichoua, S.: A variable neighborhood search algorithm for the vehicle routing problem with simultaneous pickup and delivery (2014)

An Evolutionary Variable Neighborhood Descent for Addressing an Electric VRP Variant

Dhekra Rezgui[1]([✉]), Hend Bouziri[2], Wassila Aggoune-Mtalaa[3],
and Jouhaina Chaouachi Siala[4]

[1] Institut Supérieur de Gestion, University of Tunis,
41 Rue de la liberté, 2000 Le Bardo, Tunisia
dhekra.rezgui@live.fr
[2] ESSECT, LARODEC Laboratory, ISG, University of Tunis, Tunis, Tunisia
[3] Luxembourg Institute of Science and Technology,
5 Avenue des Hauts-Fourneaux, 4362 Esch-sur-Alzette, Luxembourg
[4] Institut des Hautes Etudes Commerciales, Carthage-2016,
Carthage University, Tunis, Tunisia

Abstract. Variable neighborhood searches and evolutionary techniques
have shown their effectiveness when dealing with many combinatorial
optimisation problems. This study proposes to combine these two techniques for addressing the routing problem using electric and modular
vehicles. This is a recent problem that aims to overcome recharging battery constraints while maintaining a certain performance regarding to the
fleet cost and the traveled distance. An experimental study on benchmark
instances is provided to show the relevance of the proposed algorithm.

Keywords: Evolutionary algorithm · BCRC crossover ·
Variable neighborhood descent · Modular electric vehicles

1 Introduction

During the last decade, a strong strand of research has been devoted to design
green supply chains [3,25] and optimize their related distribution networks in
order to limit greenhouse gas emissions [4,18]. In this paper, we address a new
type of green vehicle routing problems using innovative electric vehicles. Indeed,
we consider a special variant of the electric Vehicle Routing Problem (e-VRP)
in which the studied vehicles are modular. That is to stay that each vehicle
is composed of a cabin module where the driver sits and one or more payload
modules for the freight. The main advantage of this configuration is that the
payload modules can be detached when necessary. For instance in certain urban
areas where the streets are too narrow, the vehicle can keep on running with
a limited number of modules which reduces its total length. The modules can
be also released for energy purpose. If a recharging terminal is available at a
customer location, one of the modules can recharge its battery whereas the

A. Sifaleras et al. (Eds.): ICVNS 2018, LNCS 11328, pp. 216–231, 2019.
https://doi.org/10.1007/978-3-030-15843-9_17

rest of the vehicle continues its tour. The vehicle which will later pick up this module will benefit from the additional electric charge. The payload modules can be dropped off also to enable the rest of the vehicle saving time and respecting the delays to deliver the customers. These types of vehicles exist currently as prototypes and should be commercialized in the near future.

To deal with our problem, we propose a method that exploits the best optimization strategies currently dedicated to the VRP with time windows (VRPTW) and its variants. Indeed, our assumption is that the VRP variants need efficient combinations between evolutionary methods and local search procedures, since many studies on fitness landscapes of VRPs show that they have the shape of big valleys with basin of attraction as stated in the work of [16] or that of [15]. Therefore, we look for a combination of the evolutionary schema which has the major advantage to be population based with an efficient and rapid variable neighborhood descent. An experimental study is conducted to test the solution approach and prove its effectiveness for modular electric vehicles.

The paper is organized as follows: In Sect. 2, we define in detail our problem and we present in Sect. 3 an overview of the relevant literature on metaheuristics for routing problems similar to our variant. In Sect. 4, we describe in detail the evolutionary based variable neighborhood descent algorithm and we show how genetic operators are adapted to our case. The computational study is presented and discussed in Sect. 5 and we conclude the paper in Sect. 6 by giving some perspectives for future works.

2 Problem Description

A fleet of modular electric vehicles is composed of vehicles which are a combination of modules autonomous in terms of consumption and electric charging. The vehicles may have a varying number of attached payload modules during their tours. It is different from other fleets in the sense that the payload modules which have started at the depot behind a cabin module can arrive back to the depot at the end of the distribution period with another vehicle.

Figure 1 describes a solution of a freight distribution problem using electric modular vehicles in a urban environment. This solution is detailed as follows. First, the white vehicle leaves the depot with three modules, serves the first customer at 11 a.m. and releases the last module at this location. Thereafter, it visits the next customer who has to be served at 12 a.m, and releases there another payload module. In this case, the vehicle with only one module moves to the next customer. There, the vehicle has to serve this customer before 2 p.m. When it reaches its destination, it retrieves a module left by the blue vehicle earlier in the day. Indeed, the blue vehicle began its tour with three modules and served customer 6 at 8.00 a.m and released there a module. Then it served customers 7, 8, 9 and 3 with the remaining two modules. The vehicles left then one module at customer 3 and went back to the depot. The white vehicle picked up the module at customer 3 which remained in charge. This leads to an additional battery charge for the white vehicle which had not to

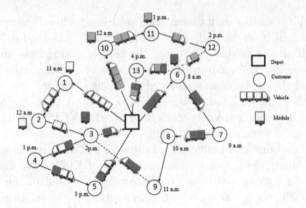

Fig. 1. Example of a vehicle routing problem using three electric modular vehicles. (Color figure online)

wait during the charging of the blue module. Then, the white vehicle pursues its tour with two payload modules. It goes and deliver goods to another customer before 1 p.m. Finally, it returns to the depot with the remaining modules after delivering the last customer at 3 p.m. At the end of the studied period, all the modules (cabin modules and payload modules) are back to the depot. The last route corresponds to the green vehicle which released also some modules along its tour and retrieved a module left by the blue vehicle at customer 6 before serving its last customer and returning to the depot.

The problem of routing a fleet of electric modular vehicles is a relatively new problem [20,21]. It extends the well-known Vehicle Routing Problem with Time Windows (VRPTW) in the sense that its objective is to minimize the total routing costs while taking into consideration the time windows for serving the customers, the capacity limit of the vehicles but also the modular property of the vehicles and the electric recharging. This problem was formulated as a Mixed Integer Linear Program (MILP) [22]. Several assumptions are associated with the recharging:

(i) All the vehicles are fully charged when they leave the depot.
(ii) Vehicles may recharge their battery only at customer locations in order to continue a tour.
(iii) The vehicle recharges at a customer location if the battery charge level drops below a given threshold.
(iv) The recharging time depends on the state of charge when arriving at the recharging station.
(v) If the vehicle has to recharge after the service time is finished, a penalty is added to the objective function.

The objective function is composed of three different components: the acquisition cost of the Electric Modular Vehicles (EMVs) used, the total travel cost and the recharging cost. Let V be the set of vehicles and C the set of customers. This objective function is defined as follows:

$$\sum_{k \in V} \sum_{j \in C} cf^k x_{0j}^k + \sum_{k \in V} \sum_{i,j \in C, i \neq j} c_{ij}^k x_{ij}^k + \sum_{k \in V} \sum_{i \in C} c_r r_i^k \qquad (1)$$

where cf^k is the fixed acquisition cost of a vehicle of type k, x_{0j}^k indicates whether the vehicle k is used or not, r_i^k is a binary decision variable indicating the recharging of vehicle k at customer i, x_{ij}^k is a binary decision variable indicating that a vehicle of type k travels from customer i to customer j, c_{ij}^k corresponds to the travel cost of a vehicle of type k traversing the pair of customers (i, j) and c_r is the recharging cost. For the detailed mathematical formulation reflecting all the constraints, please refer to [22].

3   Related Works

Our problem is a variant of the VRP which was introduced by Dantzig and Ramser [10]. It is concerned with distributing goods to a given set of customers with known demands by a homogeneous fleet of vehicles beginning from a localized depot in order to fulfill all the requests at a minimum cost. Among several variants of the VRP, researchers tend to pay close attention to the concept of VRP with Time Windows, in which, a customer has to be served in a certain time slot. As the VRPTW is a NP-hard combinatorial optimization problem, different heuristic solution techniques have been explored in order to provide near-optimal or optimal solutions for many real-life instances. However, Braÿsy and Gendreau [5] showed that despite a huge diversity of metaheuristic approaches, the genetic algorithm is still one of the most competitive heuristic providing promising results. More recently, a great attention has been devoted to the use of the Variable Neighborhood Search to deal with VRPTW variants, [13].

Pratically, we can note the work of Chen et al. [8] which proposes an adaptation of the VNS, called VNS-C that produces promising results in minimizing the total traveled distance. Another interesting work is due to [9] which proposes an iterated variable neighborhood descent that avoids the search to be trapped in local optima. A perturbation strategy is proposed to restart the search in other areas in the search space. Ferreira et al. [12] considered recently, the Vehicle Routing Problem (VRP) with Multiple Time Windows, where the customers have one or more time windows during which they can be visited. The authors propose a Variable Neighborhood Search heuristic where all the computational effort is spent on searching for feasible solutions.

Another work which extends the VRPTW appears in [23]. It considers electric constraints and introduces the Electric Vehicle Routing Problem with Time Windows and Recharging Stations (E-VRPTW). The authors propose a hybrid heuristic, including a variable neighborhood search (VNS) algorithm combined with a tabu search (TS) heuristic. Their results prove that the hybrid heuristic is able to determine efficient vehicle routes making use of the available recharging stations. Recently, Bruglieri et al. [6], addressed also a problem of serving a set of customers, within fixed time windows, by using Electric Vehicles (EVs). The authors proposed a new Variable Neighborhood Search Branching (VNSB), an approach that combines the VNS algorithm with the Local Branching one. Numerical results on benchmark instances clearly show that the VNSB is suitable to detect good quality solutions in a satisfactory computational time. More recently, Bruglieri et al. [7] developed a Three-Phase Matheuristic (TPM) for the E-VRPTW which in average outperforms the Variable Neighborhood Search Branching (VNSB), previously proposed.

Beside these works, Hiermann et al. [14] proposed a hybrid heuristic which combines an embedded local search with an Adaptive Large Neighborhood Search (ALNS). This approach has been used to analyze environmental issues in the context of routing problems using heterogeneous fleets. As a result, their experiments showed the competitiveness of the proposed approach as compared with the state of the art methods for solving the two E-VRPTW and Fleet Size and Mix VRPTW problems. Van Duin et al. [28] also studied an Electric Fleet Size and Mix Vehicle Routing Problem with Time Windows (E-FSMVRPTW). They used as a solution method a sequential insertion heuristic to solve a real instance with two different types of electric trucks.

The main contribution of this paper is that it hybridizes the genetic algorithm with a variable neighborhood descent to solve a variant of a VRPTW using heterogeneous fleets of electric modular vehicles. This idea of hybridizing the Genetic algorithm (GA) with a variant of the VNS was also followed by Baniamerian et al., [2] who used efficiently a Modified Variable Neighborhood Search (MVNS) involving four shaking procedures and two neighborhood structures for the VRP with cross-docking. The present paper deals with a VRPTW involving electric constraints as well as a modular structure of the vehicles as in [1]. In their work the resolution approach operated in two stages, one for the allocation of the modules to the customers and a second for the fusion of the routes. Our work differs from that preceding one in the sense that the problem is modeled differently and the solution method operates in a single stage by hybridizing the genetic algorithm using a problem-specific crossover with the Variable Neighborhood Descent.

4   The Proposed Evolutionary Variable Neighborhood Method

As it was stated before, we aim to use a method which tries to exploit the benefits of the best optimization strategies dedicated to the e-VRP and its variants. To achieve this, we choose to combine the evolutionary schema that have the major advantage to be population based with an efficient and rapid variable neighborhood descent.

Our evolutionary variable neighborhood algorithm is presented in Algorithm 1. It starts by constructing the graph that corresponds to the trips T to serve C customers. The initial population is then constructed and a set of

---

Algorithm 1. EVND Algorithm.

1: **Input**: T: List of Trips; C: Customers; $MaxIter$; Pc: Probability of the crossover
2: **Output**: Feasible solution
3: Generate an initial population (Pop) ;
4: **while** Termination criterion ($MaxIter$) not reached **do**
5:      $Parent1 \leftarrow$ Select(Pop);
6:      $Parent2 \leftarrow$ Select(Pop);
7:      $Child \leftarrow$ Crossover ($Parent1, Parent2, Pc$);
8:      $M \leftarrow$ Variable Neighborhood Descent($Child$);
9:       Population Updating (Pop, M);
10: **end while**
11: **Return** Best individual in Pop;

---

generations are performed using genetic operators, namely: the selection and the crossover. The mutation operator is then replaced by the VND procedure.

The selection operator ranks individuals of the current population according to their fitness value calculated as the sum of the total travel cost, acquisition cost and recharging cost, see Eq. (1). Then, a tournament selection is used to select parents for the crossover operator. After that, the variable neighborhood method is applied on the child provided by the crossover operator. The resulting feasible solution M will replace the worst individual in the population by the *Population Updating* routine. When the termination criterion is verified, namely the maximum number of iterations $MaxIter$ is reached, the algorithm returns the best individual in the current population. In the following, we explain the algorithm components.

4.1   Representation of a Solution

In this work, we choose to represent vehicles tours as sequences of customers separated by zeros. This is presented in Fig. 2, where three vehicles serve thirteen customers.

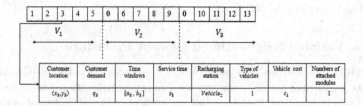

Fig. 2. Solution representation.

In this figure, the integers 1 to 13 correspond to the customers while the index 0 is dedicated to the depot. Route 1 starts at the depot, visits customers 1, 2,

3, 4, 5 and goes back to the depot. Route 2 starts from the depot to customers 6, 7, 8 and 9. Finally, route 3 begins from the depot and visits the last four customers 10, 11, 12 and 13. In our example, three types of vehicles (V_1, V_2, V_3) for the fleet are considered with different costs (c_1, c_2, c_3). Furthermore, to each customer located at a node $i \in N$ correspond a positive demand q_i, a service time s_i, and a time window $[a_i, b_i]$.

4.2   Initial Population

To generate an initial population, we propose a greedy algorithm that constructs at each step the nearest customer that verifies all the constraints related to the recharging and the time windows. Then the solution is perturbed by the 2-opt procedure to generate *popSize* solutions. If a resulting solution is unfeasible, an additional vehicle is added to respect time windows and recharging constraints.

4.3   Best Cost Route Crossover Operator

One of the particular and important features of the techniques involving the genetic algorithm is the choice of the crossover operator. In fact, the crossover is the process that allows in nature the production of chromosomes which partially inherit the characteristics of the parents in order to produce an offspring from two chosen individuals. Indeed, the crossover has to handle the problem features "to transmit" good properties of the parents to their children. In the literature of the VRP with time windows we can note the work of Moura [17] which has focused on the performance of an operator called the Best Cost Route Crossover (BCRC). This operator was originally introduced for the VRPTW by Ombuki et al. [19]. The performance of the BCRC can be explained by two factors. First, this crossover tests at each step the feasibility of the obtained solution and produces always solutions that handle time windows constraints. Secondly, this operator looks for best combinations to produce an offspring.

In our context and since our problem is highly constrained as compared with the VRPTW, we choose to apply the BCRC operator with a probability of occurrence Pc. We also check the feasibility of each move and we look for the best one as the standard BCRC does.

4.4   The Variable Neighborhood Descent Procedure

The Variable Neighborhood Search (VNS) is a popular metaheuristic that has proven its performance when dealing with many optimisation problems including the Vehicle Routing Problem and its variants [11]. It is based on varying the neighborhood structure during the search process. One simple and fast variant of the VNS is the Variable Neighborhood Descent (VND) [13]. It combines the rapidity of steepest descent procedures and the agility of the VNS, since it enables the search being efficient by using several performing move operators simultaneously. Indeed, in the literature, we can observe the use of search operators that can be more performing on some instances of the problem, and less relevant on other instances. The major idea of the VNS paradigm is to vary the move operator at each step to explore better the search space.

More formally, let $N = N_1, ..., N_{max}$ be a set of neighborhood structures where each N_i is a move operator that can be applied on each solution S to obtain a neighboring solution S' by $S' = N_i(S); i = 1, ..max$.

The VND heuristic applies alternatively the neighborhood operators until no improvement is possible. In our adaptation of the VND procedure, we apply four neighborhood operators sequentially whether an improvement is reported or not. This is motivated by the fact that we have to apply a rapid routine that will replace the mutation operator that has as the objective to perform a fast switch rather than a long walk. At the end, we retain the resulting solution only if it is better than the child provided by the crossover operator. The different steps of the VND method are given in Algorithm 2.

Algorithm 2. Variable Neighborhood Descent.

1: **Input** an initial solution SI, $N = N_1, ..., N_{max}$
2: **Output** Best neighboring solution
3:     $j \leftarrow 1$;
4:     $S \leftarrow SI$;
5: **while** $j \leq max$
6:         $S \leftarrow N_i(S)$
7:         $j \leftarrow j + 1$
8: **end while**
9: repair(S);
10:         **if** $f(S) < f(SI)$ **then** $SI \leftarrow S$;
11: **Return** SI.

The VND procedure is mainly depending on the set of neighborhood structures N and on their order. The sequence of movements in N considered in this work is the following: it involves the swap, insert, exchange and finally the 2-opt procedure. Approximately, the same order is used for other VRP problems [9]. Let Sol be the set of feasible solutions. In our adaptation of the VND, a neighborhood solution is accepted only if it verifies all the constraints, which means that it belongs to Sol.

5   Experimental Setup

We performed all the tests on a laptop PC equipped with an Intel Core i3-3217U Processor clocked at 1.8 GHz with 4 GB RAM, running Windows 8.2 Professional. The EVND algorithm was implemented as a single thread code in Java. Likewise, the tests of our work were executed over the 56 Solomon instances with 100 customers [26]. The instances are divided into six different categories denoted C1, C2, R1, R2, RCl and RC2, each with 8 to 12 problems. The sets of problems differ in the geographical distribution of the customers, the width of the time windows, the service time of each customer and the vehicle capacity. The data description is available at http://w.cba.neu.edu/~msolomon/problems.htm. The results of our EVND are based on 10 independent runs for each single Solomon test instance. We suppose that each EMV must check at each customer location the State Of Charge (SOC) of the battery of the overall vehicle including the attached modules. When the SOC is below a given threshold, the vehicle remains in charge until reaching this threshold, calculated as the maximum distance between two nodes of the distribution network. The battery capacity is set to the maximum 80 kWh. Furthermore, we set the consumption rate e^k to 1.0. The recharging time h is set to the time needed for the EMV for reaching the threshold. This time depends on the current SOC of the battery when arriving at the current customer location. In the present work, the stopping criterion of the general algorithm is a fixed number of iterations $MaxIter$ which has been set to 3000.

In our experimental study, we try to show these main issues:

- the relevance of the main components of our algorithm, which is the use of the variable neighboring descent as a mean to explore locally the search space of the solution generated by the crossover operator.
- the performance of our method to treat the electric vehicles routing problem regarding best state of the art methods basically hybrid metaheuristics.

5.1   Relevance of the VND Procedure

Here, we try to assess our theoretical assumptions that the VND could give efficient results by exploring the local neighborhood using the four operators defined before. For this reason, we experiment an Evolutionary Local Search (ELS) procedure that works similarly to the EVND, but replaces the VND procedure by a local descent procedure that iteratively applies the 2-opt move on the child until an improvement is recorded. Tables 1 and 2 compare the EVND to the ELS in terms of solutions quality to obtain the best solution. The experiments carried out for evaluating the relevance of the VND were conducted with the R1 and R2 instances of the Solomon benchmark. The best results for our evolutionary algorithm against the ELS are presented, where NV is the Number of Vehicles needed and TD is the Total Distance. The column $Gap\%$ is the percentage difference of the total traveled distance between the solutions provided by the EVND and ELS algorithms. The runtime in minutes is also provided in the columns denoted CPU.

Based on the empirical results of Tables 1 and 2, we can note that the EVND is more competitive than the ELS. For instance, considering the group of instance R1 in Table 1, the results show that the EVND reduces the gap of the traveled distance with an average of 0.08%, while the number of vehicles used is always the same as for the ELS. Moreover, the average percentage improvement of EVND, considering the instance R2 in Table 2, is less than the one of ELS with an average of 0.67%. Moreover, for the group of instances R1 and R2, concerning the number of vehicles used, the two approaches require the same number of vehicles (11 in average for the group R1 and 2.63, in average for R2).

A more detailed analysis shows that the VND hybridized with the Evolutionary Algorithm works very well as compared with the ELS for the set of instances

Table 1. Best results found for the R1 problem class.

Problem	ELS			EVND			
	NV	TD	CPU	NV	TD	CPU	Gap%
R101	14	1657.16	14.67	13	1653.80	14.82	−0.20
R102	14	1484.14	15.60	14	1479.22	14.68	−0.33
R103	12	1300.77	13.30	13	1248.64	16.26	−4.00
R104	10	1006.14	16.42	10	986.42	14.54	−1.95
R105	12	1355.41	16.81	12	1354.98	15.88	−0.03
R106	11	1253.97	15.65	11	1250.08	14.21	−0.31
R107	10	1110.04	19.62	10	1102.42	16.74	−0.68
R108	10	951.84	16.31	9	945.75	17.26	−0.63
R109	11	1104.22	13.80	11	1102.68	15.64	−0.13
R110	10	1015.72	18.21	10	1118	18.89	10.66
R111	10	1093.08	16.01	10	1064.95	17.30	−2.57
R112	8	902.28	15.87	9	900.05	17.52	−2.46
Average	11	1186.23	16.02	11	1183.92	16.14	−0.08

Table 2. Best results found for the R2 problem class.

Problem	ELS			EVND			
	NV	TD	CPU	NV	TD	CPU	Gap%
R201	3	1252.41	15.85	3	1249.77	14.57	−0.21
R202	3	1170.69	15.50	3	1164.58	14.36	−0.52
R203	3	943.93	14.20	3	940.95	13.21	−0.31
R204	2	838.66	13.06	2	827.27	15.02	−1.35
R205	3	971.24	12.36	3	969.13	14.74	−0.21
R206	3	889.19	14.66	3	889.17	16.11	−0.00
R207	2	894.92	11.80	2	894.35	12.89	−0.06
R208	2	726.84	13.31	2	725.05	12.77	−0.24
R209	3	909.68	11.58	3	908.48	13.14	−0.13
R210	3	869.62	13.76	3	848.81	14.57	−2.39
R211	2	879.25	11.83	2	862.17	12.94	−1.94
Average	2.63	940.58	13.44	2.63	934.52	14.02	−0.67

of R1 where, on average, less or the same number of electric modular vehicles (EMVs) are needed to perform the tours. In addition to that, there is only one more vehicle which is needed even if the total distance is minimized (for R103, R112). Moreover, in terms of distance EVND optimizes the TD for all problem sets except R110, as compared to ELS.

It is also worth mentioning that, for the group of instances R2, the EVND has equaled the solutions when compared to ELS according to the number of vehicles. Concerning the total distance, for all the tested problem sets the EVND has improved significantly the solutions of 10 out of 11 instances. Regarding the computational times, in average, the EVND requires 16.14 min against 16.02 min for the ELS, for the group of instances R1 and 14.02 min against 13.44 min, for the group of instances R2. This is not a big difference since the EVND involves more neighborhood operators than the ELS. It is finally worth remarking that the EVND is a powerful metaheuristic that systematically exploits the idea of neighborhood search. This leads to improve significantly the solutions of 21 out of the 23 instances for the group of instances R when comparing it to the ELS.

5.2   Comparison with Some State-of-the-Art Methods

To prove the effectiveness of our method, we choose to compare our results with three related methods; the Tabu Search (TS) [24] as an efficient competitive metaheuristic, the Localized Genetic Algorithm (LGA) [27] because of its evolutionary framework and the VNS-C [8], an effective VNS procedure. All these methods are tested on the Solomon benchmarks framework. An average runtime of 16 min are required to obtain the result of a 100-customers Solomon instance with the EVND. This is the same runtime as with the TS [24]. Since the other authors did not provide the CPU time and since our problem is more constrained than the classical VRPTW, we compare only the total distance and the number of used vehicles to report the effectiveness of our EVND. The idea behind this is to assess the ability of our EVND to overcome electric battery recharging constraints by minimizing the total distance (to serve customers rapidly) and the number of EVs (since their acquisition costs are reported to be high). A thorough analysis of the results presented in Table 3, showed that the EVND approach competes significantly well against the solutions of the 56 instances, when compared to the other state of the art methods, according to the total distance. Concerning the NV criterion, the obtained results require in almost all cases, the same number of vehicles as the best one of the other methods. It is worth noting, that for the group of instances C1 and C2, our approach is able to successfully minimize the TD for all the instances with the same number of vehicles for the distribution operations. However, for the R1 and R2 classes, EVND is less performant but it is still very competitive on the two criteria NV and TD. Indeed, it provides excellent performance essentially for the R101 and the R102 instances, where the number of vehicles used dropped to 13 and 14 respectively, against 18 and 17 for the best of the other methods. Also, we note the performance of the EVND for the R109 instance for which the TD is equal to 1109.68 with 11 vehicles, against the TS which needs a total distance of 1205.27 with the same number of vehicles. Also, we can report the R112 instance for which the EVND gains 10,33% of the TD when compared to the TS method, by using the same number of vehicles. Concerning the classes RC1 and RC2, the EVND results are very competitive especially on the number of vehicles and outperforms the other methods in some cases, namely; RC101, RC203 and RC08,

Table 3. Comparison between our best results and related methods on the Solomon instances.

Problem	TS		LGA		VNS-C		EVND	
	NV	TD	NV	TD	NV	TD	NV	TD
C101	10	828.94	10	827.3	10	828.94	10	820.66
C102	10	828.94	10	827.3	10	828.94	10	823.05
C103	10	828.07	10	827.3	10	828.94	10	827.54
C104	10	824.78	10	827.3	10	825.65	10	824.12
C105	10	828.94	10	827.3	10	898.94	10	828.48
C106	10	828.94	10	827.3	10	898.94	10	826.35
C107	10	828.94	10	827.3	10	898.94	10	827.47
C108	10	828.94	10	827.3	10	898.94	10	827.31
C109	10	828.94	10	827.3	10	898.94	10	827.99
Average	10	828.38	10	827.3	10	828.65	10	825.89
C201	3	591.56	3	589.1	3	591.56	3	583.50
C202	3	591.56	3	589.1	3	591.56	3	591.56
C203	3	588.49	3	588.7	3	591.17	3	586.45
C204	3	587.71	3	588.1	3	590.6	3	581.83
C205	3	588.49	3	586.4	3	588.88	3	581.44
C206	3	588.49	3	586.0	3	588.49	3	585.43
C207	3	588.29	3	585.8	3	588.29	3	583.89
C208	3	588.32	3	585.8	3	588.32	3	588.05
Average	3	589.11	3	587.37	3	589.86	3	585.27
R101	18	1606.07	20	1640.1	19	1652.47	13	1653.80
R102	17	1447.36	18	1467.5	18	1476.06	14	1479.22
R103	13	1257.49	14	1214.0	14	1219.89	13	1248.64
R104	9	1007.39	11	992.6	11	994.85	10	986.42
R105	13	1462.39	16	1362.3	14	1381.88	12	1354.98
R106	12	1263.29	13	1243.3	13	1243.72	11	1250.08
R107	10	1080.89	11	1069.5	11	1077.24	10	1102.42
R108	9	957.04	10	943.5	10	956.22	9	945.75
R109	11	1205.27	13	1152.4	13	1157.61	11	1102.68
R110	10	1128.61	12	1070.6	12	1081.88	10	1118
R111	10	1102.07	12	1057.3	11	1087.5	10	1064.95
R112	9	1003.76	10	960.8	10	958.7	9	900.05
Average	11.75	1134.52	13.33	1181.15	13	1079.92	11	1183.92

(*continued*)

Table 3. (*continued*)

Problem	TS		LGA		VNS-C		EVND	
	NV	TD	NV	TD	NV	TD	NV	TD
R201	4	1248.49	10	1152.7	5	1190.52	**3**	1249.77
R202	3	1177.11	7	1045.4	4	1098.06	**3**	1164.58
R203	3	939.54	6	871.2	4	905	3	940.95
R204	2	822.66	5	731.3	3	766.91	2	827.27
R205	3	1005.05	7	965.1	4	964.02	**3**	969.13
R206	2	1076.74	5	887.6	3	931.01	3	889.17
R207	2	883.502	5	807.0	3	855.37	2	894.35
R208	2	730.62	4	703.4	3	708.9	**2**	725.054
R209	3	915.07	6	867.0	3	983.75	**3**	908.48
R210	3	949.52	6	944.7	4	935.01	3	848.81
R211	2	864.83	5	754.6	3	794.04	2	862.17
Average	2.63	909.73	6	848.54	3.54	921.15	**2.63**	934.52
RC101	14	1685.39	18	1662.5	15	1624.97	**14**	**1607.48**
RC102	12	1503.25	15	1480.6	13	1497.43	12	1502.27
RC103	10	1305.20	12	1286.7	11	1265.86	11	1255.57
RC104	10	1118.42	10	1136.1	10	1136.49	10	1135.52
RC105	13	1626.49	16	1549.8	14	1642.81	**13**	1555.63
RC106	11	1366.86	13	1382.7	12	1396.59	**10**	1427.95
RC107	10	1312.23	12	1215.8	11	1254.68	**10**	1238.22
RC108	10	1132.60	11	1115.5	11	1131.23	10	1134.41
Average	11.25	1381.31	13.37	1353.71	12.12	1368.75	**11.25**	1357.13
RC201	4	1394.81	10	828.93	5	1310.44	**3**	1417.36
RC202	3	1326.40	8	828.93	4	1219.49	3	1367.42
RC203	3	1066.66	6	828.93	4	957.1	3	1051.20
RC204	2	945.44	4	828.93	3	829.13	**3**	**792.11**
RC205	3	1566.16	8	828.93	5	1233.46	4	1243.81
RC206	3	1140.98	7	828.93	4	1107.4	3	1144.37
RC207	3	1055.42	6	828.93	4	1032.78	3	1020.11
RC208	3	827.58	5	827.52	3	830.06	**3**	**822**
Average	3	1165.43	6.75	828.93	4	1064.98	3.12	1107.30

where the TD and the NV are the best. All the most interesting performances of the EVND are in bold in Table 3.

Table 4 summarizes the results of the comparison between the algorithms considered in Table 3 namely the TS, the LGA, the VNS-C and EVND. The columns represent the results obtained for each algorithm whereas the lines show the average of the total traveled distance and the average number of vehicles for each class. Additionally, the cumulative traveled distance (CTD) and the cumulative number of vehicles (CNV) are provided in the last two lines, for each problem group.

Table 4. Average number of vehicles used and total traveled distance for our method against some state of the art methods.

Benchmark	TS	LGA	VNS-C	EVND
C1	828.38	827.3	828.65	**825.89**
	10	10	10	**10**
C2	589.11	587.37	589.86	**585.27**
	3	3	3	**3**
R1	1134.52	1181.15	1079.92	1183.92
	11.75	13.33	13	**11**
R2	909.73	848.54	921.15	934.52
	2.63	6	3.54	**2.63**
RC1	1381.31	1353.71	1368.75	1357.13
	11.25	13.75	12.12	**11.25**
RC2	1165.43	828.93	1064.98	1107.30
	3	6.75	4	3.12
CND	56163.49	53113.56	54738.26	56317.5
CNV	398	501	438	**390**

As shown in Table 4, the obtained results indicate in almost all cases that the proposed EVND gives promising results as compared with Schneider's TS [24], Ursani's LGA [27] and Chen's VNS-C [8], regarding the decreased numbers of vehicles, without a significant increase in the total distance. Indeed, in all the classes of problems, we observe that the results obtained by the EVND algorithm are very competitive, in terms of distance despite the lower number of vehicles used to serve the customers. This is particularly convenient for the electric vehicles context.

6   Conclusion

In this paper, we have proposed an evolutionary variable neighborhood descent algorithm applied to a vehicle routing problem with time window constraints using heterogeneous fleets of electric modular vehicles. Experimental results show the benefits of combining the evolutionary schema with the variable neighborhood descent procedure. This performance is also confirmed on classical VRPTW instances. The effectiveness of the proposed method in reducing both the traveled distance and the number of vehicles is due to the use of appropriate genetic

operators, the ability of modules to be recharged at customers and the intensification procedure performed by the variable neighborhood descent.

This work can be extended to other VRP variants especially those handling
electric constraints in emerging freight delivery distribution problems. Moreover,
we intend to enhance the variable neighborhood descent procedure by exploring
other move operators.

References

1. Aggoune-Mtalaa, W., Habbas, Z., Ait Ouahmed, A., Khadraoui, D.: Solving new urban freight distribution problems involving modular electric vehicles. IET Intell. Transp. Syst. **9**(6), 654–661 (2015)
2. Baniamerian, A., Bashiri, M., Zabihi, F.: A modified variable neighborhood search hybridized with genetic algorithm for vehicle routing problems with cross-docking. Electron. Notes Discret. Math. **66**, 143–150 (2018). 4th International Conference on Variable Neighborhood Search
3. Bennekrouf, M., Aggoune-Mtalaa, W., Sari, Z.: A generic model for network design including remanufacturing activities. Supply Chain Forum **14**(2), 4–17 (2013)
4. Boudahri, F., Aggoune-Mtalaa, W., Bennekrouf, M., Sari, Z.: Application of a clustering based location-routing model to a real agri-food supply chain redesign. In: Nguyen, N., Trawiński, B., Katarzyniak, R., Jo, G.S. (eds.) Advanced Methods for Computational Collective Intelligence. Studies in Computational Intelligence, vol. 457, pp. 323–331. Springer, Heidelberg (2013). https://doi.org/10.1007/978-3-642-34300-1_31
5. Braÿsy, O., Gendreau, M.: Vehicle routing problem with time windows, Part I: route construction and local search algorithms. Transp. Sci. **39**(1), 101–118 (2005)
6. Bruglieri, M., Pezzella, F., Pisacane, O., Suraci, S.: A variable neighborhood search branching for the electric vehicle routing problem with time windows. Electron. Notes Discrete Math. **47**, 221–228 (2015)
7. Bruglieri, M., Mancini, S., Pezzella, F., Pisacane, O., Suraci, S.: A three-phase matheuristic for the time-effective electric vehicle routing problem with partial recharges. Electron. Notes Discrete Math. **58**, 95–102 (2017). 4th International Conference on Variable Neighborhood Search
8. Chen, B., Qu, R., Bai, R., Ishibuchi, H.: A variable neighbourhood search algorithm with compound neighbourhoods for VRPTW, pp. 25–35 (2016)
9. Chen, P., Huang, H., Dong, X.: Iterated variable neighborhood descent algorithm for the capacitated vehicle routing problem. Expert Syst. Appl. **37**(2), 1620–1627 (2010)
10. Dantzig, G.B., Ramser, R.H.: The truck dispatching problem. Manag. Sci. **6**, 80–91 (1959)
11. De Armas, J., Melián-Batista, B., Moreno-Pérez, J.A., Brito, J.: GVNS for a real-world rich vehicle routing problem with time windows. Eng. Appl. Artif. Intell. **42**, 45–56 (2015)
12. Ferreira, H.S., Bogue, E.T., Noronha, T.F., Belhaiza, S., Prins, C.: Variable neighborhood search for vehicle routing problem with multiple time windows. Electron. Notes Discrete Math. **66**, 207–214 (2018). 4th International Conference on Variable Neighborhood Search
13. Hansen, P., Mladenović, N., Todosijević, R., Hanafi, S.: Variable neighborhood search: basics and variants. EURO J. Comput. Optim. **5**(3), 423–454 (2017)

14. Hiermann, G., Puchinger, J., Hartl, R.F.: The electric fleet size and mix vehicle routing problem with time windows and recharging stations. Eur. J. Oper. Res. **252**(3), 995–1018 (2016)
15. Kubiak, M.: Distance measures and fitness-distance analysis for the capacitated vehicle routing problem. In: Doerner, K.F., Gendreau, M., Greistorfer, P., Gutjahr, W., Hartl, R.F., Reimann, M. (eds.) Metaheuristics. ORSIS, vol. 39, pp. 345–364. Springer, Boston, MA (2007). https://doi.org/10.1007/978-0-387-71921-4_18
16. Merz, P., Freisleben, B.: Fitness landscape analysis and memetic algorithms for the quadratic assignment problem. IEEE Trans. Evol. Comput. **4**(4), 337–352 (2000)
17. Moura, A.: A multi-objective genetic algorithm for the vehicle routing with time windows and loading problem. In: Bortfeldt, A., Homberger, J., Kopfer, H., Pankratz, G., Strangmeier, R. (eds.) Intelligent Decision Support, pp. 187–201. Springer, Heidelberg (2008). https://doi.org/10.1007/978-3-8349-9777-7_11
18. Mtalaa, W., Aggoune, R., Schaefers, J.: CO2 emissions calculation models for green supply chain management. In: Proceedings of POMS 20th Annual Meeting (2009). http://www.pomsmeetings.org/ConfProceedings/011/FullPapers/011-0590.pdf
19. Ombuki, B., Ross, B.J., Hanshar, F.: Multi-objective genetic algorithms for vehicle routing problem with time windows. Appl. Intell. **24**, 17–30 (2006)
20. Rezgui, D., Aggoune-Mtalaa, W., Bouziri, H.: Towards the electrification of urban freight delivery using modular vehicles. In: 10th IEEE SOLI Conference, vol. 6, pp. 154–159 (2015)
21. Rezgui, D., Chaouachi Siala, J., Aggoune-Mtalaa, W., Bouziri, H.: Application of a memetic algorithm to the fleet size and mix vehicle routing problem with electric modular vehicles. GECCO (Companion) **6**, 301–302 (2017)
22. Rezgui, D., Siala, J.C., Aggoune-Mtalaa, W., Bouziri, H.: Towards smart urban freight distribution using fleets of modular electric vehicles. In: Ben Ahmed, M., Boudhir, A.A. (eds.) SCAMS 2017. LNNS, vol. 37, pp. 602–612. Springer, Cham (2018). https://doi.org/10.1007/978-3-319-74500-8_55
23. Schneider, M., Stenger, A., Goeke, D.: The electric vehicle-routing problem with time windows and recharging stations. Transp. Sci. **48**(4), 500–520 (2014)
24. Schneider, M.: The vehicle-routing problem with time windows and driver-specific times. Eur. J. Oper. Res. **250**(1), 101–119 (2016)
25. Serrano, C., Aggoune-Mtalaa, W., Sauer, N.: Dynamic models for green logistic networks design. IFAC Proc. Vol. (IFAC-PapersOnline) **46**(9), 736–741 (2013)
26. Solomon, M.M.: Algorithms for the vehicle routing and scheduling problems with time window constraints. Oper. Res. **35**, 254–265 (1987)
27. Ursani, Z., Essam, D., Cornforth, D., Stocker, R.: Localized genetic algorithm for vehicle routing problem with time windows. Appl. Soft Comput. **11**, 5375–5390 (2011)
28. Van Duin, J.H., Tavasszy, L.A., Quak, H.J.: Towards electric-urban freight: first promising steps in the electric vehicle revolution. Eur. Transp. **54**(9), 1–19 (2013)

A Variable Neighborhood Descent Heuristic for the Multi-quay Berth Allocation and Crane Assignment Problem Under Availability Constraints

Issam Krimi[1,3,4]($\boxtimes$), Afaf Aloullal[5], Rachid Benmansour[5],
Abdessamad Ait El Cadi[3], Laurent Deshayes[2], and David Duvivier[3]

[1] Complex Systems Engineering and Human Systems,
Mohammed 6 Polytechnic University, Ben Guerir, Morocco
`issam.krimi@um6p.ma`
[2] Innovation Lab for Operations,
Mohammed 6 Polytechnic University, Ben Guerir, Morocco
[3] LAMIH UMR CNRS 8201, Valenciennes University, Valenciennes, France
[4] Mohammadia School of Engineers,
Mohammed V University in Rabat, Rabat, Morocco
[5] SI2M Laboratory, INSEA Rabat, Rabat, Morocco

Abstract. In this paper, we consider the integrated Berth Allocation and Crane Assignment problem, with availability constraints and high tides restrictions, in bulk port context. We were inspired by a real case study of a port owned by our industrial partner. The objective is to minimize the total penalty of tardiness. First, we implemented a greedy heuristic to compute an initial solution. Then, we proposed a sequential Variable Neighborhood Descent (seq-VND) for the problem. In addition, we compared the efficiency of different scenarios for the seq-VND against results given by a mathematical model for the problem.

Keywords: Berth Allocation · Crane assignment ·
Variable Neighborhood Descent · Bulk port

1 Introduction

In the last few years, the maritime industry has undergone an immense metamorphism affecting the global economy [22]. Therefore, an efficient management of this industry becomes a necessity to enhance its performance. From this perspective, several researchers have studied the operational aspect of port management by investigating two different decision problems: yard side and seaside operations. One of the major seaside problems, we cite the Berth Allocation Problem (BAP). However, this problem depends, significantly, on the resource utilization and especially the loading/unloading cranes. Hence, dealing with *BAP* and

© Springer Nature Switzerland AG 2019
A. Sifaleras et al. (Eds.): ICVNS 2018, LNCS 11328, pp. 232–242, 2019.
https://doi.org/10.1007/978-3-030-15843-9_18

Crane Assignment Problem (CAP) in one integrated decision problem is more realistic. Indeed, a significant emphasis has been given to the integrated Berth Allocation and Crane Assignment Problem [1,24] but only for the container terminal context. In this paper, we investigate the multi-quay Berth Allocation and Crane Assignment Problem under the availability constraint for a bulk port. We proposed a sequential Variable Neighborhood Descent (seq-VND) to solve a real-case study given by our industrial partner.

Formally, the bulk terminal under study contains a set Q of quays with different lengths L_q for each quay $q \in Q$ and each one is equipped with a set of mobile gantries, $c \in Cr$, with a fixed loading/unloading rate d_q. Vessels arrive within the planning horizon H and each vessel $j \in V$ has an arrival date r_j and an estimated departure time Dep_j. A vessel j may request different quantities C_{jf} of more than one quality $f \in F_j$, where F_j denotes the set of qualities requested by this vessel. During the loading process, the handling time depends on several factors. First, all vessels require a pre-loading Pre and post-loading time $Post$. In addition to a draft survey t_{ds} in the case of multiple qualities. Then, the variable handling time is calculated for, each vessel, considering loaded quantities, transfer rate between storage areas and quays, and the number of gantries used for cargo loading. To simplify the problem, we consider that the transfer rate depends only on d_q. In the actual planning tool used by the industrial partner, availability constraints are not considered. However, due to preventive maintenance of gantries or to bad weather conditions, the installation may be unavailable during periods $[start_i, finish_i]$, $i \in NVA$. We denote by NVA the set of unavailability periods within the planning horizon. In addition, the vessels should leave the docks within a high tide interval $[s^p_{tide}, f^p_{tide}]$, $p \in HT$. We define HT as the set of high tides periods over the horizon H. The objective is to minimize the total cost of vessels tardiness.

The rest of the paper is organized as follows. In Sect. 2 we describe the problem and give the related work. In Sect. 3, we propose the seq-VND algorithm to solve the problem. Finally, the computational experiments, based on a real case, as well as the corresponding conclusions are summarized in the Sects. 4 and 5 respectively.

2   Literature Review

In the container terminal environment, Park and Kim [19] presented the first work that dealt, simultaneously, with the Berth Allocation and Crane Assignment. They proposed a two-phase approach based on the Lagrangian relaxation. In the first stage, they determined the berthing positions, the handling times and the number of assigned cranes in each period. Then, in the second stage, the cranes scheduling were investigated. In the same year, Imai et al. [13] studied the notion of Multi-User terminal, in which the quays are a set of berths called the discrete layout. The authors proposed a heuristic procedure to solve the problem. Meisel and Bierwirth [18] proposed a mathematical formulation integrating the berth allocation and the crane assignment under several real-word considerations. To solve the problem, they developed a constructive heuristic for generating a feasible solution, a set of local search procedures and meta-heuristics. These methods were shown efficient on a set of real-world instances. Based on this work, Iris et al. [15] studied two variants of the crane assignment: time-variant or time-invariant and proposed new mathematical formulations. Iris et al. [14] improved the mathematical formulation provided by Meisel and Bier-wirth [18], introducing valid inequalities and variable fixing methods. The same problem but with more realistic objective was investigated by Liang et al. [17]. The aim was to minimize the handling time of vessels, the waiting time and the delay. They formulated the problem with a dynamic aspect for vessel's arrival and proposed a genetic algorithm to guarantee a good solution with a reasonable time for real word instances. Giallombardo et al. [11] formulated the integrated berth allocation and crane assignment as a quadratic model and a linearized formulation. Their models minimized the total quay crane profiles in addition to the housekeeping costs generated by transshipment flows between vessels. These models had found difficulties to solve real-case instances. Hence, the authors pro-posed a Tabu Search method based on mathematical techniques. Chang et al. [5] supposed that the crane assignment is variant within the planning horizon. They used a mathematical formulation embedded in a rolling horizon algorithm to solve medium size instances. In addition, the authors hybridized a parallel genetic algorithm using several priority rule heuristics. The handling time may depend, mainly, on the crane assignment. In order to study the importance of this assumption, Blazewicz et al. [4] considered the problem as task scheduling using a set of processors. They proposed a suboptimal algorithm that obtains a feasible solution for the discrete version of the problem using the continuous one. Raa et al. [20] studied the same model but under real-world considerations as the preferred berthing position, priority of the vessel and the handling time. They proposed a MIP model and tested it on a set of real instances. However, this model is time consuming for the daily managerial decisions. Hence, the authors proposed a rolling horizon framework based on the proposed MIP. Based on con-gested artificial data, the method shows its efficiency. In addition, Vacca et al. [24] developed a new model for the tactical berth allocation where the decisions of berthing and crane assignment were studied simultaneously. They proposed a Branch and Price algorithm to solve it. In their work, several accelerating tech-niques for the master and the pricing problem were presented. The results showed that the integration approach might be efficient in different cases. Recently, Agra and Oliveira [1] studied the integrated berth allocation, quay crane assignment and scheduling problem motivated by a real case where a heterogeneous set of cranes is considered. They proposed a MIP model based on the relative position formulation (RPF). In addition, in order to enhance the bounds of the mathe-matical formulation, they suggested a new model based on the discretization of the time and space variables. These enhancements are used in the Branch and Cut method as a relaxed method in addition to the Rolling horizon heuristic. The integrating BAP and CAP was widely investigated using different heuris-tics. Yang et al. [25] proposed evolutionary algorithm with multiple stages. First, each stage solves a specific sub-problem and then, the algorithm computes an approximation solution for the problem. In addition, Cheng et al. [6] dealt with a dynamic allocation model using objective programming. The idea was to propose a rolling-horizon technique for the *BAP* and *CAP* under the objective of minimizing the total berthing location deviation, the total penalty and the energy consumption of quay cranes. To solve large instances, the authors suggested a hybrid parallel genetic algorithm (HPGA), which combines the parallel genetic algorithm (PGA) and a heuristic algorithm. Diabat and Theodorou [7] considered the scheduling aspect in the integrated problem (*BACASP*). They proposed a MIP formulation in addition to a genetic algorithm for which the authors defined problem-specific chromosome and a set of insertions and selection strategies. Then, Fu and Diabat [10] implemented a Lagrangian relaxation based for the decomposition of the problem. The authors tested their approach using several problem instances. The results were compared to those obtained by a commercial software and to another heuristic, namely a Genetic Algorithm (GA) presented in Diabat and Theodorou [7].

In the bulk port context, Barros et al. [2] were interested in the Berth Allocation Problem in bulk port under tidal and stock level restrictions. The main decision for this problem is to define when vessels should be allocated to berths and that during tidal time windows with the aim of minimizing the overall cost. In this work, berths are homogeneous and stocks must be kept in safety levels. To deal with the complexity of the problem, they proposed a Simulated Annealing-based algorithm. Umang et al. [23] presented three formulations to solve the dynamic hybrid Berth Allocation Problem. The main differences between container and bulk port operations are the account for a cargo type and the fixed equipment facilities as conveyors that may be installed in certain berths only. The authors investigated the coordination between seaside and yard side operations. The objective is to minimize the total service time of vessels. The authors developed a Generalized Set Partitioning *GSPP* model for the problem and showed its efficiently. An alternative can be used in case of GSPP fails to generate the optimal solution in reasonable time, which is adding bulk-specific components to the metaheuristic, based on squeaky wheel optimization. Rodrigues et al. [21] proposed a MIP model for the continuous Berth Allocation Problem under cargo operation restrictions to minimize the time that vessels stayed in the port. They supposed that some cargo could not be handled in certain berths along the pier. This assumption is observed in the oil and gas logistics context. Real data showed that the model can solve instances up to 147 vessels with 440 m of pier and results confirm the possibility of significant gains. In addition, Ernst et al. [9] proposed two MIP formulations for the continuous berth allocation problem in dry bulk port context. The main restriction considered is the tidal time window for the vessel departure. The authors presented several proprieties for the optimal solution and some cuts to tighten the models. The time-indexed model was shown more suitable for large instances.

3   Variable Neighborhood Descent

Variable Neighborhood Search [12] was presented as a single solution metaheuristic based on the idea of changing systematically the neighborhood structures to escape from local optima. The use of different neighborhood structures is explained by the following observations: it is not necessary to find the same local optimum for all the neighborhood structures. (i) a global optimum is a local optimum for all the neighborhood structures. (ii) despite the diversity of the neighborhood structures, for many problems the local optimums are relatively close. Basic VNS and its variants have been applied to many optimization problems [3], as scheduling, routing, etc. In this work, we investigated the Variable Neighborhood Descent to tackle the Problem.

The seq-VND implementation requires the definition of the following parameters: an initial solution, a set of neighborhood structures, a local search and an evaluation scheme.

3.1   Initial Solution

To construct an initial solution we consider the set of quays $S = \bigcup_{q \in Q} \pi_q$, where each permutation π_q contains a list of vessels that will be loaded in the quay q. We denote this solution by a quay permutation-based representation. Based on this representation, we proposed a greedy algorithm that generates a feasible starting point for the seq-VND metaheuristic. The main idea is inspired by the workstation-balancing assignment [3], in which each equipment has the same amount of work to execute. In our problem, we considered the port as a work cell and each quay represents equipment or a workstation. We define a sequence of vessels (tasks) to be loaded (executed) respecting a priority rule. We choose the First Come First Served (FCFS) rule to select the vessels. The greedy procedure is used to assign vessels to quays based on the predefined order and respecting quays length. As in production line balancing, we define a capacity $N = \left\lceil \frac{|V|}{|Q|} \right\rceil$, as the number of vessel assigned to each quay.

After matching vessels to quays, we proposed $G - FCFS$, a constructive heuristic, to generate a feasible scheduling for the solution. In our approach, we discretized both planning horizon and quays in order to respect the time-space constraints of the problem. The $G-FCFS$ generates a feasible berthing planning for the initial solution S, in which we calculate for each vessel $v \in \pi_q$ its starting time S_v, its berthing position b_v, its handling time h_v and the resulting tardiness T_v. For the evaluation scheme, the total cost of tardiness is calculated using the greedy procedure described above and the solution generated after each move.

3.2   Neighborhood Structures

The choice of the neighborhood structures used in local search methods affects, significantly, the performance of the algorithm. We proposed a set of neighborhoods based on the permutation representation. We distinguish two types of neighborhood structures: inter-quays for the moves between two distinct docks and intra-quay used inside a specific dock. On one hand, we used insert (or-opt) and swap neighborhood structures for inter-quays case. On the other hand, we used insert (or-opt), swap and reverse neighborhood structures for intra-quay case. For the sake of simplicity, a neighborhood structure is referred by the name of its corresponding operator.

- Inter-quays swap operator $\mathcal{N}_1$: this operator selects a pair of vessels in two different quays and exchanges their positions. The operator repeats this process for all the quays until all the neighborhoods have been investigated. We assume a solution of the problem $S_a = (\pi_1^a, \pi_2^a, ..., \pi_n^a)$ and one of its neighbors $S_b = (\pi_1^b, \pi_2^b, ..., \pi_n^b)$. We fix two different indices $s \neq t$. Then, we fix another two different indices $x \neq y$ from π_s^a and π_t^a. We exchange the positions: $\pi_t^{bx} = \pi_t^{ay}$, $\pi_t^{by} = \pi_s^{ax}$ and $\forall d, d \neq (x,y)$ $\pi_s^{bd} = \pi_t^{ad}$, $\forall c, c \neq (t,s)$ $\pi_c^b = \pi_c^a$.

- Inter-quays insert operator $\mathcal{N}_2$: this operator chooses two distinct quays and removes a vessels from one and inserts it in the another, respecting the space restriction. The operator repeats this process for all the quays until all the neighborhoods have been searched. We assume a solution of the problem $S_a = (\pi_1^a, \pi_2^a, ..., \pi_n^a)$ and one of its neighbors $S_b = (\pi_1^b, \pi_2^b, ..., \pi_n^b)$. For two different indices $s \neq t$, we define $\pi_s^a = (\pi_s^{a1}, \pi_s^{a2}, ..., \pi_s^{am})$ and $\pi_t^b = (\pi_t^{b1}, \pi_t^{b2}, ..., \pi_t^{bk})$. We fix a vessel y from π_s^a and we illustrate the move as the following: $\forall y \leq i \leq m-1, \pi_s^{ai} = \pi_s^{ai+1}$ and $\pi_t^{bk+1} = \pi_s^{by}$. Finally, $\forall c, c \neq (t,s)$ $\pi_c^b = \pi_c^a$.

- Intra-quay swap operator $\mathcal{N}_3$: this operator selects a pair of vessels in one fixed quay and exchanges their positions. The operator repeats this process for all the quays until all the neighborhoods have been searched. We assume a solution of the problem $S_a = (\pi_1^a, \pi_2^a, ..., \pi_n^a)$ and one of its neighbors $S_b = (\pi_1^b, \pi_2^b, ..., \pi_n^b)$. We fix a quay s. Then, we fix another two different indices $x \neq y$ from π_s^a. We exchange the positions: $\pi_s^{bx} = \pi_s^{ay}$, $\pi_s^{by} = \pi_s^{ax}$ and $\forall d, d \neq (x,y)$ $\pi_s^{bd} = \pi_s^{ad}$, $\forall c, c \neq s$ $\pi_c^b = \pi_c^a$.

- Intra-quay insert operator $\mathcal{N}_4$: this operator removes a vessel and inserts it in another position in the same quay s. The operator repeats this process for all the quays until all the neighborhoods have been searched. We assume a solution of the problem $S_a = (\pi_1^a, \pi_2^a, ..., \pi_n^a)$ and one of its neighbors $S_b = (\pi_1^b, \pi_2^b, ..., \pi_n^b)$. For $x < y$ in π_s^a, $\pi_s^{bx} = \pi_s^{ay}$, $\pi_s^{bx+1} = \pi_s^{ax}, ..., \pi_s^{by} = \pi_s^{ay-1}$. In the case of $x > y$, $\pi_s^{by} = \pi_s^{ay+1}, ..., \pi_s^{bx-1} = \pi_s^{ax}, \pi_s^{bx} = \pi_s^{ay}$.

3.3   Sequential Variable Neighborhood Descent

In this section, we present the Variable Neighborhood Descent algorithm [8]. The main idea is to use multi-neighborhood version of a local search. More precisely, we start the seq-VND algorithm by generating an initial solution S with $G - FCFS$, this solution will be improved sequentially by applying predefined sequence of neighborhoods (see Algorithm 2). Each neighborhood is explored using the so-called first improvement search strategy, i.e., as soon as an improving solution is detected it is set as the new incumbent solution and search is resumed from it. The seq-VND algorithm stops when no more improvement is possible.

Algorithm 1. seq-VND

1 k_{max}: the number of the neighborhood structures
2 Set of neighborhoods $\mathcal{N}_k, k \in \{1, .., k_{max}\}$
3 $k = 1$
4 $S \leftarrow \text{G} - \text{FCFS}()$
  while $k \leq k_{max}$ do
5   S' $\leftarrow$ First improvement Local Search $(\mathcal{N}_k, S)$
6   if $f(S') < f(S)$ then
7        $S \leftarrow S'$
8        $k \leftarrow 1$
    end
9   else
       $k \leftarrow k + 1$
    end
  end
10 Return S

Algorithm 2. First improvement Local Search

1 Input: S, $\mathcal{N}_k$
2 $Improve \leftarrow 0$
  while $S' \in \mathcal{N}_k(S)$ et $Improve \leftarrow 0$ do
    if $f(S') < f(S)$ then
3       $S \leftarrow S'$
4       $Improve \leftarrow 1$
    end
  end
5 Return S

---

4 Computational Results

In the following, we report the computational experiments that were performed for testing the seq-VND. The algorithm were implemented in C++ and executed on an Intel Core i7 2600 CPU (2.8 GHz) and 16 GB RAM. For testing purposes test instances were generated using the data obtained from our industrial partner. We differentiate small size instances ($N = 7$) and large scale instances

($N = 14$). For each type of instances we differentiate three scenarios regarding the congestion: High Congestion, Medium Congestion and Low Congestion. The overcrowding of vessels were determined by the frequency in which boats arrive. In addition, we distinguish two types of instances based on the number of cargo: one product and two-products.

In the following, we present the port terminal characteristics considered in all instances:

- The security distance between berthed vessels Sc is equal to 5 m.
- For the high tide time-windows, we consider a daily period of 6 h (given by weather agency).
- The terminal contains four quays with the respective lengths: 400, 200, 200, 180 and their respective rates.
- We fix two unavailability periods over the planning horizon, which are due to the planned maintenance and weather conditions.

Table 1. The results of seq-VNDs using different number of neighborhoods

N	Demand	Congestion	The number of neighborhood structures							
			1		2		3		4	
			Avg_sol	Avg_time	Avg_sol	Avg_time	Avg_sol	Avg_time	Avg_sol	Avg_time
14	One product	High	9480.67	0.01	60.00	0.45	60.00	0.69	60.00	1.6
		Medium	8040.00	0.02	1557.33	0.69	1557.33	0.93	1527.33	2.03
		Low	2110.00	0.03	720.00	0.34	720.00	0.67	720.00	2.07
	Multi product	High	99091.33	0.01	29940.67	1.7	29940.67	2.02	24507.33	4.09
		Medium	86586.00	0.02	23959.33	1.8	23959.33	2.05	13422.67	4.6
		Low	17355.33	0.04	2670.00	1.3	2670.00	1.6	2100.00	3.17

For each choice of size, congestion and number of products 3 random instances are generated constituting the class of instances. Therefore, in computational results we report the average results obtained over 3 instances belonging to the same class. The results on small size instances are compared against CPLEX 12.8 MIP solver used to solve MIP formulation provided in [16]. On the other hand large size instance cannot be handled by CPLEX (CPLEX does not find even a feasible solution after two hours of running time) and therefore these instances are used as a benchmark to identify the number of neighborhoods and the best order to be used within seq-VND.

Table 1 summarizes the results for different numbers of neighborhood structures used within the seq-VND. In each of these variant, the neighborhoods are examined in order $N_1, N_2...N_k$ where k is the number of used neighborhoods. We denote, as metrics, the number of vessels (N), the congestion level (*congestion*), the average objective function value (*Avg_sol*) and its corresponding average execution time in minutes (*Avg_time*).

According to the results, all instances were solved by seq-VNDs. The most effective version turns to be seq-VND using 4 neighborhoods (i.e., $k_{max} = 4$).

However, as expected since it uses the largest number of neighborhoods it is the slowest in comparison to the other variants. However, it still consumes reasonable amount time to provide final solution. It consumes no more than 5 min to solve an instance.

Table 2. The impact of neighborhoods order on the seq-VND

N	Demand	Congestion	Order							
			1		2		3		4	
			Avg_sol	Avg_time	Avg_sol	Avg_time	Avg_sol	Avg_time	Avg_sol	Avg_time
14	One product	High	60.00	1.6	30.00	1.9	60.00	1.7	180.00	1.2
		Medium	1527.33	2.03	1527.33	2.7	1527.33	2.5	1546.67	2.00
		Low	720.00	2.07	720.00	2.1	720.00	2.06	720.00	1.9
	Multi product	High	24507.33	4.09	29624.67	6.00	26752.67	5.02	26341.33	3.2
		Medium	13422.67	4.6	14916.00	5.9	14352.67	5.2	13932.67	3.9
		Low	2100.00	3.17	2100.00	4.1	2100.00	3.8	2100.00	2.7

Now, we are interested to identify the impact of the neighborhood order on Seq-VND performance. For this purpose, we implement four seq-VNDs. Each of them uses four neighborhoods implemented in different orders. Table 2 summarizes the results obtained by the sequential VND considering different neighborhood orders. We used the same metrics except "order" that means the sequence of neighborhoods used in the proposed method. In our experiments, we suggested the following orders: $order_1 = \{\mathcal{N}_3, \mathcal{N}_1, \mathcal{N}_4, \mathcal{N}_2\}$, $order_2 = \{\mathcal{N}_4, \mathcal{N}_2, \mathcal{N}_3, \mathcal{N}_1\}$, $order_3 = \{\mathcal{N}_3, \mathcal{N}_4, \mathcal{N}_1, \mathcal{N}_2\}$, $order_4 = \{\mathcal{N}_1, \mathcal{N}_2, \mathcal{N}_3, \mathcal{N}_4\}$.

Table 3. Performances of seq-VND compared to CPLEX

N	Demand	Congestion	Avg_CPLEX	Avg_time CPLEX (min)	Avg_seq-VND	Avg_time seq-VND (min)
7	One product	High	510	1.81	510	0.004
		Medium	6772.66	1.64	6772.66	0.013
		Low	0	1.23	0	0.001
	Multi product	High	50	25.71	80	0.021
		Medium	1736.66	43.34	1870	0.023
		Low	0	1.23	0	0.001

We remark that the order may have a significant impact on the solution quality. *Order*1 provides the best solutions for all cases except one, where *order*2 exhibits the best performance. In addition, it may be observed that *order*2 is the best choice if one product is demanded, while *order*1 is the best choice in the multi-product case.

After identifying, the optimal number of neighborhoods and order within seq-VND to be used, we are interested to assess the performance of such seq-VND on small size instances. The version of seq-VND that were executed uses 4 neighborhoods and follows *order*1. The obtained results are compared against CPLEX 12.8 MIP solver which succeeded to solve all instances to the optimality within one hour of execution time. The average solution values for CPLEX and seq-VND are given in columns *Avg_CPLEX* and *Avg_Seq-VND*, respectively. Analogously, corresponding average execution times are given in columns *Avg_time CPLEX* and *Avg_time seq-VND* (Table 3).

From the reported results it follows that on all instances with one product, seq-VND finds optimal solutions in very short time. In the two-product case seq-VND exhibits different performances regarding solution quality depending to the level of congestion. In the case of low congestion seq-VND finds optimal solution on all instances, on medium and high congestion instances it finds 2 out of 3 optimal solutions. Regarding CPU time consumption seq-VND consumes negligible amount of time in comparison to CPLEX. So, we may conclude that the proposed seq-VND is very competitive comparing to the CPLEX.

5 Conclusion and Perspectives

In this paper, we present a sequential Variable Neighborhood Descent based heuristics for an important seaside decision problem, which is the integrated berth allocation and crane assignment problem. The variant studied in our work is realistic. Indeed, we consider a multi-quay bulk port layout and the planning is done under several availability restrictions. In order to evaluate the proposed method, we were relying on a real case study. The preliminary experiments showed satisfactory results. Proposed seq-VND turns out to be very competitive in comparison to CPLEX 12.8 MIP solver and succeeds to solve instances that cannot be handled by it. Future work may investigate proposing some variants of *VNS* to tackle the problem.

References

1. Agra, A., Oliveira, M.: MIP approaches for the integrated berth allocation and quay crane assignment and scheduling problem. Eur. J. Oper. Res. **264**(1), 138–148 (2018)
2. Barros, V.H., Costa, T.S., Oliveira, A.C., Lorena, L.A.: Model and heuristic for berth allocation in tidal bulk ports with stock level constraints. Comput. Ind. Eng. **60**(4), 606–613 (2011)
3. Becker, C., Scholl, A.: A survey on problems and methods in generalized assembly line balancing. Eur. J. Oper. Res. **168**(3), 694–715 (2006)
4. Blazewicz, J., Cheng, T.E., Machowiak, M., Oguz, C.: Berth and quay crane allocation: a moldable task scheduling model. J. Oper. Res. Soc. **62**(7), 1189–1197 (2011)
5. Chang, D., Jiang, Z., Yan, W., He, J.: Integrating berth allocation and quay crane assignments. Transp. Res. Part E: Logist. Transp. Rev. **46**(6), 975–990 (2010)
6. Cheng, J.K., Tahar, R.M., Ang, C.L.: A system dynamics approach to operational and strategic planning of a container terminal. Int. J. Logist. Syst. Manag. **10**(4), 420–436 (2011)
7. Diabat, A., Theodorou, E.: An integrated quay crane assignment and scheduling problem. Comput. Ind. Eng. **73**, 115–123 (2014)

8. Duarte, A., Mladenović, N., Sánchez-Oro, J., Todosijević, R.: Variable neighborhood descent. In: Handbook of Heuristics, pp. 1–27 (2016)
9. Ernst, A.T., Oğuz, C., Singh, G., Taherkhani, G.: Mathematical models for the berth allocation problem in dry bulk terminals. J. Sched. **20**(5), 459–473 (2017)
10. Fu, Y.M., Diabat, A.: A lagrangian relaxation approach for solving the integrated quay crane assignment and scheduling problem. Appl. Math. Model. **39**(3–4), 1194–1201 (2015)
11. Giallombardo, G., Moccia, L., Salani, M., Vacca, I.: Modeling and solving the tactical berth allocation problem. Transp. Res. Part B: Methodol. **44**(2), 232–245 (2010)
12. Hansen, P., Mladenović, N., Pérez, J.A.M.: Variable neighbourhood search: methods and applications. Ann. Oper. Res. **175**(1), 367–407 (2010)
13. Imai, A., Sun, X., Nishimura, E., Papadimitriou, S.: Berth allocation in a container port: using a continuous location space approach. Transp. Res. Part B: Methodol. **39**(3), 199–221 (2005)
14. Iris, Ç., Pacino, D., Ropke, S.: Improved formulations and an adaptive large neighborhood search heuristic for the integrated berth allocation and quay crane assignment problem. Transp. Res. Part E: Logist. Transp. Rev. **105**, 123–147 (2017)
15. Iris, Ç., Pacino, D., Ropke, S., Larsen, A.: Integrated berth allocation and quay crane assignment problem: set partitioning models and computational results. Transp. Res. Part E: Logist. Transp. Rev. **81**, 75–97 (2015)
16. Krimi, I., Benmansour, R., Ait El Cadi, A., Deshayes, L., Duvivier, D., Elhachemi, N.: A rolling horizon approach for the multi-quay berth allocation and crane assignment problem in bulk ports. Computer and Industrial Engineering, under review
17. Liang, C., Huang, Y., Yang, Y.: A quay crane dynamic scheduling problem by hybrid evolutionary algorithm for berth allocation planning. Comput. Ind. Eng. **56**(3), 1021–1028 (2009)
18. Meisel, F., Bierwirth, C.: Heuristics for the integration of crane productivity in the berth allocation problem. Transp. Res. Part E: Logist. Transp. Rev. **45**(1), 196–209 (2009)
19. Park, Y.-M., Kim, K.H.: A scheduling method for berth and quay cranes. In: Günther, H.-O., Kim, K.H. (eds.) Container Terminals and Automated Transport Systems, pp. 159–181. Springer, Heidelberg (2005). https://doi.org/10.1007/3-540-26686-0_7
20. Raa, B., Dullaert, W., Van Schaeren, R.: An enriched model for the integrated berth allocation and quay crane assignment problem. Expert Syst. Appl. **38**(11), 14136–14147 (2011)
21. Rodriguez-Molins, M., Salido, M.A., Barber, F.: A grasp-based metaheuristic for the berth allocation problem and the quay crane assignment problem by managing vessel cargo holds. Appl. Intell. **40**(2), 273–290 (2014)
22. Stopford, M.: Maritime Economics. Routledge, Abingdon (2013)
23. Umang, N., Bierlaire, M., Vacca, I.: Exact and heuristic methods to solve the berth allocation problem in bulk ports. Transp. Res. Part E: Logist. Transp. Rev. **54**, 14–31 (2013)
24. Vacca, I., Salani, M., Bierlaire, M.: An exact algorithm for the integrated planning of berth allocation and quay crane assignment. Transp. Sci. **47**(2), 148–161 (2013)
25. Yang, C., Wang, X., Li, Z.: An optimization approach for coupling problem of berth allocation and quay crane assignment in container terminal. Comput. Ind. Eng. **63**(1), 243–253 (2012)

A Variable Neighborhood Search Approach for Solving the Multidimensional Multi-Way Number Partitioning Problem

Alexandre Frias Faria[1], Sérgio Ricardo de Souza[1(✉)],
Marcone Jamilson Freitas Souza[2], Carlos Alexandre Silva[3],
and Vitor Nazário Coelho[4]

[1] Federal Center of Technological Education of Minas Gerais, Belo Horizonte, Brazil
alexandrefrias1@hotmail.com, sergio@dppg.cefetmg.br
[2] Federal University of Ouro Preto, Ouro Preto, Brazil
marcone@iceb.ufop.br
[3] Federal Institute of Minas Gerais, Sabará, Brazil
carlos.silva@ifmg.edu.br
[4] Fluminense Federal University, Niterói, Brazil
vncoelho@gmail.com

Abstract. This paper presents an implementation of the Variable Neighborhood Search (VNS) metaheuristic for solving the optimization version of the Multidimensional Multi-Way Number Partitioning Problem (MDMWNPP). This problem consists in distributing the vectors of a given sequence into k disjoint subsets such that the sums of each subset form a set of vectors with minimum diameter. The proposed VNS for solving MDMWNPP has a good performance over instances with three and four subsets. A comparative study of results found from this proposed VNS and an implementation of Memetic Algorithm (MA) is carried out, running in the same proportional time interval. Although the average results are different, the statistical tests show that results of the proposed VNS are not significantly better than MA in a set of instances analyzed.

Keywords: Multidimensional Multi-Way Number Partitioning Problem · Variable Neighborhood Search · Number Partitioning Problem · Combinatorial optimization

1 Introduction

This paper addresses the Multidimensional Multi-Way Number Partitioning Problem (MDMWNPP), a more general version of the classical Number Partitioning Problem (NPP). This problem is related to any problems involving partitions set like Bin Packing, Machine Scheduling and Clustering, for example.

© Springer Nature Switzerland AG 2019
A. Sifaleras et al. (Eds.): ICVNS 2018, LNCS 11328, pp. 243–258, 2019.
https://doi.org/10.1007/978-3-030-15843-9_19

As it is a generalization, it is necessary to review the problems that originated it to contextualize its study. Throughout this text, a partition of a X set is a collection of mutually disjoint subsets whose union forms X. A k-partition of a set X is a partition of this set, with exactly k non-empty subsets. In this text, the subsets belonging to the partition are called parts. The notation $I_p = \{y \in \mathbb{Z} : 1 \leq y \leq p\}$ denotes the closed set of all integers between 1 and p.

The Two-Way Number Partitioning Problem (TWNPP) is a well known problem in the literature. Its purpose is to find a 2-partition of the indexes of a V sequence so that the difference between the sums of the elements of each part is minimal. This problem was listed in [7] as one of the basic NP-complete problems, and a series of equivalences between TWNPP and other NP-complete problems are also demonstrated. The first exact algorithms trivially adaptable to TWNPP were presented in [5] and [24], both proposed to solve the Knapsack Problem. In [9], two proposals are presented to transform the heuristics into exact methods of the Branch & Bound type, which are: (i) Complete Greedy Algorithm (CGA), using the Longest Processing Time heuristic (LPT) [4]; and (ii) Complete Karmarkar-Karp Algorithm (CKK), using the Differencing Method, well-known as Karmarkar-Karp Heuristic (KKH) [6]. An exact method based on CKK appears in [14]. In this case, the proposal is to increase the number of prunings in the CKK search tree using a new heuristic called the Balanced Largest Differencing Method (BLDM). Already [19] presents an improvement in the CKK search tree search using a new data structure.

The first generalization of TWNPP is the Multi-Way Number Partitioning Problem (MWNPP), which expands the number of parts in which the sequence indices V must be distributed. Given a numeric sequence V, the goal is to find a k-partition for its indexes, such that the sums of the elements of each part fits into the shortest possible interval. MWNPP is explicitly stated in [6], in which an analysis of Karmarkar-Karp Heuristics (KKH) is presented. This heuristic is focused on the idea of dividing the largest numbers into distinct parts, inserting the differences between the removed elements in the set of unallocated elements as long as this set is not empty. In [2], it is shown that MWNPP is a very difficult problem to be solved by general-purpose metaheuristics, such as Genetic Algorithms, Simulated Annealing, and others. In many cases, these methods have a worse computational cost (in terms of time and performance) when compared to HKK and even to LPT. The exact algorithms presented in [9] were already adapted for the MWNPP.

The first improvement of these works happens with the algorithm Recursive Number Partitioning (RNP), proposed by [10] working with the resolution of minor subproblems derived from MWNPP. Through successive MWNPP conversions of a $(k-1)$-partition to a k-partition, [16] propose an algorithm based on solving smaller subproblems. Currently, the state of the art for the resolution of MWNPP is the Sequential Number Partitioning (SNP) algorithm, presented in [11], and the Cached Iterative Weakening (CIW) algorithm, presented in [23], both fully analyzed in [22]. An application of VNS algorithm for solving MWNPP is described in [1].

The second generalization of TWNPP is the Multidimensional Two-Way Number Partitioning Problem (MDTWNPP). This variant considers a V sequence of vectors of dimension m instead of real numbers, as TWNPP is originally defined. Its purpose is to find a 2-partition of the set of vectors so that the vectors resulting from the sum of each part have minimized the distance induced by the infinite norm. This generalization is initially proposed in [8]. In this same article, a mathematical model in integer linear optimization for MDTWNPP is proposed and solved using CPLEX. This is the only known exact method for solving this problem up to the present moment, according to the knowledge of the authors of this current article. MDTWNPP is also addressed in [20], but in this case using population metaheuristics, such as Memetic Algorithm (MA) and Genetic Algorithm (GA), for its solution. Already [12] presents implementations of Variable Neighborhood Search (VNS) and Electromagnetism-like (EM) metaheuristics to the solution of MDTWNPP and compares the results with those presented in [8] and [20]. The results show that the EM metaheuristic performs slightly superior to the others and strongly superior to the direct solution of the exact model. Another important article addressing MDTWNPP is [21], in which this problem is solved using GRASP+Exterior Path-relinking hybrid metaheuristics. The results obtained, from the same set of instances used in [8,12,20], show the superiority of the proposed procedure.

The third generalization of TWNPP is the Multidimensional Multi-Way Number Partitioning Problem (MDMWNPP). Given a V sequence of vectors of dimension m, the goal of MDMWNPP is to determine a k-partition of vectors such that, added the elements of each part, the diameter of the resulting vectors is minimized. MDMWNPP is originally proposed in [20] with the resolution of three-way ($k = 3$) and four-way ($k = 4$) cases using Memetic Algorithm (MA). It should be stressed that this problem is still little studied in the literature and, on the other hand, is the central object of study of the current article.

This article presents an adaptation of the VNS metaheuristic proposed in [15] for the MDMWNPP solution. The results are compared with those presented in [20] using the same instances of this last article. The justification for applying VNS to MDMWNPP is the set of good results found in [12] for this metaheuristic when solving MDTWNPP.

The article is organized as follows. Section 2 presents the synthetic statement of MDMWNPP and a equivalence proof between the diameter induced by the infinite norm and the objective function introduced in [20]. Section 3 shows the operation of the proposed VNS and the particularity of its neighborhood in rings. Section 4 presents a delineation of the tests performed for the comparison between the results, while Sect. 5 criticizes the results obtained regarding the number of executions required and the form of the instances used for a really valid statistical test. Finally, Sect. 5 concludes the article and presents proposals for future work.

2   Problem Statement

2.1   Fundamental Notions

This section introduces fundamental notions concerning MDMWNPP. Let $V = \{v_i\}_{i \in I_n}$ be a set of vectors. The function $gv : P(I_n) \to \mathbb{R}^m$ receives a discrete subset of vectors in $\mathbb{R}^m$ and returns the sum of its elements. Calculate the function $gv(\cdot)$ as:

$$X \in P(I_n) \quad : \quad gv(X) = \sum_{i \in X} v_i \tag{1}$$

Referring to the l-th coordinate of $gv(X)$, the notation $gv_l(X)$ is used.

Definition 1. *Let $V = \{v_i\}_{i \in I_n}$ be a sequence of vectors such that $v_i \in \mathbb{R}^m$ and k an integer positive number. Find a k-partition of the V indexes, in the form $\{A_j\}_{j \in I_k}$, that minimizes the diameter of the multiset $\{gv(A_j)\}_{j \in I_k}$ given by:*

$$diam_\infty \left(\{A_j\}_{j \in I_k}\right) = \max_{j',j} \{\|gv(A_{j'}) - gv(A_j)\|_\infty\} \tag{2}$$

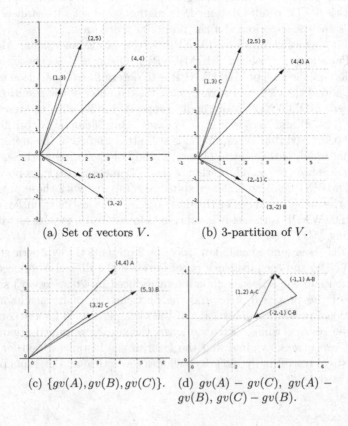

(a) Set of vectors V.   (b) 3-partition of V.

(c) $\{gv(A), gv(B), gv(C)\}$.   (d) $gv(A) - gv(C)$, $gv(A) - gv(B)$, $gv(C) - gv(B)$.

Fig. 1. Representation of $V = \{(1,3), (4,4), (3,-2), (2,5), (2,-1)\}$ from Example 1.

Example 1. Let $V = \{(1,3), (4,4), (3,-2), (2,5), (2,-1)\}$ be a set of vectors, as shown in Fig. 1(a). This example comes from [20]. Figure 1(b) shows an optimal 3-partition for V. The value of the objective function, from Eq. (2), is $\max_l\{|gv_l(A) - gv_l(C)|, |gv_l(A) - gv_l(B)|, |gv_l(C) - gv_l(B)|\} = 2$.

2.2 Analysis of the Objective Function

The objective function for MDMWNPP proposed in this article, presented in Eq. (2), appears to be different from that originally introduced in [20], given by:

$$f(\{A_j\}_{j \in I_k}) = \max_l \left\{ \left| \max_{j'} gv_l(A_{j'}) - \min_j gv_l(A_j) \right| \right\} \tag{3}$$

In fact, these two functions represent different ways for calculating the diameter of a set of vectors. In the following, an analysis of these two objective functions is presented.

First, it is possible to remove the module from the Expression (3) without any loss, since $\max_{j'} gv_l(A_{j'}) - \min_j gv_l(A_j) \geq 0$ in any case. The idea is to denote the objective function only as the set diameter, allowing a clear interpretation that applies to all variants of the NPP problem (TWNPP, MWNPP, MDTWNPP and MDMWNPP) listed in the current article.

Proposition 1 and Corollary 1 are applied in the demonstration of equivalence of the two expressions in Proposition 2.

Proposition 1. *Let I be a limited real interval. Then:*

$$\max_{x,y \in I} |x - y| = \max_{z \in I} z - \min_{z \in I} z \tag{4}$$

Proof. Since $x, y \in I$, then:

$$\min_{z \in I} z \leq x \leq \max_{z \in I} z \tag{5}$$

$$\min_{z \in I} z \leq y \leq \max_{z \in I} z \tag{6}$$

Manipulating these two expressions, the result is:

$$\min_{z \in I} z - \max_{z \in I} z \leq x - y \leq \max_{z \in I} z - \min_{z \in I} z \tag{7}$$

That is:

$$|x - y| \leq \max_{z \in I} z - \min_{z \in I} z \tag{8}$$

Therefore:

$$\max_{x,y \in I} |x - y| \leq \max_{z \in I}(z) - \min_{z \in I}(z) \tag{9}$$

Equality occurs for the maximum value. This is verified by fixing $x = \max_{z \in I} z$ and $y = \min_{z \in I} z$.

Corollary 1. *Let I be a limited real interval and consider a discrete sequence $\{a_i\}_{i \in I_n} \subset I$. Then:*

$$\max_{i,j \in I_n} |a_i - a_j| = \max_{i \in I_n} a_i - \min_{j \in I_n} a_j \qquad (10)$$

Proposition 2. *The distance induced by the infinite norm is given by:*

$$diam_\infty(\{gv(A_j)\}_{j \in I_k}) = \max_{l \in I_m} \left\{ \max_{j \in I_k} gv_l(A_j) - \min_{j' \in I_k} gv_l(A_{j'}) \right\} \qquad (11)$$

Proof. From Definition 1:

$$\begin{aligned}
diam_\infty(\{gv(A_j)\}_{j \in I_k}) &= \max_{j,j' \in I_k} \|gv(A_j) - gv(A_{j'})\|_\infty \\
&= \max_{j,j' \in I_k} \max_{l \in I_m} |gv_l(A_j) - gv_l(A_{j'})| \\
&= \max_{l \in I_m} \max_{j,j' \in I_k} |gv_l(A_j) - gv_l(A_{j'})| \qquad (12)
\end{aligned}$$

Then, by Corollary 1:

$$\max_{l \in I_m} \max_{j,j' \in I_k} |gv_l(A_j) - gv_l(A_{j'})| = \max_{l \in I_m} \left\{ \max_{j \in I_k} gv_l(A_j) - \min_{j' \in I_k} gv_l(A_{j'}) \right\} \qquad (13)$$

and the proof is finished.

Example 2. Consider the sequence:

$$\begin{aligned}
V = \{&(14, 48, 23), (87, 61, 48), (76, 14, 23), (24, 25, 33), (84, 13, 49), (25, 48, 78), \\
&(56, 14, 73), (55, 21, 20), (16, 13, 86), (74, 55, 31)\}
\end{aligned}$$

with $n = 10$ and $m = 3$. For $k \in \{3, 4, 5\}$, Table 1 shows a feasible (non-optimal) partition and the optimal partition of this sequence V when solving MDTWNPP, with the associated values of objective function. It is worth mentioning that V has a larger dimension than that associated to sequence shown in Example 1. To represent the partitions, the classic coding in [18] is used: a sequence $(s_i)_{i \in I_n}$, where $s_i \in I_k$, indicating the part to which the vector v_i belongs. This notation has some extra details that will be explained in Sect. 3.

Table 1. Example 2: feasible solutions vs optimal solutions

k	Feasible	Obj. val.	Optimal	Obj. val.
3	[1, 1, 1, 2, 3, 3, 2, 3, 1, 2]	54	[1, 2, 1, 2, 3, 3, 2, 1, 1, 3]	22
4	[1, 2, 3, 3, 3, 2, 4, 4, 1, 1]	81	[1, 2, 3, 1, 1, 3, 2, 4, 4, 4]	44
5	[1, 2, 3, 1, 1, 3, 4, 5, 5, 4]	59	[1, 2, 3, 2, 1, 3, 4, 4, 5, 5]	51

In this problem, it should be emphasized not only that the calculation of the objective function is particularly complicated, but also how difficult it is to perform the complete search in the whole space of feasible solutions. According to [25], the number of possibilities for each value of k in Example 2 is given by the Stirling Numbers: $S(10,3)_{k=3} = 9330$, $S(10,4)_{k=4} = 34105$ and $S(10,5)_{k=5} = 42525$, respectively.

3 Proposed Algorithm

The proposed Variable Neighborhood Search (VNS) algorithm works with real-location moves of an element between parts of the partition $\{A_j\}_{j \in I_k}$. The move $m_{i,j}$ means that an index element $i \notin A_j$ leaves the part where it is and goes to the index part j. This move derives from the enumeration algorithm presented in [18]. It is able to generate all k-partitions of a set.

Consider $s' = \{A'_j\}_{j \in I_k}$ e $s = \{A_j\}_{j \in I_k}$. The neighborhoods $N_1(s)$, $N_2(s)$ e $N_3(s)$ are given by:

$$N_1(s) = \{s' \; : \; s' \leftarrow s \oplus m_{i,j}, \; \forall (i,j) \in I_n \times I_k\} \tag{14}$$

$$N_2(s) = \{s' \; : \; s' \leftarrow s \oplus m_{i,j} \oplus m_{i',j'}, \; \forall (i,j) \in I_n \times I_k\} \tag{15}$$

$$N_3(s) = \{s' \; : \; s' \leftarrow s \oplus m_{i,j} \oplus m_{i',j'} \oplus m_{i'',j''}, \; \forall (i,j) \in I_n \times I_k\} \tag{16}$$

Therefore, these neighborhoods are formed, respectively, by compositions of one, two and three distinct moves $m_{i,j}$. Thus, if $i \in A_l$, a reallocation move $\{A_1, \ldots, A_l, \ldots, A_j, \ldots, A_k\} \oplus m_{i,j}$ leads to $\{A_1, \ldots, A_l - \{i\}, \ldots, A_j \cup \{i\}, \ldots, A_k\}$. The neighborhoods are such that $N_l(s) \cap N_j(s) = \emptyset, \quad \forall i \neq j$.

The encoding used is a vector of size n whose entries are numbers from 1 to k. There are restrictions to these moves, in the form of the following rules:

(i) The index 1 of v_1 must always be in the part 1;
(ii) If v_i is in a part with a single element, the motion $m_{i,j}$ can not be applied;
(iii) A part j will always have at least a i' index less than any index contained in the part $j + 1$. This holds for all $j \in I_{k-1}$.

Example 3. Consider the two encodings below:

$$[1,1,1,3,2], \quad [1,1,1,2,3]$$

Note that the first vector does not satisfy the rule (iii) while the second vector satisfies. It is possible to make a move by following rules (i) and (ii) as:

$$[1,1,1,2,3] \oplus m_{2,2} = [1,2,1,2,3]$$

but not:

$$[1,1,1,2,3] \oplus m_{2,3} = [1,3,1,2,3]$$

since $[1,3,1,2,3] = [1,2,1,3,2]$, representing the 3-partition $\{v_1, v_3\}, \{v_2, v_5\}, \{v_4\}$.

Algorithm 1. Codification of solutions s

1: V, k, n, m
2: $s : s_i$ ▷ vector $(s_i)_{i \in I_n}$ where $s_i \in I_k$
3: $b : b_i$ ▷ vector $(b_i)_{i \in I_n}$ where $b_1 = 1$ and $b_i = \min\{\max_{2 \le h \le i}\{s_h + 1\}, k\}$
4: $card : card_j$ ▷ cardinality of the parts
5: $sum : sum_j : sum_{jl}$ ▷ $sum_j = gv(A_j)$ and $sum_{jl} = gv_l(A_j)$

Algorithm 2. Objective function

1: **function** $f(s)$
2:     $r_1 \leftarrow \max_j sum_{j1} - \min_j sum_{j1}$
3:     $obj \leftarrow r_1$
4:     **for** $l \in I_m \setminus \{1\}$ **do**
5:         $r_l \leftarrow \max_j sum_{jl} - \min_j sum_{jl}$
6:         **if** $r_l > obj$ **then**
7:             $obj \leftarrow r_l$
8:         **end if**
9:     **end for**
10:     **return** obj
11: **end function**

By rule (iii), the leader element of j can not be passed to a part $j' > j$, because in this case the same solution would have many encodings. The leader element in j can only be moved to $j' < j$ if the second lowest index in j is smaller than all indices of $j + 1$.

These conditions are described in [17], which presents the most efficient known codings to make enumerations of combinatorial structures. There is also a demonstration of the bijection between the set of k-partitions and the coding presented in [18].

With this encoding, the neighborhood size $N_r(s)$ depends on the solution s. In Example 3, there is $N_1([1, 1, 1, 2, 3]) = \{[1, 2, 1, 2, 3], [1, 1, 2, 2, 3]\}$ but, also, $N_1([1, 2, 1, 2, 3]) = \{[1, 1, 1, 2, 3], [1, 2, 2, 2, 3], [1, 2, 3, 2, 3], [1, 2, 1, 3, 3], [1, 2, 1, 1, 3]\}$.

These cardinality differences accumulate as r increases its value to 2 or 3. It is only possible to limit the cardinality of neighborhoods by upper bounds to show that they are polynomials. Thus, consider s being a vector $(s_i)_{i \in I_n}$ and $s_i \in I_k$ representing a partition:

$$|N_r(s)| < \left(\sum_{i \in I_n \setminus \{1\}} \max_{2 \le h \le i} \{s_h\} \right)^r \le \left(\frac{k(2n - k - 1)}{2} \right)^r \qquad (17)$$

The upper bound shown in Expression (17) is not tight, that is, there is no case where equality occurs, but its expression is compact and already shows that the search space of the neighborhoods used is limited polynomially since $r \le 3$.

The proposed VNS, described in the Algorithm 8, follows the guidelines of [15]. The coding of the solution s is given by Algorithm 1, which holds information essential for the manipulation of the search space and to save computational operations. Instance data can be accessed directly from solution s.

The objective function calculation is done by Algorithm 2. This method is equivalent to the implementation of function (3). The computational cost is

Algorithm 3. Initial solution algorithm

```
1: function LPT(V, k)
2:     for j ∈ I_k do
3:         L_j = 0
4:     end for
5:     l ← rand(I_m)                                    ▷ Select one coordinate in I_m for all vectors of V
6:     for i ∈ I_n do
7:         s_i = arg min_j L_j                          ▷ build the k-partition
8:         L_{s_i} = L_{s_i} + v_{il}                   ▷ Update sums of the parts
9:     end for
10:    return s
11: end function
```

$\mathcal{O}(\max\{n, km\})$, being $n - k$ operations necessary to obtain the vector sum in Algorithm 1 and $(\frac{3}{2}k - 2)(m - 1)$ the number of operations of Algorithm 2.

The initial solution, found by Algorithm 3, is given by an adaptation of the algorithm proposed in [3]. This is a greedy method that fixes a coordinate l and applies a greedy allocation of vectors v_i in the part L_j less loaded at each iteration. A k-partition resulting from this method will be the initial solution of the proposed VNS.

Algorithm 4. Movement of one element

```
1: function move(i, s)
2:     s' ← s
3:     if card_{s_i} = 1 then                           ▷ rule 1
4:         return s'
5:     end if
6:     h ← s_i                                          ▷ part of element i
7:     for j ∈ I_{b_i} \ {h} do
8:         s_i ← j
9:         sum_j ← sum_j + v_i                          ▷ v_i move out from h to j
10:        sum_h ← sum_h - v_i
11:        if f(s) < obj then
12:            obj ← f(s)
13:            s' ← s                                    ▷ The new best solution
14:        end if
15:    end for
16:    return s'
17: end function
```

Algorithm 4 returns the best of all possible valid moves of a vector v_i between the possible parts. This procedure is used to enumerate all neighbors in the structures $N_1(s)$, $N_2(s)$ and $N_3(s)$.

Algorithms 5, 6 and 7 show the implementations of the local search method Best Improvement for neighborhoods $N_1(s)$, $N_2(s)$ and $N_3(s)$, respectively. These algorithms explore all the neighbors of a solution s and return the one with the lowest objective function value. The computational complexity of each of them is upperly limited by Expression (17).

Algorithm 5. Best improvement for $N_1(s)$

1: **function** $best1(s)$
2:     $obj \leftarrow f(s)$ ▷ Save objective value of the current solution
3:     **for** $i \in I_n$ **do**   ▷ For each v_i the $move()$ is applied
4:         $s' \leftarrow move(i, s)$
5:         **if** $f(s') < obj$ **then**
6:             $obj \leftarrow f(s')$
7:             $s'' \leftarrow s'$ ▷ The new best solution
8:         **end if**
9:     **end for**
10:     **return** s''
11: **end function**

Algorithm 6. Best improvement for $N_2(s)$

1: **function** $best2(s)$
2:     $obj \leftarrow f(s)$
3:     **for** $i_1 \in I_{n-1}$ **do**
4:         $s' \leftarrow move(i_1, s)$
5:         **for** $i_2 \in I_n \setminus \{I_{i_1+1}\}$ **do**   ▷ For each tuple (v_{i_1}, v_{i_2}) the $move()$ is applied
6:             $s'' \leftarrow move(i_2, s')$
7:             **if** $f(s'') < obj$ **then**
8:                 $obj \leftarrow f(s'')$
9:                 $s''' \leftarrow s''$
10:             **end if**
11:         **end for**
12:     **end for**
13:     **return** s'''
14: **end function**

Algorithm 7. Best improvement for $N_3(s)$

1: **function** $best3(s)$
2:     $obj \leftarrow f(s)$
3:     **for** $i_1 \in I_{n-2}$ **do**
4:         $s' \leftarrow move(i_1, s)$
5:         **for** $i_2 \in I_{n-1} \setminus \{I_{i_1+1}\}$ **do**
6:             $s'' \leftarrow move(i_2, s')$
7:             **for** $i_3 \in I_n \setminus \{I_{i_2+1}\}$ **do**   ▷ For each tuple $(v_{i_1}, v_{i_2}, v_{i_3})$ $move()$ is applied
8:                 $s''' \leftarrow move(i_3, s'')$
9:                 **if** $f(s''') < obj$ **then**
10:                     $obj \leftarrow f(s''')$
11:                     $s^{(4)} \leftarrow s'''$
12:                 **end if**
13:             **end for**
14:         **end for**
15:     **end for**
16:     **return** $s^{(4)}$
17: **end function**

Algorithm 8 shows the proposed VNS metaheuristic for solving MDTWNPP. The input data are the initial solution, determined by Algorithm 3, and the limit value for runtime. The perturbation in the current solution is a valid random move $m_{i,j}$. The local search uses Best Improvement method to select the neighbor that causes the greatest decrease of objective function and updates it as a current solution, if it is worse than the global solution so far. This local search enumerates the neighbors of a solution using the classical enumeration methods presented in [17] and [18].

Algorithm 8. Adapted VNS algorithm

```
1: function VNS(s, Time)                              ▷ Initial solution and time limit
2:     r ← 1                                                         ▷ Initial N_r(s).
3:     s' ← s
4:     while t < Time do                              ▷ Stopping criterion by the time limit
5:         choose s' ∈ N_r(s) at valid random                              ▷ Shake
6:         if r = 1 then
7:             s' ← best1(s')
8:         else if r = 2 then
9:             s' ← best2(s')
10:        else
11:            s' ← best3(s')
12:        end if
13:        if f(s') < f(s) then                            ▷ Neighborhood exchange
14:            s ← s'
15:            r = 1
16:        else
17:            r ← 1 + (r mod 3)
18:        end if
19:        count time t
20:    end while
21:    return s, f(s)
22: end function
```

4   Experimental Results

The proposed VNS algorithm was implemented in C++ language. The computational tests were performed on a computer with Intel Core i7-3770 CPU, 3.4 GHz with 8 cores, 32 GB RAM and Ubuntu 16.04 64-bit operating system using version 3.8 of the clang compiler.

Algorithm 8 uses only a single core for its execution. Of the 8 processor cores, only 4 are used simultaneously in sets of distinct instances. The instances used for the experiments are the same used in the articles [8,12,20,21]. The main comparison is with the latest experiments in [20], where MDMWNPP is solved with a Memetic Algorithm.

The goal of the computational experiments of this paper is to compare the result, i.e., the objective function values found by the algorithm, in a same time interval in seconds. However, there is a difference in computational processing capacity between the processor used in the current article and the processor used in [20]. The difference between the single-core performance of the processors used is approximately $\frac{3765}{1109}$, as shown in [13][1,2], which provides benchmarks for processors. In consequence, it is fairer that the experiments in this paper use $\frac{1}{3}$ of the average computational time used in [20] as the time limit. Thus, the values of computacional time shown in Tables 2 and 3 reflect this adjustment factor and are therefore equivalent.

The measures for the comparison of results are based on the average of ten executions. With this average, the relative error measure is given by:

$$Gap(B, A) = \frac{z(A) - z(B)}{z(A)}.100\% \qquad (18)$$

---

[1] https://browser.geekbench.com/processors/748.
[2] https://browser.geekbench.com/processors/309.

This measure shows that the response of the algorithm B is less than that of the algorithm A, when $Gap(B, A) > 0$, where $z(A)$ and $z(B)$ are their respective objective function values on average.

Tables 2 and 3 show the results used in comparing the two methods for $k = 3$ and $k = 4$, respectively. The instances have the form n_ma, following the pattern of Definition 1. Table 4 shows the mean value, standard deviation and p-value of a 95% confidence paired t-test, calculated from the results of the column "Avg. Sol". The hypothesis formulation of the test performed is:

$$\begin{cases} H_1 : f_{VNS} < f_{MA} \\ H_0 : f_{VNS} \geq f_{MA} \end{cases} \tag{19}$$

For a descriptive analysis of the results, we can observe, in these two tables, the column $Gap(VNS, MA)$. In Table 2 there are eight instances with mark "no", i.e., the MA algorithm was better than Algorithm 8. On the other hand, in Table 3, the number of times the MA beats the Algorithm 8 is six. The difference between the MA and VNS results are, on average, 3359.92 for $k = 3$ and 3509.21 for $k = 4$.

Table 2. Results of [20] vs results from Algorithm 8 for $k = 3$

Instance	MA		VNS		Comparison	
$k = 3$	Avg. sol	Avg. time	Avg. sol	Avg. time	Gap (VNS, MA)	Better
50_2a	86.2	182.45	130.98	60.94	51.95%	No
50_3a	334.4	202.34	1045.94	67	−212.78%	No
50_4a	3382.5	673.83	2044.01	223.98	39.57%	Yes
50_5a	4125.8	781.82	14266.4	259.97	−245.79%	No
50_10a	37521.6	1189.38	38136.1	395.96	−1.64%	No
50_15a	56015.2	1212.27	68396.1	403.91	−22.10%	No
50_20a	102652	1235.22	92299.1	410.92	10.09%	Yes
100_2a	178.1	342.39	50.7	113.98	71.53%	Yes
100_3a	531.3	428.38	2886.44	141.96	−443.28%	No
100_4a	867.5	673.22	8374.57	223.96	−865.37%	No
100_5a	6224.5	834.62	11527.3	277.97	−85.19%	No
100_10a	47004.8	1436.08	37146.2	477.94	20.97%	Yes
100_15a	96827.3	2073.76	61427.6	690.94	36:56%	Yes
100_20a	113112.5	2564.38	84093.4	853.91	25.66%	Yes

Table 4 reports that there is no significant statistical difference between the averages of the results of the algorithms in the set of tested instances when $k = 3$ and $k = 4$. Even if the average difference between the algorithms is positive, the large standard deviation in the results does not allow to reject the

Table 3. Results of [20] vs results from Algorithm 8 for $k = 4$

Instance	MA		VNS		Comparison	
$k = 4$	Avg. sol	Avg. time	Avg. sol	Avg. time	Gap (VNS, MA)	Better
50_2a	391.4	342.37	321.69	113.66	17.81%	Yes
50_3a	678.3	536.24	687.75	177.95	−1.39%	Yes
50_4a	836.7	1023.39	3185.66	340.95	−280.74%	No
50_5a	1094.4	1243.28	19979.4	413.87	−1725.60%	No
50_10a	42005.6	1647.76	57025.2	548.83	−35.76%	No
50_15a	56034.7	1843.92	91035.6	613.82	−62.46%	No
50_20a	123627.8	2135.48	108404	710.83	12.31%	Yes
100_2a	687.2	564.02	99.28	187.94	85.55%	Yes
100_3a	1213.7	847.37	5765.02	281.9	−375.00%	No
100_4a	1924.6	922.14	12219.9	306.9	−534.93%	Yes
100_5a	8356.8	1972.39	19056.7	656.78	−128.04%	No
100_10a	75034.8	2819.32	58992.6	938.71	21.38%	Yes
100_15a	122892.7	3193.84	76239.6	1063.82	37.96%	Yes
100_20a	174981.6	4392.01	107619	1463.61	38.50%	Yes

Table 4. Statistical analysis with paired sample for $k \in \{3, 4\}$.

Measures	$k = 3$	$k = 4$
Normality	1.63%	6.34%
$p - value_{t-test}$	19.01%	**31.39%**
$p - value_{wilcox-test}$	**47.58%**	54.84%
$\mu_{MA} - \mu_{VNS}$	3359.92	3509.21
$\sigma_{MA,VNS}$	13840.88	26436.14

null hypothesis. The t-test assumes that the data are normally distributed. The Shapiro-Wilk test verifies this condition. The results of the column "Avg. Sol." with $k = 3$ do not satisfy the normality assumption; therefore, the Wilcoxon test was used. For $k = 4$, the t-test can be used.

Table 5 shows the obtained results using the proposed VNS algorithm for solving MDMWNPP to $k \in \{5, 6\}$, i.e., the Multidimensional Five-way and Six-way Multidimensional Number Partitioning Problems, considering the same instances used to solve the cases in which $k \in \{3, 4\}$. The results were obtained considering ten executions for each instance with maximum computational time equal to 1800 s. It is important to highlight that the cases for $k \in \{5, 6\}$ have not been solved previously in any other article, at least according to the knowledge of the authors of the current article.

Table 5. Results for $k \in \{5, 6\}$ with Algorithm 8

VNS	k = 5		k = 6	
Instance	Avg. sol	Avg. time	Avg. sol	Avg. time
50_2a	194.77	1799.31	374.48	1799.7
50_3a	1498.14	1799.96	3073.79	1799.93
50_4a	6678.27	1799.95	9607.94	1799.93
50_5a	11388.1	1799.92	17682.3	1799.93
50_10a	66263.7	1799.49	76327.7	1799.56
50_15a	93023.4	1799.62	93350.7	1799.71
50_20a	113159	1799.58	131340	1799.83
100_2a	618.52	1799.49	393.75	1799.78
100_3a	1229.68	1799.77	2988.6	1799.87
100_4a	11480.8	1799.86	18304.9	1799.77
100_5a	23440.7	1799.45	27570.8	1799.74
100_10a	63401.4	1799.94	76073.5	1799.82
100_15a	98647.3	1799.89	106270	1799.85
100_20a	127963	1799.99	125540	1799.9

5   Conclusion

This article presents a proposal to adapt the VNS metaheuristic to the solution of the Multidimensional Multi-Way Number Partitioning Problem (MDMWNPP). This problem is a generalization of the classical Number Partition Problem (NPP), in which it is assumed that each element of the sequence is a vector and, in addition, k-partitions of the sequence are performed, for $k \geq 2$. Despite this attractive and challenging formulation, this problem remains little studied in the literature. For the purposes of validation of the obtained results, a comparison is made with the only algorithm found in the literature proposed directly to solve the addressed problem, according to the knowledge of the authors of the current article. The proposed VNS algorithm, shown in Algorithm 8, was tested using the same instances used in [20], in which MDMWNPP is solved using Memetic Algorithm. The results are satisfactory as to the quality of the proposed VNS by comparing only the averages and the $gap()$ between the results of the two algorithms, according to Tables 2 and 3. In the statistical analysis, it is not possible to conclude that there is a significant difference between the VNS and the MA in the instances with $k \in \{3, 4\}$.

Specific difficulties were found to support better statistical analysis, such as the low number of executions, and the fact that instances used in this work are dependent on one another. As future work, we intend to apply the VNS metaheuristic combined with mathematical programming formulations for the solution of MDMWNPP, using, as a test basis, a new group of uniformly distributed generated instances.

Acknowledgements. The authors would like to thank the CAPES Foundation, the Brazilian Council of Technological and Scientific Development (CNPq), the Minas Gerais State Research Foundation (FAPEMIG), the Federal Center of Technological Education of Minas Gerais (CEFET-MG), and the Federal University of Ouro Preto (UFOP) for supporting this research.

References

1. Faria, A.F., de Souza, S.R., Silva, C.A.: Variable neighborhood descent applied to multi-way number partitioning problem. Electron. Notes Discret. Math. **66**, 103–110 (2018). https://doi.org/10.1016/j.endm.2018.03.014. 5th International Conference on Variable Neighborhood Search
2. Gent, I.P., Walsh, T.: Analysis of heuristics for number partitioning. Comput. Intell. **14**(3), 430–451 (1998)
3. Graham, R.L.: Bounds for certain multiprocessing anomalies. Bell Syst. Tech. J. **XLV**(9), 1563–1581 (1966)
4. Graham, R.L.: Bounds on multiprocessing timing anomalies. SIAM J. Appl. Math. **17**(2), 416–429 (1969)
5. Horowitz, E., Sahni, S.: Computing partitions with applications to the knapsack problem. J. ACM (JACM) **21**(2), 277–292 (1974)
6. Karmarkar, N., Karp, R.M.: The differencing method of set partition. Report UCB/CSD 81/113, Computer Science Division, University of California, Berkeley, CA (1982)
7. Karp, R.M.: Reducibility among combinatorial problems. In: Miller, R.E., Thatcher, J.W., Bohlinger, J.D. (eds.) Proceedings of a Symposium on the Complexity of Computer Computations. The IBM Research Symposia, pp. 85–103. Springer, Boston (1972). https://doi.org/10.1007/978-1-4684-2001-2_9
8. Kojić, J.: Integer linear programming model for multidimensional two-way number partitioning problem. Comput. Math. Appl. **60**(8), 2302–2308 (2010)
9. Korf, R.E.: A complete anytime algorithm for number partitioning. Artif. Intell. **106**(2), 181–203 (1998)
10. Korf, R.E.: Multi-way number partitioning. In: Proceedings of the 21st International Joint Conference on Artifical Intelligence (IJCAI 2009), pp. 538–543 (2009)
11. Korf, R.E., Schreiber, E.L., Moffitt, M.D.: Optimal sequential multi-way number partitioning. In: Proceedings of the International Symposium on Artificial Intelligence and Mathematics (ISAIM-2014) (2014)
12. Kratica, J., Kojic, J., Savic, A.: Two metaheuristic approaches for solving multidimensional two-way number partitioning problem. Comput. Oper. Res. **46**, 59–68 (2014)
13. Labs, P.: Geekbench, May 2018. https://browser.geekbench.com/
14. Mertens, S.: A complete anytime algorithm for balanced number partitioning (1999). http://arxiv.org/abs/cs.DS/9903011
15. Mladenović, N., Hansen, P.: Variable neighborhood search. Comput. Oper. Res. **24**(11), 1097–1100 (1997)
16. Moffitt, M.D.: Search strategies for optimal multi-way number partitioning. In: Proceedings of the Twenty-Third International Joint Conference on Artificial Intelligence, pp. 623–629. AAAI Press (2013)

17. Nijenhuis, A., Wilf, H.S.: Combinatorial Algorithms: For Computers and Calculators. Academic Press, Cambridge (2014)
18. Orlov, M.: Efficient generation of set partitions. Technical report, Faculty of Engineering and Computer Sciences, University of Ulm (2002). http://www.cs.bgu.ac.il/~orlovm/papers/partitions.pdf
19. Pedroso, J.P., Kubo, M.: Heuristics and exact methods for number partitioning. Eur. J. Oper. Res. **202**(1), 73–81 (2010)
20. Pop, P.C., Matei, O.: A memetic algorithm approach for solving the multidimensional multi-way number partitioning problem. Appl. Math. Modell. **37**(22), 9191–9202 (2013)
21. Rodriguez, F.J., Glover, F., García-Martínez, C., Martí, R., Lozano, M.: GRASP with exterior path-relinking and restricted local search for the multidimensional two-way number partitioning problem. Comput. Oper. Res. **78**, 243–254 (2017)
22. Schreiber, E.L.: Optimal Multi-Way Number Partitioning. Ph.D. thesis, University of California Los Angeles (2014)
23. Schreiber, E.L., Korf, R.E.: Cached iterative weakening for optimal multi-way number partitioning. In: Proceedings of the Twenty-Eighth Annual Conference on Artificial Intelligence (AAAI-2014), Quebec City, Canada (2014)
24. Schroeppel, R., Shamir, A.: A $T = O(2^{n/2})$, $S = O(2^{n/4})$ algorithm for certain NP-complete problems. SIAM J. Comput. **10**(3), 456–464 (1981)
25. Sloane, N.: On-Line Encyclopedia of Integer Sequences (1991). https://oeis.org/

A General Variable Neighborhood Search with Mixed VND for the multi-Vehicle multi-Covering Tour Problem

Manel Kammoun[1](✉), Houda Derbel[1], and Bassem Jarboui[2]

[1] MODILS, FSEGS, Route de l'aéroport km 4,
3018 Sfax, Tunisia
kamounemanel@gmail.com, derbelhouda@yahoo.fr
[2] Emirates College of Technology, Abu Dhabi, United Arab Emirates
bassem_jarboui@yahoo.fr

Abstract. The well-known Vehicle Routing-Allocation Problem (VRAP) receives recently more attention than the classical routing problems. This article deals with a special case of the VRAP named the multi-vehicle multi-Covering Tour Problem (mm-CTP-p). More precisely, the mm-CTP-p is a generalized variant of the multi-vehicle Covering Tour Problem (m-CTP-p). In both problems, the objective is to find a minimum length set of vehicle routes while satisfying the total demands by visiting vertices by the route or covering vertices which does not included in any route. But, in the m-CTP-p, the demand of a vertex can be satisfied with only one coverage whereas in the mm-CTP-p, a vertex must be covered several times to be completely served. Indeed, a vertex is covered if it lies within a specified distance of at least one vertex of a route. We develop a General Variable Neighborhood Search algorithm (GVNS) with a mixed Variable Neighborhood Descent (mixed-VND) method to solve the problem. Experiments were conducted using benchmark instances from the literature. Extensive computational results on mm-CTP-p problems show the performance of our method.

Keywords: Vehicle Routing-Allocation Problems · multi-Covering · Local search · Variable Neighborhood Descent

1 Introduction

Since 1980, several researchers investigate in the area of transportation due to its economics and social importance in our life. The vehicle routing problem (VRP) is a central problem in the family of routing problems. It is a NP-hard problem since it is not easy to find the optimal solution especially for large size problems [7]. The main objective is to find the optimal set of routes that satisfies the demand of all customers such that each route begins and ends at the same depot and the routing cost is minimized. However, in many real world routing problems, we do not need to visit all the customers or more precisely we cannot reach all the

© Springer Nature Switzerland AG 2019
A. Sifaleras et al. (Eds.): ICVNS 2018, LNCS 11328, pp. 259–273, 2019.
https://doi.org/10.1007/978-3-030-15843-9_20

customers because of some constraints which can prevent the vehicles from serving them. This particularity in routing problems represents an interesting area of study. The assumption of letting unvisited or isolated customers was considered by [1] while they introduce the Vehicle Routing-Allocation Problems (VRAP). Two concepts are provided: the covered vertices and the covering distance. The first represents unvisited customers which their demands should be allocated to a visited one on the vehicle route and the second measures the travel distance between a covered vertex and a visited one. The covering tour problem (CTP) is a special case of the VRAP where a subset of vertices must be visited and covering constraints must be respected. In other word, the objective of the CTP is to determine the shortest Hamiltonian cycle using a set of optional locations such that every vertex in the unreachable locations is within a predefined distance from the cycle. In this work we consider a particular case of the m-CTP called the m-CTP-p in which the number of vertices on a route (excluding the depot) is less than a given value p. More precisely we deal with a generalized variant of the m-CTP-p named the multi-vehicle multi-Covering Tour Problem (mm-CTP-p) where multiple covering is necessary. In the following we present a description of the problem and we propose a VNS approach to solve the generalized variant of the m-CTP-p.

2   Problem Description

The m-CTP-p is defined by [3] as an undirected graph $G = (V \cup W, E)$, where $V \cup W$ is the vertex set and E is the edge set. We consider different kinds of locations: V, T and W. Let V the set of vertices that can be visited and W the set of vertices that must be covered with up to m vehicles. $T \subseteq V$ is the set of vertices that must be visited including the depot where identical vehicles are located. We consider that each vertex of V has a unit demand and each vehicle has a capacity p. We note that we do not need to visit all vertices of V with the exception of the vertices of T to satisfy the demand of each customer. The demand of each customer could be satisfied in two different ways: either by visiting the customer along the tour or by covering it. Covered vertices must be within a predefined distance d from the tour. So, we focus to find a minimum length set of vehicle routes with respecting the number of vertices in each route and satisfying the following constraints: (i) each vehicle route starts and ends at the depot, (ii) each vertex of T belongs to exactly one route while each vertex of $V \backslash T$ belongs to at most one route, (iii) each vertex of W must be covered by a route, (iv) the number of vertices on a route (excluding the depot) is less than a given value p while the constraint on the length of each route is relaxed. In the VRP families, a routing decision must be made. As for the VRAP families an additional decision was introduced: "the allocation decision". In addition, in some cases, the allocation decision involves the location decision such in the postal example where the design of good postal collection routes is related to the allocation and the location decision. In fact, we need to fix the location of collection points so that each customer should be served by their nearest collection points. Consequently, the VRAP combines routing, location and allocation decisions. In this paper, we considered only the routing and allocation decisions. The m-CTP-p was generalized in [9]. They proposed an extended graph $\bar{G}$ based on the original graph as follows: Let S_{v_i} the vertex i from the route S_v belonging to the solution S. In the new graph $\bar{G}$ a copy S_{v_n} of the depot S_{v_0} was considered. Let $\bar{G} = (\bar{V} \cup W, \bar{E}_1 \cup E_2)$ where $\bar{V} = V \cup \{S_{v_n}\}$ and $V' = \bar{V} \setminus \{S_{v_0}, S_{v_n}\}$, $\bar{E}_1 = E_1 \cup \{(S_{v_i}, S_{v_n}), S_{v_i} \in V'\}$ and $E_2 = \{(S_{v_i}, S_{v_j}), S_{v_i} \in V \setminus T, S_{v_j} \in W\}$. Each customer $W_j \in W$ should be covered at least u_j times by visited S_{v_j} vertices of V. The mm-CTP-p is an NP-hard problem as it is reduced to a m-CTP-p when $u_j = 1$ (see Fig. 1). For each covered vertex $w_j \in W$, $u_j \in [1, min(3, nb_j)]$ where nb_j represents the maximal number of vertices in V which can cover w_j.

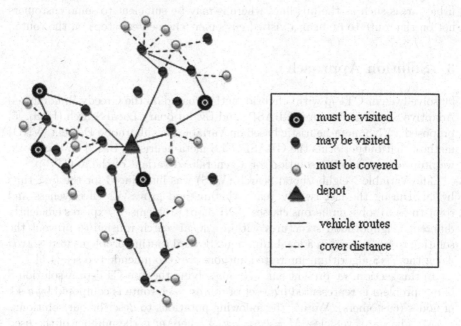

Fig. 1. An illustrative example of the mm-CTP-p.

Three variants of the mm-CTP-p are introduced in the literature [2]. In the first one each customer can be visited by the routes at most once. Whereas in the two other variants, there is the possibility of visiting each customer more than once. But for the second variant, it is not allowed to revisit the same customer immediately only if the tour visits another node. In other words, to revisit a customer S_{v_i}, the tour S_v has to visit another node before it can return to S_{v_i}. As for the third variant, each customer can be visited several times subsequently by the route. Similar definitions of the three variants above can be found in [9] where they reduce the last two variants to the first one by proposing some appropriate graph transformations. In the new graphs, they add copies co_i for each node $S_{v_i} \in V \setminus T$ where $co_i = max_{w_j \in W_i} u_j - 1$. For the two variants two graphs are built in a similar way except the length of the new edges that was defined in a different way. In the second variant, they affect a very large number to the length of edges whose two endpoints in $C_i(C_i = \{S_{v_i}, co_i\})$ to prevent from revisiting the node S_{v_i}. Whereas, in the third variant they allow revisit this node and the length of edges linking a copy of node S_{v_i} to a node S_{v_k} is set zero. The VRAP is applied in many real world routing situations. Particularly, when only the more urgent or profitable requests are served at a given time because of the unavailability of some resources. For example, such problem emerges in rural health care when a route is designed for serving the urgent requests and giving the essential medical care services. In fact, if the demands of some customers are too large, customers must be served several times not only once in order to respect the capacity constraint of each vehicle. The VRAP is also applied in urban areas such as the bus lines where it may be sufficient to some customers not on the route to be near a visited customer (the nearest stop) on the route.

3   Solution Approach

[3] solved the m-CTP-p with a hybrid method based on the Greedy Randomized Adaptive Search Procedure (GRASP) and Evolutionary Local Search (ELS). [6] proposed a VNS meta-heuristic based on Variable Neighborhood Descent (VND) method that outperforms the GRASP/ELS meta-heuristic. Based on this work, we propose a promoting solution for a generalized variant of the m-CTP-p.

The Variable Neighborhood Search (VNS) was introduced for the first time by [8]. During the last twenty years, this method proves its effectiveness and robustness to solve numerous classes of NP-hard problems. It explores efficiently different neighborhood structures within a local search routine to prevent the solution to be trapped in a local minimum. Several routing problems were solved using the VNS algorithm, interested authors are recommended to see [4,5].

In this section, we present our VNS for solving the mm-CTP-p. A solution S of our problem is represented by a set of routes, each route is composed by a set of nodes (customers). We use the following notations to describe our solution:

M: The set of vehicles, $M = \{v_1, ..., v_m\}$ where m is the number of the used vehicles.

S: The set of m routes, $S = \{S_1, ..., S_m\}$

S_v: The set of $N(v)$ nodes visited by the vehicle v, $S_v = \{S_{v_1}, ..., S_{v_{N(v)}}\}$ where each element S_{v_k} of S_v represent the kth node visited by vehicle v and $N(v) \in \bar{V}$.

As already mentioned, the VNS method uses different neighborhood structures in the local search.

Let $N_k, k = 1, ..., k_{max}$ be the set of neighborhoods used in our GVNS algorithm and $N_k(S)$ be the neighbors of a solution S via a neighborhood structure N_k. In the following, we present four neighborhood structures N_1, N_2, N_3, N_4 used in our algorithm to obtain S', a neighbor of a solution S.

- $N_1(S) = \{S' : S' = \{S'_1, .., S'_v, .., S'_m\}$ and $S'_v = Insert(S_v, k, i) \forall v \in M, k \in N(v)\}$. This neighborhood consists on removing a customer i from a route S_v and inserting it in a new position k in the same route or into another route $S_{v'}$ This move can change the number and the order of customers in those routes.
- $N_2(S) = \{S' : S' = \{S'_1, .., S'_v, .., S'_m\}$ and $S'_v = Swap(S_v, k, S_{v'k'})$ and $S'_{v'} = Swap(S_{v'}, k', S_{vk}) \forall v, v' \in M, k \in N(v), k' \in N(v')\}$. This move consists on exchanging two customers, from the same route or from different routes. In this move we change a node at position k by a new node i in the route S_v.
- $N_3(S) = \{S' : S' = \{S'_1, .., S'_v, .., S'_m\}$ and $S'_v = Drop(S_v, k) \forall v \in M, k \in N(v)\}$. In this neighborhood structures we select a route S_v from the solution S and select a customer at position k from this route. Then, a new route S'_v is obtained by removing the selected node from S_v.
- $N_4(S) = \{S' : S' = \{S'_1, .., S'_v, .., S'_m\}$ and $S'_v = Insert(Drop(S_v, k), k, i) \forall v \in M, k \in N(v), i \in V \setminus N(v)\}$. This move consists on replacing a customer in a route S_v by another node which does not belong to the solution.

The main steps of our VNS algorithm for solving the mm-CTP-p are given in Algorithm 1. First, we determine the stopping criteria and the maximum number of neighborhood structures. Second, we generate initial feasible solution. Then, we aim to improve this solution based on three phases: Shaking, Local search and Move or Not. In the shaking phase, we perturb the solution by using the neighborhood structure N_4. In Local search phase, we propose a mixed-VND for each visited vertex. We note that before applying some swap move in order to improve the solution we must check if the selected vertex i from the set of visited vertices I represents a redundant visited node or not. If that is the case, we use the neighborhood structure N_3 to drop it from the solution and we continue with the next node from the set I. Finally, in the last phase we decide to move or not.

4   Computational Results

The work of [3] is based on benchmark instances $KroA100, KroB100, KroC100$ and $KroD100$ from $TSPLIB$ to build instances for m-CTP-p. Let $nb_{total} = |V| + |W| = 100$ vertices and the tests are run for $|V| = \lceil 0.25 nb_{total} \rceil$, $\lceil 0.5 nb_{total} \rceil$, $|T| = 1$ and $\lceil 0.20n \rceil$ where W is composed of remaining vertices ($|W| = 100 - |V|$) Large instances were generated in the same way using $KroA200$ and $KroB200$ of $TSPLIB$ with $|nb_{total}| = 200$. This set of instances was used later by [9] to generate instances for the mm-CTP-p. As a result, we have 92 instances, where each one is labeled as follows: $X - |T| - |V| - |W| - |p|$, X represents the name of $TSPLIB$ instance and the values taken by V, W and T are explained above while the value of p is set to $\{4, 5, 6, 8\}$.

Algorithm 1. GVNS for the mm-CTP-p

1: Initialize : max number of iteration $iter_{max} = 1000$ and the number of neighbor-hood structures $k_{max} = 1$;
2: Let I be the set of vertices in an initial solution S ;
3: Insert randomly the vertices from T in I ;
4: **repeat**
5:    select randomly a vertex i from V to cover a maximum number of vertices in $V \cup W$;
6:    Insert i in I ;
7: **until** all vertices of W will be covered ;
8: iter=1 ;
9: Begin
10: **while** $(iter \leq iter_{max})$ **do**
11:   k=1 ;
12:   **repeat**
13:     Remove randomly f vertices from $I \setminus T$; /* Shaking phase*/
14:     **repeat**
15:       Insert randomly a vertex from $V \setminus I$ that maximizes the number of covered vertices in $V \cup W$;
16:     **until** a feasible solution $S' \in N_4(S)$ is reached ;
17:     **repeat**
18:       Let $R = V \setminus I$ the remaining set of vertices ; /* Local search phase*/
19:       Select $i \in I$;
20:       Let $J_i \subset R$;
21:       **for** $j \in J_i$ **do**
22:         Swap move between i and j to get $S"$;
23:         Let $N_l, l = 1, \ldots, l_{max}$;
24:         l=1;
25:         **repeat**
26:           Find the best neighbor $S_1 \in N_l(S")$;
27:           **if** $f(S_1) < f(S")$ **then**
28:             $S" \leftarrow S_1$;
29:             l=1;
30:           **else**
31:             l= l + 1;
32:           **end if**
33:         **until** $l = l_{max}$;
34:         **if** $f(S_1) < f(S')$ **then**
35:           $S' \leftarrow S_1$;
36:           Update I ;
37:           Go to line 18 ;
38:         **end if**
39:       **end for**
40:     **until** No possible improvement ;
41:     **if** $f(S_1) < f(S)$ **then**
42:       $S \leftarrow S_1$; /* Move or Not phase*/
43:       k= 1 ;
44:     **else**
45:       k= k + 1;
46:     **end if**
47:   **until** $k = k_{max}$;
48:   iter = iter + 1 ;
49: **end while**;
50: End ;

The algorithm described in the previous section was coded in C++ programming language and is run on a computer with an Intel Core i 5-4200U and 2.3 Ghz processor and 6 GB memory.

In this section, a computational study is performed to evaluate the performance of our algorithm compared with the best and the average known solutions. Tables 1 and 2 summarize the results of two meta-heuristics from the literature, namely GA-VLG and GRASP-ELS, developed by [9] and our proposed method for the mm-CTP-p. The column headings are as follows: **Data** is the name of instance, **Best** and **Avg.** columns presents the best and the average cost of the solution over 10 runs respectively. The column **Time** is the total running time in seconds and the column **Gap$_{UB}$** shows the deviation between the value of the solution given by the meta-heuristics developed in the literature and the value obtained by our proposed approach.

To evaluate the performance of our methods we calculate the Gap_{UB} between each meta-heuristic from the literature and our proposed algorithm.

Let $Meta$ be the solution value given by a meta-heuristic and UB be the solution value of our algorithm, the percentage deviation (Gap_{UB}) is computed as follows:

$$Gap_{UB} = 100.\frac{Meta - UB}{UB} \tag{1}$$

It is worthy to mention that the GVNS is run 10 times for each instance to better observe their variance. In Tables 1 and 2, three criteria are used to evaluate the quality of the solution provided by our GVNS algorithm: the best cost solution (column **Best**) over 10 runs, the average solution (Column **Avg.**) over 10 runs and the total running time in seconds of 10 runs (column **Time**).

Table 2 detailed the variance of solution costs of each instance over 10 runs. Based on this results we investigate the stability of our GVNS algorithm and compared with GRASP-ELS and GA-VLG meta-heuristics. Table 3 summarizes this result where the column headings are as follows: **SameCost** column presents the number of problem instances that an algorithm (GVNS or GRASP-ELS or GA-VLG) has better variance than the others while they provide the same best cost. A comparison between our GVNS and GRASP-ELS is presented where **GVNSBetter** and **GRASP − ELSBetter** columns are the same as **SameCost** column, GVNS provides better solutions than GRASP-ELS does or vice versa, respectively. Also in the same way, we compare the variance of our GVNS algorithm with the GA-VLG algorithm where each of them has better best cost than the other does(Columns **GVNSBetter** and **GA − VLGBetter**).

Table 3 show that our GVNS method has 16 times better (smaller) variance than GRASP-ELS and 6 times better variance than GA-VLG and 5 times better variance than both of them where this three methods (GVNS, GRASP-ELS, GA-VLG) provide the same best cost. It is clear from this table that, when this three methods achieve the same best cost, our GVNS algorithm is much more stable especially more than GRASP-ELS. Our GVNS algorithm find better solution cost and variance than GRASP-ELS in six instances. The GVNS achieve smaller variance than the GA-VLG in only one instance where the GA-VLG has better

Table 1. Comparaison of best cost solution on mm-CTP-p

Data	GVNS	GRASP-ELS		GA-VLG	
	Best	Best	Gap_{UB}	Best	Gap_{UB}
A1-1-25-75-4-250	17774	17774	0	17774	0
A1-1-25-75-5-250	15793	15793	0	15793	0
A1-1-25-75-6-250	14628	14628	0	14628	0
A1-1-25-75-8-250	12590	12590	0	12590	0
A1-1-50-50-4-250	21473	21473	0	21473	0
A1-1-50-50-5-250	18680	18680	0	18680	0
A1-1-50-50-6-250	17481	17481	0	17481	0
A1-1-50-50-8-250	14380	14380	0	14380	0
A1-10-50-50-4-250	25340	25340	0	25340	0
A1-10-50-50-5-250	21712	21712	0	21712	0
A1-10-50-50-6-250	20125	20125	0	20125	0
A1-10-50-50-8-250	17603	17603	0	17603	0
A1-5-25-75-4-250	13082	13082	0	13082	0
A1-5-25-75-5-250	11969	11969	0	11969	0
A1-5-25-75-6-250	11746	11746	0	11746	0
A1-5-25-75-8-250	9081	9081	0	9081	0
A2-1-100-100-4-250	25026	25051	0.099	25026	0
A2-1-100-100-5-250	21626	21626	0	21626	0
A2-1-100-100-6-250	19108	19119	0.057	19108	0
A2-1-100-100-8-250	16209	16226	0.104	16209	0
A2-1-50-150-4-250	23601	23601	0	23601	0
A2-1-50-150-5-250	20439	20439	0	20439	0
A2-1-50-150-6-250	18410	18410	0	18410	0
A2-1-50-150-8-250	15502	15565	0.406	15502	0
A2-10-50-150-4-250	25702	25702	0	25702	0
A2-10-50-150-5-250	21503	21503	0	21503	0
A2-10-50-150-6-250	20250	20250	0	20250	0
A2-10-50-150-8-250	16676	16676	0	16676	0
A2-20-100-100-4-250	38074	38074	0	38074	0
A2-20-100-100-5-250	32583	32646	0.193	32583	0
A2-20-100-100-6-250	28490	28490	0	28490	0
A2-20-100-100-8-250	24593	24615	0.089	24593	0
B1-1-25-75-4-250	17417	17417	0	17417	0
B1-1-25-75-5-250	15891	15891	0	15891	0
B1-1-25-75-6-250	14260	14260	0	14260	0

(continued)

Table 1. (*continued*)

Data	GVNS	GRASP-ELS		GA-VLG	
	Best	Best	Gap_{UB}	Best	Gap_{UB}
B1-1-25-75-8-250	11538	11538	0	11538	0
B1-1-50-50-4-250	19966	19966	0	19966	0
B1-1-50-50-5-250	17113	17113	0	17113	0
B1-1-50-50-6-250	15989	15989	0	15989	0
B1-1-50-50-8-250	14027	14027	0	14027	0
B1-10-50-50-4-250	20075	20075	0	20075	0
B1-10-50-50-5-250	17986	17986	0	17986	0
B1-10-50-50-6-250	15924	15924	0	15924	0
B1-10-50-50-8-250	13672	13672	0	13672	0
B1-5-25-75-4-250	17079	17079	0	17079	0
B1-5-25-75-5-250	15110	15110	0	15110	0
B1-5-25-75-6-250	14707	14707	0	14707	0
B1-5-25-75-8-250	11319	11319	0	11319	0
B2-1-100-100-4-250	40974	40974	0	40974	0
B2-1-100-100-5-250	34848	34848	0	34848	0
B2-1-100-100-6-250	30849	30829	−0.064	30849	0
B2-1-100-100-8-250	25804	25804	0	25804	0
B2-1-50-150-4-250	23288	23288	0	23288	0
B2-1-50-150-5-250	20039	20039	0	20039	0
B2-1-50-150-6-250	18046	18046	0	18046	0
B2-1-50-150-8-250	15668	15668	0	15668	0
B2-10-50-150-4-250	25967	25967	0	25967	0
B2-10-50-150-5-250	22359	22359	0	22359	0
B2-10-50-150-6-250	19792	19792	0	19792	0
B2-10-50-150-8-250	17106	17106	0	17106	0
B2-20-100-100-4-250	53590	53590	0	53590	0
B2-20-100-100-5-250	45209	45209	0	45209	0
B2-20-100-100-6-250	39184	39184	0	39184	0
B2-20-100-100-8-250	32513	32513	0	32512	−0.003
C1-1-25-75-4-250	13012	13012	0	13012	0
C1-1-25-75-5-250	11666	11666	0	11666	0
C1-1-25-75-6-250	9820	9820	0	9820	0
C1-1-25-75-8-250	9818	9818	0	9818	0
C1-1-50-50-4-250	20294	20294	0	20294	0
C1-1-50-50-5-250	17378	17378	0	17378	0

(*continued*)

Table 1. (*continued*)

Data	GVNS	GRASP-ELS		GA-VLG	
	Best	Best	Gap_{UB}	Best	Gap_{UB}
C1-1-50-50-6-250	16365	16365	0	16365	0
C1-1-50-50-8-250	13900	13900	0	13900	0
C1-10-50-50-4-250	26931	26931	0	26931	0
C1-10-50-50-5-250	23544	23544	0	23544	0
C1-10-50-50-6-250	20818	20818	0	20818	0
C1-10-50-50-8-250	18154	18154	0	18154	0
C1-5-25-75-4-250	13738	13738	0	13738	0
C1-5-25-75-5-250	13575	13575	0	13575	0
C1-5-25-75-6-250	10826	10826	0	10826	0
C1-5-25-75-8-250	10556	10556	0	10556	0
D1-1-25-75-4-250	18127	18127	0	18127	0
D1-1-25-75-5-250	15972	15972	0	15972	0
D1-1-25-75-6-250	14532	14532	0	14532	0
D1-1-25-75-8-250	12700	12700	0	12700	0
D1-1-50-50-4-250	23275	23275	0	23275	0
D1-1-50-50-5-250	20402	20402	0	20402	0
D1-1-50-50-6-250	18072	18072	0	18072	0
D1-1-50-50-8-250	14930	14930	0	14930	0
D1-10-50-50-4-250	30390	30390	0	30390	0
D1-10-50-50-5-250	26284	26284	0	26284	0
D1-10-50-50-6-250	23646	23646	0	23646	0
D1-10-50-50-8-250	19986	19986	0	19986	0
D1-5-25-75-4-250	18464	18464	0	18464	0
D1-5-25-75-5-250	15767	15767	0	15767	0
D1-5-25-75-6-250	14851	14851	0	14851	0
D1-5-25-75-8-250	12705	12705	0	12705	0

solution cost. We note that our method is faster than the other for almost all instances but it need a more time when $|V| = 100$. Finally, the obtained results show that the GVNS performs better than the other meta-heuristics according to the different criteria on the mm-CTP-p. We observe also that our GVNS gives better solutions with smaller running time for almost all instances.

Table 2. Comparaison of average solution on mm-CTP-p

Data	GVNS			GRASP-ELS				GA-VLG			
	Avg.	σ^2	Time	Avg.	σ^2	Time	Gap_{UB}	Avg.	σ^2	Time	Gap_{UB}
A1-1-25-75-4-250	17774	0	0.862	17774	0	56.77	0	17774	0	178.34	0
A1-1-25-75-5-250	15793	0	1.282	15793	0	61.86	0	15793	0	177.52	0
A1-1-25-75-6-250	14628	0	1.142	14628	0	59.14	0	14628	0	178.32	0
A1-1-25-75-8-250	12590	0	1.463	12590	0	60.39	0	12590	0	179.16	0
A1-1-50-50-4-250	21473	0	5.727	21473	0	860.54	0	21473	0	283.15	0
A1-1-50-50-5-250	18680	0	4.09	18680	0	898.8	0	18680	0	287.94	0
A1-1-50-50-6-250	17481	0	46.369	17481	0	944.39	0	17481	0	284.99	0
A1-1-50-50-8-250	14380	0	22.764	14380	0	965.42	0	14380	0	278.43	0
A1-10-50-50-4-250	25340	0	4.705	25340	0	1165.99	0	25340	0	298.6	0
A1-10-50-50-5-250	21712	0	17.387	21712	0	1133.15	0	21712	0	309.51	0
A1-10-50-50-6-250	20125	0	11.819	20125	0	1144.12	0	20125	0	300.33	0
A1-10-50-50-8-250	17603	0	5.969	17603	0	1253.44	0	17603	0	308.99	0
A1-5-25-75-4-250	13082	0	1.909	13082	0	20.83	0	13082	0	159.75	0
A1-5-25-75-5-250	11969	0	2.507	11969	0	21.74	0	11969	0	159.81	0
A1-5-25-75-6-250	11746	0	2.648	11746	0	21.1	0	11746	0	162.48	0
A1-5-25-75-8-250	9081	0	3.483	9081	0	21.46	0	9081	0	155.13	0
A2-1-100-100-4-250	25026	0	14205.364	25058.2	60.96	3656.09	0.128	25033.6	134.84	538.48	0.03
A2-1-100-100-5-250	21656.8	20.163	19119.834	21677.3	292.41	4140.76	0.094	21669.1	629.89	717.98	0.056
A2-1-100-100-6-250	19108	0	4062.644	19180.2	7823.96	4026.62	0.377	19108	0	565.29	0
A2-1-100-100-8-250	16228	7.214	21948.215	16241	235.8	4051.89	0.08	16266.4	3803.84	564.42	0.236
A2-1-50-150-4-250	23601	0	59.764	23613.6	635.04	798.37	0.053	23601	0	533.38	0
A2-1-50-150-5-250	20439	0	269.389	20443.2	158.76	835.08	0.02	20483.4	3591.84	617.65	0.217
A2-1-50-150-6-250	18410	0	91.372	18410	0	829.21	0	18410	0	493.03	0

(continued)

Table 2. (*continued*)

Data	GVNS			GRASP-ELS				GA-VLG			
	Avg.	σ^2	Time	Avg.	σ^2	Time	Gap_{UB}	Avg.	σ^2	Time	Gap_{UB}
A2-1-50-150-8-250	15502	0	7.391	15593.3	145.41	768.58	0.588	15502	0	371.09	0
A2-10-50-150-4-250	25702	0	7.021	25712.4	432.64	1072.54	0.04	25702	0	380.91	0
A2-10-50-150-5-250	21503	0	9.551	21503	0	1046.22	0	21503	0	369.09	0
A2-10-50-150-6-250	20250	0	13.22	20250	0	1126.13	0	20250	0	353.1	0
A2-10-50-150-8-250	16676	0	9.136	16676	0	1091.21	0	16676	0	354.46	0
A2-20-100-100-4-250	38074	0	117.162	38104.4	316.04	16115.14	0.079	38078.6	84.64	704.69	0.012
A2-20-100-100-5-250	32583	0	2801.434	32680.9	872.09	16736.41	0.3	32634.2	5179.16	825.72	0.157
A2-20-100-100-6-250	28568.6	64.176	25043.3	28576	3811.6	18798.71	0.025	28490	0	683.06	−0.275
A2-20-100-100-8-250	24600.5	11.456	21527.085	24652.9	555.49	16746.77	0.213	24605.1	351.09	901.37	0.018
B1-1-25-75-4-250	17417	0	11.963	17417	0	71.63	0	17417	0	194.08	0
B1-1-25-75-5-250	15891	0	6.564	15891	0	77.48	0	15891	0	183.65	0
B1-1-25-75-6-250	14260	0	5.286	14260	0	70.93	0	14260	0	186.37	0
B1-1-25-75-8-250	11538	0	15.754	11538	0	72.92	0	11538	0	188.12	0
B1-1-50-50-4-250	19966	0	8.464	19966	0	555.05	0	19966	0	280.26	0
B1-1-50-50-5-250	17113	0	5312.75	17179.1	10915.89	573.71	0.386	17113	0	328.04	0
B1-1-50-50-6-250	15989	0	45.122	15999.5	785.45	534.8	0.065	15989	0	292.18	0
B1-1-50-50-8-250	14027	0	43.377	14027	0	540.37	0	14027	0	296.37	0
B1-10-50-50-4-250	20075	0	4.624	20075	0	735.56	0	20075	0	277.25	0
B1-10-50-50-5-250	17986	0	6.415	17986	0	789.64	0	17986	0	307.1	0
B1-10-50-50-6-250	15924	0	1.803	15924	0	803.43	0	15924	0	258.94	0
B1-10-50-50-8-250	13672	0	6.665	13705.6	4515.84	703.8	0.245	13672	0	267.91	0
B1-5-25-75-4-250	17079	0	1.656	17079	0	54.82	0	17079	0	201.98	0
B1-5-25-75-5-250	15110	0	1.211	15110	0	59.68	0	15110	0	190.72	0
B1-5-25-75-6-250	14707	0	1.392	14707	0	62.32	0	14707	0	192.43	0

(*continued*)

Table 2. (*continued*)

Data	GVNS			GRASP-ELS				GA-VLG			
	Avg.	σ^2	Time	Avg.	σ^2	Time	Gap_{UB}	Avg.	σ^2	Time	Gap_{UB}
B1-5-25-75-8-250	11319	0	1.493	11319	0	60.69	0	11319	0	194.38	0
B2-1-100-100-4-250	40974	0	198.112	40993.1	741.69	20287.35	0.046	41001.5	3025.05	821.72	0.067
B2-1-100-100-5-250	34848	0	4904.273	34856.3	34.61	21132.51	0.023	34848	0	883.7	0
B2-1-100-100-6-250	30849	0	15885.026	30880.1	1715.69	21999	0.1	30894.3	996.81	856.52	0.146
B2-1-100-100-8-250	25804	0	13342.465	25914.1	3048.29	20826.46	0.426	25820	256	993.06	0.062
B2-1-50-150-4-250	23288	0	5.545	23288	0	881.96	0	23288	0	339.12	0
B2-1-50-150-5-250	20039	0	20.227	20039	0	866.44	0	20039	0	332.39	0
B2-1-50-150-6-250	18046	0	47.292	18046	0	891.85	0	18046	0	345.81	0
B2-1-50-150-8-250	15668	0	40.007	15668	0	959.18	0	15668	0	313.84	0
B2-10-50-150-4-250	25967	0	19.954	25967	0	1452.23	0	25967	0	346.35	0
B2-10-50-150-5-250	22359	0	19.741	22359	0	1421.98	0	22359	0	334.17	0
B2-10-50-150-6-250	19792	0	17.002	19792	0	1539.92	0	19792	0	348.38	0
B2-10-50-150-8-250	17106	0	18.184	17106.1	0.09	1386.24	0.0005	17106	0	361.92	0
B2-20-100-100-4-250	53590	0	3650.731	53591.9	32.49	38501	0.003	53590	0	763.91	0
B2-20-100-100-5-250	45209	0	1152.306	45209	0	42990.69	0	45213.4	174.24	743.29	0.009
B2-20-100-100-6-250	39184	0	13073.578	39194.2	560.16	41914.83	0.026	39184	0	712.49	0
B2-20-100-100-8-250	32513	0	8897.264	32524.2	1128.96	38976.69	0.034	32531	1444	861.18	0.055
C1-1-25-75-4-250	13012	0	1.203	13012	0	31.86	0	13012	0	160.63	0
C1-1-25-75-5-250	11666	0	1.421	11666	0	31.39	0	11666	0	159.93	0
C1-1-25-75-6-250	9820	0	1.127	9820	0	30	0	9820	0	156.82	0
C1-1-25-75-8-250	9818	0	1.24	9818	0	31.94	0	9818	0	159.01	0
C1-1-50-50-4-250	20294	0	4.778	20294	0	574.6	0	20294	0	258.98	0
C1-1-50-50-5-250	17378	0	6.102	17378	0	619.47	0	17378	0	268.75	0

(*continued*)

Table 2. (*continued*)

Data	GVNS			GRASP-ELS				GA-VLG			
	Avg.	σ^2	Time	Avg.	σ^2	Time	Gap_{UB}	Avg.	σ^2	Time	Gap_{UB}
C1-1-50-50-6-250	16365	0	4.726	16365	0	636.53	0	16365	0	265.5	0
C1-1-50-50-8-250	13900	0	12.698	13900	0	616.37	0	13900	0	260.33	0
C1-10-50-50-4-250	26931	0	3.017	26931	0	937.78	0	26931	0	291.93	0
C1-10-50-50-5-250	23544	0	11.571	23544	0	1075.82	0	23544	0	412.64	0
C1-10-50-50-6-250	20818	0	2.727	20818	0	1001.74	0	20818	0	331.56	0
C1-10-50-50-8-250	18154	0	8.028	18158.8	34.56	980.82	0.026	18154	0	292.64	0
C1-5-25-75-4-250	13738	0	1.412	13738	0	35.89	0	13738	0	168.41	0
C1-5-25-75-5-250	13575	0	3.05	13575	0	34.92	0	13575	0	175.35	0
C1-5-25-75-6-250	10826	0	2.164	10826	0	37.02	0	10826	0	166.63	0
C1-5-25-75-8-250	10556	0	1.437	10556	0	34.4	0	10556	0	168.97	0
D1-1-25-75-4-250	18127	0	1.003	18127	0	35.35	0	18127	0	175.32	0
D1-1-25-75-5-250	15972	0	1.016	15972	0	36.79	0	15972	0	175.92	0
D1-1-25-75-6-250	14532	0	1.457	14532	0	39.3	0	14532	0	175.72	0
D1-1-25-75-8-250	12700	0	1.544	12700	0	36.71	0	12700	0	174.48	0
D1-1-50-50-4-250	23275	0	13.041	23275	0	716.26	0	23275	0	271.06	0
D1-1-50-50-5-250	20402	0	80.018	20402	0	719.32	0	20402	0	275.12	0
D1-1-50-50-6-250	18072	0	32.065	18072	0	741.83	0	18072	0	257.36	0
D1-1-50-50-8-250	14930	0	4.998	14930	0	684.95	0	14930	0	249.68	0
D1-10-50-50-4-250	30390	0	2.026	30390	0	1407.15	0	30390	0	308.98	0
D1-10-50-50-5-250	26284	0	3.713	26284	0	1509.47	0	26284	0	331.55	0
D1-10-50-50-6-250	23646	0	8.203	23646	0	1433.92	0	23646	0	304.1	0
D1-10-50-50-8-250	19986	0	55.444	19986	0	1404.4	0	19986	0	323.79	0
D1-5-25-75-4-250	18464	0	1.107	18464	0	21.99	0	18464	0	177.63	0
D1-5-25-75-5-250	15767	0	1.028	15767	0	21.86	0	15767	0	176.24	0
D1-5-25-75-6-250	14851	0	1.125	14851	0	21.89	0	14851	0	180.31	0
D1-5-25-75-8-250	12705	0	1.328	12705	0	20.65	0	12705	0	183.84	0

Table 3. Stability of our GVNS algorithm

Methods	Criteria		GVNS/GRASP-ELS		GVNS/GA-VLG	
	Same cost		GVNS Better	GRASP-ELS Better	GVNS Better	GA-VLG Better
GVNS	GRASP-ELS	16	6	1	0	1
	GA-VLG	6				
	GRASP-ELS and GA-VLG	5				
GRASP-ELS	GVNS	0	0	0	-	-
	GA-VLG	4				
	GVNS and GA-VLG	0				
GA-VLG	GVNS	1	-	-	0	0
	GRASP-ELS	13				
	GVNS and GRASP-ELS	1				

5   Conclusions

In this paper, we solve a generalized variant of the well-know multi-Vehicle Covering Tour Problem (m-CTP-p) called the multi-vehicle multi-Covering Tour Problem (mm-CTP-p) where some customers must be covered multiple times. We develop a GVNS algorithm to solve this variant. Our proposed approach solved small and large size instances and improved best known solutions. In addition, our results show that our algorithm provides better solutions than that of [9] in all solved instances in term of time.

References

1. Beasley, J., Nascimento, E.: The vehicle routing-allocation problem: a unifying framework. Top **4**(1), 65–86 (1996)
2. Golden, B., Naji-Azimi, Z., Raghavan, S., Salari, M., Toth, P.: The generalized covering salesman problem. INFORMS J. Comput. **24**(4), 534–553 (2012)
3. Hà, M.H., Bostel, N., Langevin, A., Rousseau, L.: An exact algorithm and a metaheuristic for the multi-vehicle covering tour problem with a constraint on the number of vertices. Eur. J. Oper. Res. **226**, 211–220 (2013)
4. Hansen, P., Mladenović, N., Pérez, J.A.M.: Variable neighbourhood search: methods and applications. Ann. Oper. Res. **175**(1), 367–407 (2010)
5. Hansen, P., Mladenović, N., Todosijević, R., Hanafi, S.: Variable neighborhood search: basics and variants. EURO J. Comput. Optim. **5**(3), 423–454 (2017)
6. Kammoun, M., Derbel, H., Ratli, M., Jarboui, B.: A variable neighborhood search for solving the multi-vehicle covering tour problem. Electron. Notes Discrete Math. **47**, 285–292 (2015)
7. Lenstra, J.K., Kan, A.R.: Complexity of vehicle routing and scheduling problems. Networks **11**(2), 221–227 (1981)
8. Mladenović, N., Hansen, P.: Variable neighborhood search. Comput. Oper. Res. **24**(11), 1097–1100 (1997)
9. Pham, T.A., Hà, M.H., Nguyen, X.H.: Solving the multi-vehicle multi-covering tour problem. Comput. Oper. Res. **88**, 258–278 (2017)

A Hybrid Firefly - VNS Algorithm for the Permutation Flowshop Scheduling Problem

Andromachi Taxidou[1], Ioannis Karafyllidis[1], Magdalene Marinaki[1], Yannis Marinakis[1(✉)], and Athanasios Migdalas[2]

[1] School of Production Engineering and Management, Technical University of Crete, Chania, Greece
machi_taxidou@hotmail.com, Io_Karafi@hotmail.com, magda@dssl.tuc.gr, marinakis@ergasya.tuc.gr
[2] Industrial Logistics, Luleå Technical University, 97187 Luleå, Sweden
athmig@ltu.se

Abstract. In this paper a Permutation Flowshop Scheduling Problem is solved using a hybridization of the Firefly algorithm with Variable Neighborhood Search algorithm. The Permutation Flowshop Scheduling Problem (PFSP) is one of the most computationally complex problems. It belongs to the class of combinatorial optimization problems characterized as NP-hard. In order to find high quality solutions in reasonable computational time, heuristic and metaheuristic algorithms have been used for solving the problem. The proposed method, Hybrid Firefly Variable Neighborhood Search algorithm, uses in the local search phase of the algorithm a number of local search algorithms, 1-0 relocate, 1-1 exchange and 2-opt. In order to test the effectiveness and efficiency of the proposed method we used a set of benchmark instances of different sizes from the literature.

Keywords: Permutation Flowshop Scheduling Problem ·
Firefly algorithm · Variable Neighborhood Search

1 Introduction

The flowshop scheduling problem was proposed by Johnson [6] and it is a widely studied scheduling problem. In this problem, there is a set of jobs (task or items) that must be processed from a set of machines in the same order. All jobs are independent and must start their process from the same starting machine. The machines are constantly available. The purpose of the process is to find the sequence for the processing of the jobs in the machines in order to optimize a given criterion. In the literature, the most frequently used criterion is the minimization of the maximum completion time (makespan) [6].

The Firefly Algorithm (FA), proposed by Yang [24], is a nature inspired algorithm and it depends on the glowing behavior of fireflies. Each firefly has

© Springer Nature Switzerland AG 2019
A. Sifaleras et al. (Eds.): ICVNS 2018, LNCS 11328, pp. 274–286, 2019.
https://doi.org/10.1007/978-3-030-15843-9_21

brightness and attracts other fireflies in a predefined way. Each firefly is attracted by the one firefly which is the brightest of all in a specific neighborhood. It is obvious from the available literature that FA can be well applied to various different optimization fields and problems. Moreover, FA is an algorithm that has been used by many scientists so far [10].

The goal of the research presented in this paper is to solve the Permutation Flowshop Scheduling Problem using a hybrid Firefly - Variable Neighborhood Search algorithm (HFVNS). Three local search algorithms are incorporated in the VNS algorithm, the 1-0 relocate, 2-opt, 1-1 exchange. In order to test the efficiency of the algorithm we test the algorithm in 120 classic benchmark instances from the literature. Also, for comparison reasons and in order to see how each of the local search contributes in the effectiveness of the algorithm we use two different versions of VNS as it will be described later and we test the firefly algorithm using each one of the local search algorithms used in the VNS scheme independently. Thus, we compared, using the same number of function evaluations and the same parameters, five different versions of the hybrid firefly algorithm, three of them using in the local search phase only one local search algorithm and the other two using two different versions of VNS.

The structure of the paper is as follows. In Sect. 2 the Permutation Flowshop Scheduling Problem is presented and the formulation of the problem is given. In Sect. 3 a sort description of the classic firefly algorithm is given while in Sect. 4 the proposed algorithm is presented and analyzed in detail. In Sect. 5 the computational results of the algorithm are presented and analyzed, finally, in the last section some conclusions are presented and future directions based on this research are given.

2 Permutation Flowshop Scheduling Problem

The Permutation Flowshop Scheduling Problem (PFSP) consists of n jobs and m machines. Each job is processing in all machines with the same order. The jobs and the machines match one by one. Moreover, each job cannot be terminated or interrupted before the predefined termination time. No jobs depend on other jobs and are ready to start the process at the predefined time zero. The set up times can be omitted. The machines are all the time available. The aim is to find a sequence for the processing of the jobs in the machines in order to optimize one given criterion, usually the minimization of the maximum completion time (makespan).

In the permutation flowshop scheduling problem (PFSP) [14,17], solutions are represented by the permutation of n jobs, i.e., $\pi = \{\pi_1, \pi_2, ..., \pi_n\}$. Each job is composed of m operations, and every operation is performed by a different machine. Thus, given the processing time p_{jk} for the job j on the machine k (these times are fixed, known in advance and non-negative), the PFSP is to find the best permutation of jobs $\pi^* = \pi_1^*, \pi_2^*, ..., \pi_n^*$ to be processed on each machine subject to the makespan criterion. Let $C(\pi_j, m)$ denote the completion time of the job π_j on the machine m. Then, given the job permutation π, the completion time for the n-job, m-machine problem is calculated as follows:

$$C(\pi_1, 1) = p_{\pi_1,1} \tag{1}$$
$$C(\pi_j, 1) = C(\pi_{j-1}, 1) + p_{\pi_j,1}, j = 2, ..., n \tag{2}$$
$$C(\pi_1, k) = C(\pi_1, k-1) + p_{\pi_1,k}, k = 2, ..., m \tag{3}$$
$$C(\pi_j, k) = \max\{C(\pi_{j-1}, k), C(\pi_j, k-1) + p_{\pi_j,k}\},$$
$$j = 2, ..., n, k = 2, ..., m \tag{4}$$

So, the makespan of a permutation π can be formally defined as the completion time of the last job π_n on the last machine m, i.e.:

$$C_{max}(\pi) = C(\pi_n, m). \tag{5}$$

Therefore, the PFSP with the makespan criterion is to find the optimal permutation π^* in the set of all permutations Π such that:

$$C_{max}(\pi^*) \leq C(\pi_n, m) \text{ for each permutation } \pi \text{ belonging to } \Pi. \tag{6}$$

The computational complexity of the PFSP has been proved to be NP-hard by [3]. Due to this fact, the solution procedure for the PFSP is often either heuristic or metaheuristic. A number of heuristic and metaheuristic algorithms have been developed in the past for this problem:

- Iterated Greedy Algorithm [19]
- Hybrid Genetic Algorithms [1,18,22,23,27]
- Tabu Search [4,12]
- Differential Evolution [14]
- Hybrid Particle Swarm Optimization [7,8,11,25,26]
- Ant Colony Optimization Algorithms [2,15,16]
- Hybrid Artificial Bee Colony Algorithm [9].

3   Classic Firefly Algorithm

The basic form of the Firefly Algorithm was proposed by Yang [24]. It simulates the mating process based on the glow of each firefly. It is a metaheuristic nature inspired algorithm. It follows the same principle as algorithms like the Particle Swarm Optimization Algorithm. In nature there are about 2000 species of fireflies and most of them produce a short and rhythmic glow. This glow is produced after a bioluminescence process and serves as a mean of communicating between fireflies for mating and warning in case of an imminent danger. Bioluminescence is characterized by a specific generation of light at various wavelengths emitted by various living organisms, often called incorrectly phosphorescence [24]. There are three basic features in order to develop and describe the algorithm:

1. Fireflies are attracted to each other regardless of the gender which means that no mutant operator is required.

2. The sharing of information between fireflies is proportional to their attractiveness, more specifically, is proportional to their distance, so the closer the two fireflies are, the more likely it is to mate with each other. If there is no firefly that is brighter than the firefly we are testing, then, the firefly moves randomly in the space.
3. The value of the objective function depends on the glow of each firefly (firefly is the solution of the problem). In case of maximization (minimization) problems, the brighter the firefly is, the higher (lower) is the value of the objective function [13].

There are two important parameters in the firefly algorithm: Changing the glow of the firefly and formulating the attractiveness. However, for convenience we can assume that the attractiveness of a firefly is proportional to the intensity of the light that is emitted to the other fireflies, so it depends on the value of the objective function. Glow I is proportional to the objective function f at point x with $I(x) \propto f(x)$. The brightness varies and is inversely proportional to the distance between the two fireflies and is equal to:

$$I = \frac{I_s}{d^2} \tag{7}$$

where I_s is the glow at the source and d is the distance between the two fireflies and is symbolized as $r_{ij} = d(x_i, x_j)$. The glow is given by the equation:

$$I = I_0 e^{-\gamma r_{ij}} \tag{8}$$

where I_0 is the initial glow and γ is the stable parameter indicating the absorbance of the glow. The r_{ij} is the Euclidean distance between the two fireflies and is given by the equation:

$$r_{ij} = \sqrt{(x_i - x_j)^2 + (y_i - y_j)^2} \tag{9}$$

The attractiveness β of the firefly is proportional to the light intensity and is equal to:

$$\beta = \beta_0 e^{-\gamma r_{ij}^2} \tag{10}$$

where β_0 is the attractiveness when $d = 0$. One of the basic factors of the algorithm is the movement of each firefly. The movement of a firefly i located at position x_i and attracted by another firefly j which shines more than i and is at position x_j is determined by the equation:

$$x_i(t + 1) = x_i(t) + \beta_0 e^{-\gamma r_{ij}^2} (x_j(t) - x_i(t)) + \alpha \epsilon_i \tag{11}$$

The first term is the current position of the firefly while the second term shows the intensity with which a firefly will see a glower firefly located in the near area and the third term is the random movement that a firefly makes when there are no brighter fireflies than it in the near area. The variable α is a random number. As γ approaches zero, the glow $\beta = \beta_0$ is stable and does not depend from the distance. As γ increases the attractiveness of each firefly decreases which means that no firefly can see the others and all of them move randomly in the area [13, 20].

4 Hybrid Firefly Variable Neighborhood Search Algorithm

4.1 General Structure of the Algorithm

In the proposed algorithm we use a hybridization of the firefly algorithm with a Variable Neighborhood Search algorithm. Initially we used the classic firefly algorithm in the way that it was proposed by Yang [24] and it is described in the previous section of this research. However, as the problem is a combinatorial optimization problem and we have to transform each of the solutions from the continuous space in order to apply the main equation of the firefly algorithm to discrete space and vice versa it was observed that the differences in the solutions from the one iteration to the other were very small due to the movement equation of the fireflies which lead to a fast convergence to a not a very good solution. Thus, we have tried a number of different equations for the movement equation of the fireflies and we concluded that the most effective equation for the current problem is the one given in the following:

$$x_i(t+1) = rand\, x_i(t) + \beta_0 e^{-\gamma r_{ij}^2}(x_j(t) - x_i(t)) + \alpha \epsilon_i \qquad (12)$$

A pseudocode of the whole procedure is presented in the following:

Initialization
Select the number of fireflies
Generate the initial solution in discrete space
Evaluate the fitness function of each firefly
Apply Variable Neighborhood Search in each firefly
Keep Best solution
Main Phase
Do until the maximum number of iterations has not been reached:
    Calculate the distance between two fireflies
    Calculate each firefly's glow
    Decide whether the firefly is going to make the move
    Evaluate the new makespan for each solution
    Apply variable neighborhood search in each new solution
    Update the best solution for each firefly
    if a new best solution is found **then**
        Update the best firefly
    endif
Enddo
Return the best firefly (the best solution).

4.2 Variable Neighborhood Search

A Variable Neighborhood Search (VNS) [5] algorithm is applied in order to optimize the particles. The basic idea of the method is the successive search in a number of neighborhoods of a solution. In VNS, with the term neighborhood it is meant different number of local search algorithms. The search is applied either with random or with a more systematical manner in order the solution to escape from a local minimum. This method takes advantage of the fact that different local search algorithms will lead to different local minimums [11].

The local search algorithms that are incorporated into the VNS scheme are the 1-0 relocate, 1-1 exchange and 2-opt. In all algorithms the choice of the job that it will be relocated (1-0 relocate) or the two jobs that they will be exchanged (1-1 exchange) or the two jobs that all the jobs between them will be reversed (2-opt) are selected randomly.

In this paper two different versions of local search were applied. The first one is a sequential VNS where each local search algorithm is run for a number of predefined numbers and, then, another local search is selected until all (three) local search algorithms are selected and run for the predefined number of iterations. The other one is a parallel VNS where in each iteration a step from each one of the local search algorithms is realized and the one, that better improves the solution is selected. If none of the local search algorithms find a better solution we proceed with the current solution and the selection of the moves for any local search algorithm is performed randomly in the iteration different moves are selected.

5 Computational Results

The algorithm was tested on the 120 benchmark instances of Taillard [21]. In these instances, there are different sets having 20, 50, 100, 200 and 500 jobs and 5, 10 or 20 machines. There are 10 problems inside every size set. In total there are 12 sets and these are: 20×5 (i.e. 20 jobs and 5 machines), 20×10, 20×20, 50×5, 50×10, 50×20, 100×5, 100×10, 100×20, 200×10, 200×20 and 500×20.

The parameters of the proposed algorithm are selected after thorough testing. A number of different alternative values were tested and the ones selected are those that gave the best computational results concerning both the quality of the solution and the computational time needed to achieve this solution. The efficiency of the HFVNS algorithm is measured by the quality of the produced solutions. The quality is given in terms of the relative deviation from the best known solution, that is $\omega = \frac{(c_{HFVNS} - c_{BKS})}{c_{BKS}}\%$, where c_{HFVNS} denotes the cost of the solution found by HFVNS and c_{BKS} is the cost of the best known solution.

In Table 1, the results for the 120 Taillard benchmark instances are presented. More specifically, in this table the best known results from the literature (BKS), the best results produced from HFVNS (BS) (the parallel version of VNS) and the quality (ω) for each of the 120 instances are presented. As it can be seen, the algorithm finds very satisfactory results as it finds solutions with quality less than 1% from the BKS in 19 instances and solutions with quality between 1% and 5% from the BKS in 50 instances. The instances can be divided in 5 different categories based on the number of jobs. The first category contains the

Table 1. Results of HFVNS in Taillard benchmark instances for the PFSP

Problem	BKS	BS	ω	Problem	BKS	BS	ω	Problem	BKS	BS	ω
20 × 5	1278	1284	0.47	50 × 10	2991	3182	6.39	100 × 20	6202	6799	9.63
20 × 5	1359	1360	0.07	50 × 10	2867	3059	6.70	100 × 20	6183	6774	9.56
20 × 5	1081	1098	1.57	50 × 10	2839	3040	7.08	100 × 20	6271	6850	9.23
20 × 5	1293	1309	1.24	50 × 10	3063	3222	5.19	100 × 20	6269	6792	8.34
20 × 5	1235	1243	0.65	50 × 10	2976	3165	6.35	100 × 20	6314	6881	8.98
20 × 5	1195	1200	0.42	50 × 10	3006	3162	5.19	100 × 20	6364	6949	9.19
20 × 5	1239	1249	0.81	50 × 10	3093	3257	5.30	100 × 20	6268	6909	10.23
20 × 5	1206	1213	0.58	50 × 10	3037	3179	4.68	100 × 20	6401	7037	9.94
20 × 5	1230	1255	2.03	50 × 10	2897	3069	5.94	100 × 20	6275	6851	9.18
20 × 5	1108	1121	1.17	50 × 10	3065	3244	5.84	100 × 20	6434	6905	7.32
20 × 10	1582	1586	0.25	50 × 20	3850	4091	6.26	200 × 10	10862	11162	2.76
20 × 10	1659	1694	2.11	50 × 20	3704	3997	7.91	200 × 10	10480	10992	4.89
20 × 10	1496	1525	1.67	50 × 20	3640	3905	7.28	200 × 10	10922	11287	3.34
20 × 10	1377	1401	1.74	50 × 20	3720	4005	7.66	200 × 10	10889	11143	2.33
20 × 10	1419	1455	2.54	50 × 20	3610	3904	8.14	200 × 10	10524	10957	4.11
20 × 10	1397	1424	1.93	50 × 20	3681	3953	7.39	200 × 10	10329	10802	4.58
20 × 10	1484	1508	1.62	50 × 20	3704	4003	8.07	200 × 10	10854	11249	3.64
20 × 10	1538	1569	1.62	50 × 20	3691	3996	8.26	200 × 10	10730	11176	4.16
20 × 10	1593	1624	1.95	50 × 20	3743	4030	7.67	200 × 10	10438	10906	4.48
20 × 10	1591	1615	1.51	50 × 20	3756	4035	7.43	200 × 10	10675	11125	4.22
20 × 20	2297	2326	1.22	100 × 5	5493	5521	0.51	200 × 20	11195	12243	9.36
20 × 20	2099	2119	0.86	100 × 5	5268	5284	0.30	200 × 20	11203	12387	10.57
20 × 20	2326	2360	1.37	100 × 5	5175	5236	1.18	200 × 20	11281	12389	9.82
20 × 20	2223	2258	1.48	100 × 5	5014	5044	0.60	200 × 20	11275	12365	9.67
20 × 20	2291	2323	1.26	100 × 5	5250	5307	1.09	200 × 20	11259	12279	9.06
20 × 20	2226	2265	1.62	100 × 5	5135	5161	0.51	200 × 20	11176	12280	9.88
20 × 20	2273	2326	2.33	100 × 5	5246	5291	0.86	200 × 20	11360	12469	9.76
20 × 20	2200	2230	1.27	100 × 5	5094	5145	1.00	200 × 20	11334	12399	9.40
20 × 20	2237	2266	1.16	100 × 5	5448	5510	1.14	200 × 20	11192	12333	10.19
20 × 20	2178	2209	1.42	100 × 5	5322	5372	0.94	200 × 20	11288	12363	9.52
50 × 5	2724	2729	0.18	100 × 10	5770	6025	4.42	500 × 20	26059	28174	8.12
50 × 5	2834	2871	1.16	100 × 10	5349	5600	4.69	500 × 20	26520	28648	8.02
50 × 5	2621	2648	1.03	100 × 10	5676	5907	4.07	500 × 20	26371	28446	7.87
50 × 5	2751	2782	1.13	100 × 10	5781	6099	5.50	500 × 20	26456	29126	10.09
50 × 5	2863	2888	0.87	100 × 10	5467	5774	5.62	500 × 20	26334	28415	7.90
50 × 5	2829	2847	0.64	100 × 10	5303	5500	3.71	500 × 20	26477	28522	7.72
50 × 5	2725	2758	1.21	100 × 10	5595	5784	3.38	500 × 20	26389	28210	6.90
50 × 5	2683	2715	1.19	100 × 10	5617	5847	4.09	500 × 20	26560	28615	7.74
50 × 5	2552	2577	0.98	100 × 10	5871	6076	3.49	500 × 20	26005	28242	8.60
50 × 5	2782	2784	0.07	100 × 10	5845	6059	3.66	500 × 20	26457	28384	7.28

Table 2. Comparisons of the results (average qualities) of HFVNS with the other four algorithms in Taillard benchmark instances for the PFSP

Problem	Firefly + 1-0 relocate	Firefly + 1-1 exchange	Firefly + 2opt	Firefly + SVNS	Firefly + PVNS (HFVNS)
20 × 5	1.90	2.66	3.36	2.12	0.90
20 × 10	4.40	5.00	6.03	4.81	1.69
20 × 20	4.06	4.47	5.58	3.86	1.40
50 × 5	2.58	2.58	3.02	2.04	0.85
50 × 10	9.67	10.23	11.28	9.38	5.86
50 × 20	12.22	12.77	13.72	12.24	7.61
100 × 5	2.20	2.22	2.43	1.95	0.81
100 × 10	7.69	7.66	8.42	7.49	4.26
100 × 20	13.48	13.62	14.60	13.35	9.16
200 × 10	6.86	6.73	6.64	6.58	3.85
200 × 20	13.61	13.84	14.28	13.42	9.72
500 × 20	10.56	10.42	10.55	10.22	8.03

instances in which the number of jobs is equal to 20 and the number of machines varies between 5 to 20. In these instances, the quality of the solution is between 0.07 and 2.53. In the second category, the number of jobs is equal to 50 and the number of machines varies between 5 to 20. In these instances, the quality of the solution is between 0.07 and 8.26. In the third category, the number of jobs is equal to 100 and the number of machines varies between 5 to 20. In these instances, the quality of the solution is between 0.30 and 10.22. In the fourth category, the number of jobs is equal to 200 and the number of machines varies between 10 to 20. In these instances, the quality of the solution is between 2.33 and 10.56. And, finally, in the last category, the number of jobs is equal to 500 and the number of machines is equal to 20. In these instances, the quality of the solution is between 6.90 and 10.09.

As it has, already, been mentioned the instances can be divided in 12 sets based on the number of jobs and the number of machines. Thus, in Table 2 for each of the 12 sets we have averaged the quality of the 10 corresponding instances that belong to each one of the 12 sets. This is the way that most researchers that study the Permutation Flowshop Scheduling Problem present their results. In this Table, except of the results of the proposed algorithm, the results of the other four algorithms described previously are presented. In this Table in columns 2 to 4 the results of the firefly algorithm using a single local search (1-0 relocate, 1-1 exchange and 2opt, respectively) are presented, in column 5 the results of the firefly hybridized with a sequential VNS are given, while in the last column the results of the proposed algorithm (hybrid firefly with a parallel VNS - HFVNS) are given.

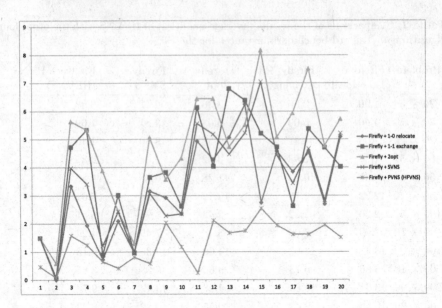

Fig. 1. Quality of solutions of the five algorithms for the 1–20 benchmark instances of Taillard

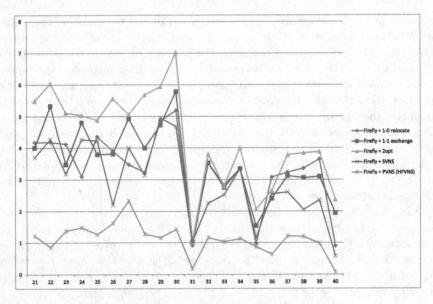

Fig. 2. Quality of solutions of the five algorithms for the 21–40 benchmark instances of Taillard

From the results that they are presented in this Table we can see that the proposed algorithm performs better than the algorithms that use only one local search algorithm which is expected as the VNS algorithm is a more sophisticated

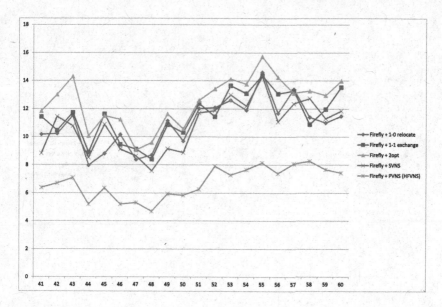

Fig. 3. Quality of solutions of the five algorithms for the 41–60 benchmark instances of Taillard

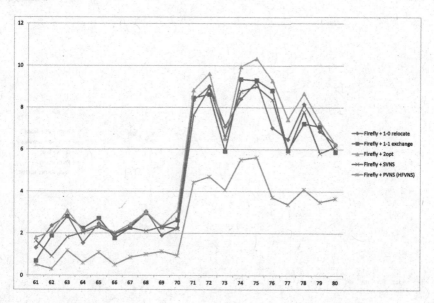

Fig. 4. Quality of solutions of the five algorithms for the 61–80 benchmark instances of Taillard

and powerful algorithm. The interesting part of this Table is the large differences between the qualities produced by the proposed algorithm and the qualities produced by the hybridization of the firefly algorithm with the sequential VNS

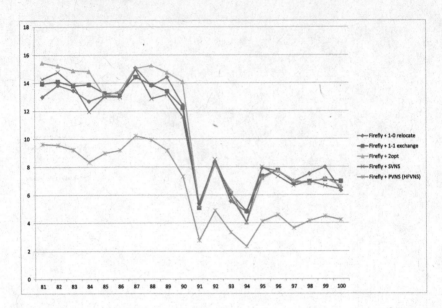

Fig. 5. Quality of solutions of the five algorithms for the 81–100 benchmark instances of Taillard

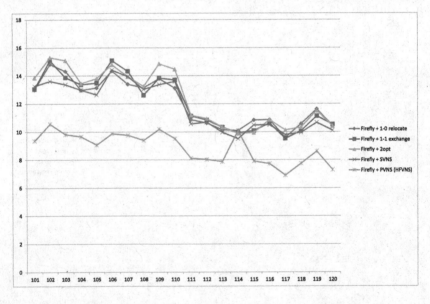

Fig. 6. Quality of solutions of the five algorithms for the 101–120 benchmark instances of Taillard

algorithm. This is happening because we kept the same number of iterations and function evaluations for all algorithms and the parallel version of the VNS produce in this implementation more effective moves from the one local optimum to the other than the moves produced by the sequential version of the VNS when both are hybridized with the firefly algorithm.

In Figs. 1, 2, 3, 4, 5 and 6, the quality for all algorithms in the 120 instances is presented. We preferred to present the results in 6 different figures and not in a single one as there are 120 instances and 5 algorithms and thus, the figures were not very clear. In each figure the results of 20 instances are presented. More precisely in Fig. 1 the results of the 20 first instances are given, while in Fig. 2 the results of the next 20 instances are given and so on. As it can be seen from these figures the results of the proposed algorithm are better from all the other algorithms used in the comparisons.

6   Conclusions

In this paper, a new algorithm based on the Firefly algorithm for the solution of the Permutation Flowshop Scheduling Problem is presented. This algorithm is a hybridization of the Firefly algorithm with the Variable Neighborhood Search algorithm. We compared the algorithm with different local search algorithms and with another VNS version. The algorithm was tested in 120 benchmark instances that are usually used in the literature and gave very good results. The computational results were satisfactory. In the future, this algorithm will be used for the solution of other NP-hard combinatorial optimization problems.

References

1. Chen, S.-H., Chang, P.-C., Cheng, T.C.E., Zhang, Q.: A self-guided genetic algorithm for permutation flowshop scheduling problems. Comput. Oper. Res. **39**, 1450–1457 (2012)
2. Gajpal, Y., Rajendran, C.: An ant-colony optimization algorithm for minimizing the completion-time variance of jobs in flowshops. Int. J. Prod. Econ. **101**(2), 259–272 (2006)
3. Garey, M.R., Johnson, D.S., Sethi, R.: The complexity of flowshop and jobshop scheduling. Math. Oper. Res. **1**, 117–129 (1976)
4. Grabowski, J., Wodecki, M.: A very fast tabu search algorithm for the permutation flow shop problem with makespan criterion. Comput. Oper. Res. **31**, 1891–1909 (2004)
5. Hansen, P., Mladenovic, N.: Variable neighborhood search: principles and applications. Eur. J. Oper. Res. **130**, 449–467 (2001)
6. Johnson, S.: Optimal two-and-three stage production schedules with setup times included. Naval Res. Logistics Q. **1**, 61–68 (1954)
7. Liao, C.-J., Tseng, C.-T., Luarn, P.: A discrete version of particle swarm optimization for flowshop scheduling problems. Comput. Oper. Res. **34**, 3099–3111 (2007)
8. Liu, B., Wang, L., Jin, Y.-H.: An effective PSO-based memetic algorithm for flow shop scheduling. IEEE Trans. Syst. Man Cybern.-Part B: Cybern. **37**(1), 18–27 (2007)
9. Liu, Y.-F., Liu, S.-Y.: A hybrid discrete artificial bee colony algorithm for permutation flowshop scheduling problem. Appl. Soft Comput. **13**, 1459–1463 (2013)

10. Ma, Y., Zhao, Y., Wu, L., He, Y., Yang, X.-S.: Navigability analysis of magnetic map with projecting pursuit-based selection method by using firefly algorithm. Neurocomputing **159**, 288–297 (2015)
11. Marinakis, Y., Marinaki, M.: Particle swarm optimization with expanding neighborhood topology for the permutation flowshop scheduling problem. Soft Comput. **17**(7), 1159–1173 (2013)
12. Nowicki, E., Smutnicki, C.: A fast tabu search algorithm for the permutation flowshop problem. Eur. J. Oper. Res. **91**, 160–175 (1996)
13. Osaba, E., Yang, X.S., Diaz, F., Onieva, E., Masegosa, A.D., Perallos, A.: A discrete firefly algorithm to solve a rich vehicle routing problem modelling a newspaper distribution system with recycling policy. Soft Comput. **21**(18), 5295–5308 (2017)
14. Pan, Q.-K., Tasgetiren, M.F., Liang, Y.-C.: A discrete differential evolution algorithm for the permutation flowshop scheduling problem. Comput. Ind. Eng. **55**, 795–816 (2008)
15. Rajendran, C., Ziegler, H.: Ant-colony algorithms for permutation flowshop scheduling to minimize makespan/total flowtime of jobs. Eur. J. Oper. Res. **155**(2), 426–438 (2004)
16. Rajendran, C., Ziegler, H.: Two ant-colony algorithms for minimizing total flowtime in permutation flowshops. Comput. Ind. Eng. **48**(4), 789–797 (2005)
17. Ruiz, R., Maroto, C.: A comprehensive review and evaluation of permutation flowshop heuristics. Eur. J. Oper. Res. **165**, 479–494 (2005)
18. Ruiz, R., Maroto, C., Alcaraz, J.: Two new robust genetic algorithms for the flowshop scheduling problem. Omega **34**, 461–476 (2006)
19. Ruiz, R., Stützle, T.: A simple and effective iterated greedy algorithm for the permutation flowshop scheduling problem. Eur. J. Oper. Res. **177**, 2033–2049 (2007)
20. Srikakulapu, R., Vinatha, U.: Combined approach of firefly algorithm with travelling salesmen problem for optimal design of offshore wind farm. In: IEEE Power and Energy Society General Meeting 2017, pp. 1–5 (2017)
21. Taillard, E.: Benchmarks for basic scheduling problems. Eur. J. Oper. Res. **64**, 278–285 (1993)
22. Tseng, L.-Y., Lin, Y.-T.: A hybrid genetic local search algorithm for the permutation flowshop scheduling problem. Eur. J. Oper. Res. **198**, 84–92 (2009)
23. Tseng, L.-Y., Lin, Y.-T.: A genetic local search algorithm for minimizing total flowtime in the permutation flowshop scheduling problem. Int. J. Prod. Econ. **127**, 121–128 (2010)
24. Yang, X.-S.: Nature-Inspired Metaheuristic Algorithms, 2nd edn. Luniver Press, London (2010)
25. Zhang, C., Sun, J., Zhu, X., Yang, Q.: An improved particle swarm optimization algorithm for flowshop scheduling problem. Inf. Process. Lett. **108**, 204–209 (2008)
26. Zhang, C., Ning, J., Ouyang, D.: A hybrid alternate two phases particle swarm optimization algorithm for flow shop scheduling problem. Comput. Ind. Eng. **58**, 1–11 (2010)
27. Zobolas, G.I., Tarantilis, C.D., Ioannou, G.: Minimizing makespan in permutation flow shop scheduling problems using a hybrid metaheuristic algorithm. Comput. Oper. Res. **36**, 1249–1267 (2009)

Studying the Impact of Perturbation Methods on the Efficiency of GVNS for the ATSP

Christos Papalitsas[1]([⊠]) [iD], Theodore Andronikos[1],
and Panagiotis Karakostas[2] [iD]

[1] Department of Informatics, Ionian University, 7 Tsirigoti Square, Corfu, Greece
{c14papa,andronikos}@ionio.gr
[2] Department of Applied Informatics, University of Macedonia,
Thessaloniki, Greece
pkarakostas.tm@gmail.com

Abstract. In this work we examine the impact of three shaking procedures on the performance of a GVNS metaheuristic algorithm for solving the Asymmetric Travelling Salesman Problem (ATSP). The first shaking procedure is a perturbation method that is commonly used in the literature as intensified shaking method. The second one is a quantum-inspired shaking method, while the third one is a shuffle method. The shaped GVNS schemes are tested with both first and best improvement and with a time limit of one and two minutes. Experimental analysis shows that the first two methods perform equivalently and much better than the shuffle approach, when using the best improvement strategy. The first method also outperforms the other two when using the first improvement strategy, while the second method produces results that are closer to the results of the third in this case.

Keywords: Metaheuristics · VNS · GVNS · Optimization · TSP · aTSP · Perturbation comparisons · Performance study

1 Introduction

A whole class of problems of practical importance can be reduced to Combinatorial Optimization (CO) problems. In such problems one typically searches for a solution from a discrete finite set of feasible solutions that achieves the minimization (or maximization) of a cost function and at the same time satisfies certain given constraints. One of the most famous CO problems is the Travelling Salesman Problem (TSP). Solving the TSP amounts to finding the minimum cost route so that the salesman starts from a specific node and returns to this node after passing from all other nodes once. The TSP was first expressed mathematically by Hamilton and Kirkman [29]. A *cycle* in a graph is a closed path beginning and ending at the same node and visiting all other nodes exactly once. A cycle containing all vertices of a graph is called *Hamiltonian*. Hence, TSP is

© Springer Nature Switzerland AG 2019
A. Sifaleras et al. (Eds.): ICVNS 2018, LNCS 11328, pp. 287–302, 2019.
https://doi.org/10.1007/978-3-030-15843-9_22

the problem of finding the shortest Hamiltonian cycle. TSP, which is NP-hard, features prominently in many fields such as operational research and theoretical computer science. Practically, TSP seeks the optimal way one can visit all the cities, return to the starting point, and minimize the cost of the tour.

TSP is usually formulated in terms of a complete graph $G = (V, A)$, where $V = \{v_1, v_2, \ldots, v_n\}$ is the set of nodes and $A = \{(v_i, v_j) : v_i, v_j \in V$ and $v_i \neq v_j\}$ is the set of the directed edges or arcs. Each arc is associated with a weight c_{ij} representing the cost (or the distance) of moving from node i to node j. If c_{ij} is equal to c_{ji}, the TSP is symmetric (sTSP), otherwise it is called asymmetric (aTSP). The fact that TSP is NP-hard means that there is no known polynomial-time algorithm for finding an optimal solution regardless of the size of the problem instance [22], Therefore, in an effort to improve the computational time, it is a commonly accepted practice the sacrifice of the optimality of the solution by adopting heuristic and metaheuristic approaches [15,17,25].

The main contribution of this paper is the thorough investigation of different perturbation strategies for the GVNS schema and their impact on the quality of the solutions. We have made a performance analysis and we have concluded that in fact different shaking methods provide different sets of solutions. Our performance analysis has specifically focused on the asymmetric TSP. The main rationale behind our decision is that known solvers such as Concorde are not designed to solve asymmetric TSP benchmarks. For that reason we have chosen to make a comprehensive analysis based on aTSP benchmarks.

This paper is organized as follows. Related work is presented in Sect. 2; in Sect. 3 we introduce the Variable Neighborhood Search procedure, we describe the shaking strategies we investigate and present the three different GVNS variations we use for this performance analysis. Section 4 contains the experimental results of our performance analysis, which are presented in a series of Tables that demonstrate the performance of each specific implementation. Finally, conclusions and ideas for future work are given in Sect. 5.

2   Related Work

Researchers have always tried to solve real world problems using methods and techniques from CO problems. Recently, a new trend has gained momentum: researchers have been striving to enhance conventional optimization methods by introducing principles from unconventional methods of computation in the hope that they would prove superior to traditional approaches. For example, Dey et al. [5] proposed several novel techniques which they called quantum inspired Ant Colony Optimization, quantum inspired Differential Evolution and quantum inspired Particle Swarm Optimization, respectively, for Multi-level Colour Image Thresholding. These techniques find optimal threshold values at different levels of thresholding for colour images.

Variable Neighborhood Search (VNS) based solutions have been applied to route planning problems. Sze et al. proposed a hybrid adaptive variable neighborhood search algorithm for solving the capacitated vehicle routing problem (capacitated VRP) [26]. A two level VNS heuristic has been developed in order to tackle the clustered VRP by Defryn and Sorensen [4]. In [10] a VNS approach for the solution of the recently introduced Swap-Body VRP is proposed. Curtin et al. made an extensive comparative study of well known methods and ready-to-use software and they concluded that no software or classic method can guarantee an optimal solution to the TSP problems that model GIS problem with more than 25 nodes [3]. Papalitsas et al. proposed a GVNS approach for the TSP with Time windows [17] and a quantum inspired GVNS (qGVNS) for solving the TSP with Time Windows [19]. Furthermore, qGVNS was successfully applied to the real world routing problem of garbage collectors at [18].

A new quantum inspired Social Evolution algorithm was proposed by hybridizing a well-known Social Evolution algorithm with an emerging quantum-inspired evolutionary one. The proposed QSE algorithm was applied to the 0-1 knapsack problem and the performance of the algorithm was compared to various evolutionary, swarm and quantum inspired evolutionary algorithmic variants. Pavithr and Gursaran claim that the performance of the QSE algorithm is better than or at least comparable to the different evolutionary algorithmic variants it was tested against [21].

Fang et al. proposed a decentralized form of quantum-inspired particle swarm optimization with a cellular structured population for maintaining population diversity and balancing global and local search [6]. Zheng et al. conducted an interesting study by applying a novel Hybrid Quantum Inspired Evolutionary Algorithm to a permutation flow-shop scheduling problem. They proposed a simple representation method for the determination of job sequence in the permutation flow-shop scheduling problem based on the probability amplitude of qubits [32].

Lu et al. designed a quantum inspired space search algorithm in order to solve numerical optimization problems. In their algorithm, the feasible solution is decomposed into regions in terms of quantum representation. The search progresses from one generation to the next, while the quantum bits evolve gradually to increase the probability of region selection [13]. Wu et al. in [31] proposed a novel approach using a quantum inspired algorithm based on game-theoretic principles. In particular, they reduced the problem they studied to choosing strategies in evolutionary games. Quantum games and their strategies seem very promising, offering enhanced capabilities over classic ones [7]. Moreover, the proposed method can also be applied to other classes of real world problems, such as optimization on localization in hospitals, smart cities, as well as smart parking systems etc.

Tsiropoulou et al. applied RFID technologies to tag-to-tag communication paradigms in order to achieve improved energy-efficiency and operational effectiveness [27]. Liebig et al. presented a system for trip planning that consolidates future traffic threats [12]. Specifically, this system measures traffic flow in areas with low sensor coverage by using a Gaussian Process Regression. Many studies also deal with the optimization of localization and positioning of doctors and nurses in hospitals and health care organizations [28,30].

3 Solution Method

3.1 Variable Neighborhood Search

Variable Neighborhood Search (VNS) is a metaheuristic for solving combinatorial and global optimization problems, proposed by Mladenovic and Hansen [8,9,14]. The main idea of this framework is the systematic neighborhood change in order to achieve an optimal (or a close-to-optimal) solution [16]. VNS and its extensions have proven their efficiency in solving many combinatorial and global optimization problems [11].

Each VNS heuristic consists of three parts. The first one is a shaking procedure (diversification phase) used to escape local optimal solutions. The next one is the neighborhood change move, in which the following neighborhood structure that will be searched is determined; during this part, an approval or rejection criterion is also applied on the last solution found. The third part is the improvement phase (intensification) achieved through the exploration of neighborhood structures through the application of different local search moves. Variable Neighborhood Descent (VND) is a method in which the neighborhood change procedure is performed deterministically. General Variable Neighborhood Search (GVNS) is a VNS variant where the VND method is used as the improvement procedure. GVNS has been successfully tested in many applications, as several recent works have demonstrated [23,24].

In this work a GVNS method is applied for solving the ATSP, using the pipe-VND scheme (keep searching in the same neighborhood as improvements occur) as its improvement phase.

3.2 Neighborhood Structures

Three local search operators are considered for exploring different solutions:

- **1-0 Relocate.** This move removes node i from its current position in the route and re-inserts it after a selected node b.
- **2-Opt.** The 2-Opt move breaks two arcs in the current solution and reconnects them in a different way.
- **1-1 Exchange.** This move swaps two nodes in the current route.

All three neighborhood structures ($l_{max} = 3$) are incorporated in a pipe-VND scheme, as illustrated in Sect. 3.2.

3.3 Shaking Methods

In order to avoid local optimum traps, three different shaking procedures are examined. These perturbation methods are the following:

Shake_1. This diversification method randomly selects one of the predefined neighborhood structures and applies it k times ($1 < k < k_{max}$) in the current solution. The method is summarized in Algorithm 2.

Algorithm 1. pipe-VND

```
1: procedure PVND(N, l_max)
2:     l = 1
3:     while l <= l_max do
4:         select case(l)
5:             case(1) : S' ← 1-0 Relocate(S)
6:             case(2) : S' ← 2-Opt(S)
7:             case(3) : S' ← 1-1 Exchange(S)
8:         end select
9:         if f(S') < f(S) then
10:            S ← S'
11:        else
12:            l = l + 1
13:        end if
14:    end while
15:    return S
16: end procedure
```

Algorithm 2. Shake_1

```
procedure SHAKE_1(S, k_max)
    l = random_integer(1, l_max)

    for k ← 1, k_max do
        select case(l)
        case(1)
        S' ← 1-0 Relocate(S)
        case(2)
        S' ← 2-Opt(S)
        case(3)
        S' ← 1-1 Exchange(S)
        end select
    end for
    return S'
```

Shake_2 [20]. In each call of this shaking method, a number of required qubits are generated by a quantum register (e.g., $N \leq 2^n$, where n is the number of nodes in problem) and they produce the corresponding components, according to the problem's dimension. These components must be equal or greater than the number of the nodes in the tour. Then, all the required components are placed in a $1 \times n$ vector. In addition, each one of the selected components corresponds to a node of the current solution. The components are to used as a flag for each node of the incumbent solution (note: components can be $0 \leq C \leq 1$). Because of the matching between components and nodes in a tour, sorting the first vector affects the order in the solution vector and, consequently, drives the exploration effort to another point in the search space. The pseudocode of this shaking procedure is given in Algorithm 3.

Algorithm 3. Shake_2

---

procedure SHAKE_2(S, n)
    $NQubits \leftarrow$ **QuantumRegister**(n)

    Compute the components based to the qubits.

    Save the n components in the vector $QCompVector$.

    Matching each element in the $QCompVector$ with a node in S.

    Descending sorting on $QCompVector$ produces S'.

    Recalculate the cost of the new S'.
return S'

Shake_3. This shaking method acts like a shuffle method. In each call the position of each node in the new route is selected randomly. The method is given in Algorithm 4.

Algorithm 4. Shake_3

---

procedure SHAKE_3(S)
$S' \leftarrow Shuffle(S)$
return S'

3.4   GVNS Schemes

For each shaking method a GVNS scheme is formed. More specifically, the GVNS_1 contains Shake_1 as its shaking method, GVNS_2 uses Shake_2 in order to diversify solutions, and GVNS_3 adopts the perturbation method Shake_3. The initial solution is produced by the Nearest Neighbor heuristic in all GVNS schemes. The pseudocode for all three GVNS approaches is given in Algorithms 5, 6 and 7, respectively.

Algorithm 5. GVNS_1

---

procedure GVNS_1(S, k_{max}, max_time)
    while $time \leq max\_time$ **do**

        $S^* = $ Shake_1(S, k_{max})

        $S' = pVND(S^*)$

        if $f(S') < f(S)$ **then**
            $S \leftarrow S'$
        end if

    end while

    return S

Algorithm 6. GVNS_2

---

  procedure GVNS_2(S, n, max_time)

    while $time \leq max\_time$ **do**

       $S^* = \text{Shake\_2}(S, n)$

       $S' = pVND(S^*)$

       if $f(S') < f(S)$ **then**
         $S \leftarrow S'$
       end if

    end while

    return S

Algorithm 7. GVNS_3

---

  procedure GVNS_3(S, max_time)

    while $time \leq max\_time$ **do**

       $S^* = \text{Shake\_3}(S)$

       $S' = pVND(S^*)$

       if $f(S') < f(S)$ **then**
         $S \leftarrow S'$
       end if

    end while

    return S

It should be mentioned that the neighborhoods in all three GVNS methods are computed using both the first and best improvement strategy.

4 Computational Analysis

4.1 Computing Environment and Parameter Settings

The aforementioned methods were implemented in Fortran and were executed on a laptop PC running Windows 10 Home 64-bit with an Intel Core i7-6700 CPU at 2.6 GHz and 16 GB RAM. The compilation of codes was done using Intel Fortran 64 compiler XE with optimization option/O3. The maximum execution time limit was set to $max\_time = 60$ s or $max\_time = 120$ s and the maximum number of the random jumps in the Shake_1 was experimentally set to $k_{max} = 3$.

4.2   Computational Results

This section presents the computational results for the different perturbation strategies for each class of experiments. All experiments ran 5 times and the average value of all runs was computed.

Tables 1 and 2 contain the aggregated experimental results. Specifically, they contain the benchmark name, the optimal value (zOpt), the cost of the three GVNS variations (GVNS_1, GVNS_2 and GVNS_3) and the GAPs from the optimal value. Table 1 depicts GVNS using first improvement and execution time of 1 min. Table 2 shows GVNS using best improvement and execution time of 1 min. The cost of each GVNS variation is the average of the 5 runs for each problem. The GAP is computed as follows: given the outcome x, its gap from the optimal value OV is given by the formula $\frac{x-OV}{OV}$. The gap is widely used in the field of Optimization to measure how close a particular solution is to the optimal. The data from the experiments demonstrate that both GVNS_1 and GVNS_2 outperform GVNS_3 in most cases. Recall that GVNS_3 is based on a suffle-wise perturbation strategy. For example, for benchmark ftv47 we can see that GVNS_1's cost is 1821, GVNS_2's is 1992 and GVNS_3's is 2101. GVNS_1 and GVNS_2 both outperform GVNS_3 and are also near the optimal (1778).

Table 1. Perturbation impact on FI for 1 min runs

Instance	zOpt	GVNS_1	GVNS_2	GVNS_3	GAP_1	GAP_2	GAP_3
br17.atsp	39	39	39	39	0.00	0.00	0.00
ft53.atsp	6905	7189	7328	7737	4.11	6.13	12.05
ft70.atsp	38673	39782	40691	40537	2.87	5.22	4.82
ftv33.atsp	1286	1318	1339	1450	2.49	4.12	12.75
ftv35.atsp	1473	1484	1499	1596	0.75	1.77	8.35
ftv38.atsp	1530	1546	1585	1579	1.05	3.59	3.20
ftv44.atsp	1613	1651	1760	1797	2.36	9.11	11.41
ftv47.atsp	1778	1821	1992	2101	2.42	12.04	18.17
ftv55.atsp	1608	1666	1985	1912	3.61	23.45	18.91
ftv64.atsp	1839	1961	2382	2395	6.63	29.53	30.23
ftv70.atsp	1950	2136	2557	2484	9.54	31.13	27.38
ftv170.atsp	2755	3487	3923	3923	26.57	42.40	42.40
kro124p.atsp	36230	39024	43187	40259	7.71	19.20	11.12
p43.atsp	5620	5620	5623	5658	0.00	0.05	0.68
rbg323.atsp	1326	1516	1563	1626	69.61	17.27	102.11
rbg358.atsp	1163	1347	1437	1404	80.40	22.27	136.89
rbg403.atsp	2465	2535	2587	2565	9.78	4.42	11.76
rbg443.atsp	2720	2814	2859	2814	3.46	5.11	3.46
ry48p.atsp	14422	14549	14901	14738	0.88	3.32	2.19
Average	**6599.74**	**6920.26**	**7328.37**	**7190.21**	**12.33**	**12.64**	**24.10**

Table 2. Perturbation impact on BI for 1 min runs

Instance	zOpt	GVNS_1	GVNS_2	GVNS_3	GAP_1	GAP_2	GAP_3
br17.atsp	39	39	39	39	0.00	0.00	0.00
ft53.atsp	6905	7043	7135	7674	2.00	3.33	11.14
ft70.atsp	38673	39507	40206	40539	2.16	3.96	4.83
ftv33.atsp	1286	1289	1286	1379	0.23	0.00	7.23
ftv35.atsp	1473	1476	1473	1533	0.20	0.00	4.07
ftv38.atsp	1530	1538	1541	1599	0.52	0.72	4.51
ftv44.atsp	1613	1632	1644	1728	1.18	1.92	7.13
ftv47.atsp	1778	1792	1816	1940	0.79	2.14	9.11
ftv55.atsp	1608	1642	1665	2012	2.11	3.54	25.12
ftv64.atsp	1839	1908	1986	2193	3.75	7.99	19.25
ftv70.atsp	1950	2110	2157	2346	8.21	10.62	20.31
ftv170.atsp	2755	3341	3852	3923	21.27	39.82	42.40
kro124p.atsp	36230	36501	37076	38195	0.75	2.34	5.42
p43.atsp	5620	5620	5620	5627	0.00	0.00	0.12
rbg323.atsp	1326	1486	1539	1633	107.77	107.77	107.77
rbg358.atsp	1163	1307	1409	1437	136.89	136.89	136.89
rbg403.atsp	2465	2510	2547	2554	11.76	11.76	11.76
rbg443.atsp	2720	2765	2824	2844	1.65	3.16	4.56
ry48p.atsp	14422	14480	14498	14659	0.40	0.12	1.64
Average	**6599.74**	**6736.11**	**6858.58**	**7044.95**	**15.88**	**17.69**	**22.28**

Table 3. Perturbation impact on FI for 2 min runs

Instance	zOpt	GVNS_1	GVNS_2	GVNS_3	GAP_1	GAP_2	GAP_3
br17.atsp	39	39	39	39	0.00	0.00	0.00
ft53.atsp	6905	7024	7498	7752	1.72	8.59	12.27
ft70.atsp	38673	39615	40827	40505	2.44	5.57	4.74
ftv33.atsp	1286	1330	1370	1454	3.42	6.53	13.06
ftv35.atsp	1473	1482	1519	1604	0.61	3.12	8.89
ftv38.atsp	1530	1547	1618	1576	1.11	5.75	3.01
ftv44.atsp	1613	1628	1839	1812	0.93	14.01	12.34
ftv47.atsp	1778	1787	2020	2097	0.51	13.61	17.94
ftv55.atsp	1608	1668	2012	1912	3.73	25.12	18.91
ftv64.atsp	1839	1951	2484	2476	6.09	35.07	34.64
ftv70.atsp	1950	2165	2571	2484	11.03	31.85	27.38
ftv170.atsp	2755	3412	3923	3923	23.85	42.40	42.40
kro124p.atsp	36230	39344	44243	40849	8.60	22.12	12.75
p43.atsp	5620	5620	5628	5657	0.00	0.14	0.66
rbg323.atsp	1326	1499	1576	1586	14.33	17.12	107.77
rbg358.atsp	1163	1329	1410	1406	16.34	21.93	136.89
rbg403.atsp	2465	2509	2586	2547	2.27	4.10	11.76
rbg443.atsp	2720	2808	2849	2811	3.24	4.74	3.35
ry48p.atsp	14422	14475	14936	14708	0.37	3.56	1.98
Average	**6599.74**	**6906.95**	**7418.32**	**7220.95**	**5.29**	**13.96**	**24.78**

Table 1 presents the data for the different GVNS variations for 1 min run on First Improvement. Table 2 shows the results of the GVNS variations for 1 min run on Best Improvement. The results of Table 2 lead to the same conclusion, i.e., that both GVNS_1 and GVNS_2 outperform GVNS_3 in most cases. Table 3 shows the results of the GVNS variations for 2 min run on First Improvement. The results of Table 3 demonstrate that GVNS_1 outperform GVNS_2 and GVNS_3 in most cases. However, the main difference from the results of Table 1 and Table 2 is that now the behavior of GVNS_2 is closer to GVNS_3's. Table 4 shows the results of the GVNS variations for 2 min run on First Improvement. The results of Table 4 corroborate the conclusion of Tables 1 and 2, i.e., that both GVNS_1 and GVNS_2 outperform GVNS_3 in most cases.

Table 4. Perturbation impact on BI for 2 min runs

Instance	zOpt	GVNS_1	GVNS_2	GVNS_3	GAP_1	GAP_2	GAP_3
br17.atsp	39	39	39	39	0.00	0.00	0.00
ft53.atsp	6905	7043	7207	7773	2.00	4.37	12.57
ft70.atsp	38673	39358	40230	40588	1.77	4.03	4.95
ftv33.atsp	1286	1286	1290	1370	0.00	0.31	6.53
ftv35.atsp	1473	1474	1475	1509	0.07	0.14	2.44
ftv38.atsp	1530	1538	1555	1599	0.52	1.63	4.51
ftv44.atsp	1613	1636	1664	1731	1.43	3.16	7.32
ftv47.atsp	1778	1787	1837	1903	0.51	3.32	7.03
ftv55.atsp	1608	1640	1686	2012	1.99	4.85	25.12
ftv64.atsp	1839	1914	2032	2217	4.08	10.49	20.55
ftv70.atsp	1950	2038	2189	2342	4.51	12.26	20.10
ftv170.atsp	2755	3351	3918	3923	21.63	42.21	42.40
kro124p.atsp	36230	36379	37378	37915	0.41	3.17	4.65
p43.atsp	5620	5620	5620	5625	0.00	0.00	0.09
rbg323.atsp	1326	1473	1531	1610	10.71	16.14	107.77
rbg358.atsp	1163	1292	1405	1435	9.29	20.55	136.89
rbg403.atsp	2465	2498	2547	2553	1.30	3.25	11.76
rbg443.atsp	2720	2771	2822	2842	1.88	3.75	4.49
ry48p.atsp	14422	14468	14464	14678	0.32	0.29	1.78
Average	**6599.74**	**6716.05**	**6888.89**	**7034.95**	**3.28**	**7.05**	**22.16**

4.3 Statistical Analysis on Computational Results

This section presents the statistical tests which were performed on the computational results in order to evaluate the performance of the three different GVNS methods. Different statistical tests are applied to different data structures. Particularly, statistical analysis methods can be divided on parametric and non parametric tests. The first category examines normal variables whereas the other methods concern non-normal variables [2].

Initially, the application of a normality test showed that the numerical data did not follow the normal distribution. As a consequence, we applied the Kruskal-Wallis test for checking the existence of a statistically significant difference between the methods. In this test receiving a p-value less than 0.05 means that there is statistically significant difference between the three methods.

Table 5. Kruskal-Wallis rank sum test

	X^2	df	p-value
FI_1min	6.8689	2	0.0322
FI_2mins	9.0314	2	0.0109
BI_1min	9.2739	2	0.0097
BI_2mins	9.6658	2	0.008

In Table 5 we can see that for all cases p-value is less than 0.5. So we can conclude that for all cases there is a significant statistical difference.

Table 6. KPairwise comparisons using Wilcoxon signed rank test.

FI_1min		
	GVNS1	GVNS2
GVNS2	0.00064	
GVNS3	0.00064	0.6701
FI_2mins		
	GVNS1	GVNS2
GVNS2	0.00064	
GVNS3	0.00064	0.4488
BI_1min		
	GVNS1	GVNS2
GVNS2	0.00109	
GVNS3	0.00064	0.00064
BI_2mins		
GVNS2	0.00109	
GVNS3	0.00064	0.00064

In Table 6, we examine the experiment results in pairs for each case of run. Once again if any of values in Table 6 are less 0.05 then there is significant difference between the two methods. For example, for the Best Improvement 1 min run the GVNS_1 has significant difference on performance from GVNS_2 since the value is 0.00064. However, at the First improvement 1 min run the GVNS_2 and GVNS_3 schemes have no significant differences on their performance.

As a result of this Kruskal-Wallis statistical analysis, we have the following four box plots. Each one depicts either First Improvement or Best Improvement for one minute, as well as for two minutes runs. In particular, one can observe in the figures below the median value of each method. This enables us to conclude which method gives the best results (Fig. 1).

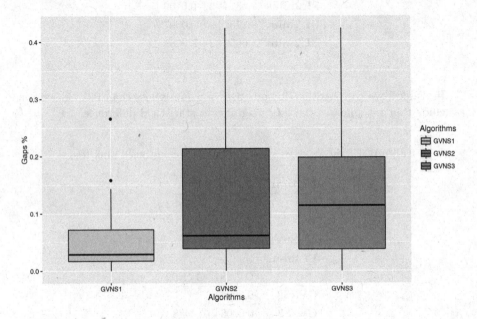

Fig. 1. Box plot for FI 1 min

In all cases GVNS_1 outperforms the other two GVNS schemes. However, by checking the medians at the box plots we can also conclude that the GVNS_2 method performs significantly better on Best improvement, producing results that are "close" to the results of GVNS_1, while on the First improvement is only slightly better than GVNS_3. GVNS_3 exhibits the worst performance in all cases according to median analysis of the box plots (Figs. 2, 3 and 4).

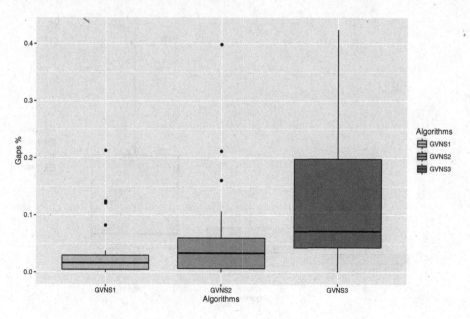

Fig. 2. Box plot for BI 1 min

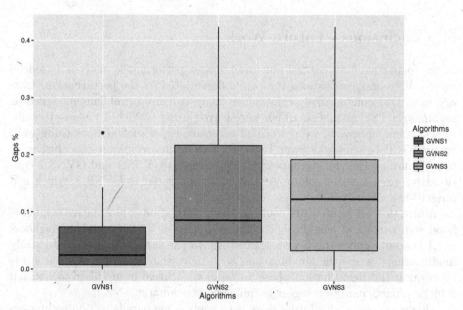

Fig. 3. Box plot for FI 2 min

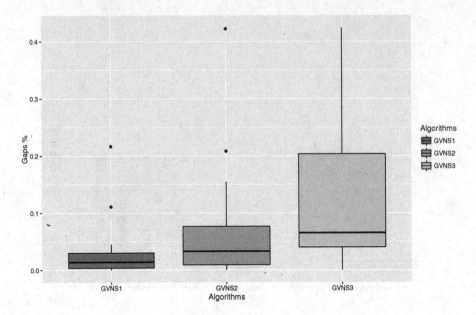

Fig. 4. Box plot for BI 2 min

5   Conclusions - Future Work

In this paper we have carried out an extensive performance analysis based on three GVNS implementations that are differentiated by the perturbation strategy used. Our comparative performance analysis ivolved problems modelled as asymmetric TSP instances, which were solved using GVNS. To assess the efficiency of our approach, we performed extensive experimental tests using well-known aTSP benchmarks from TSPLIB. The results were quite conclusive, as they confirmed that for asymmetric TSP instances GVNS_1 and GVNS_2 consistently provide better solutions in all cases compared to GVNS_3, which is a random-like strategy.

A direction for future work could be the investigation of alternative neighborhood structures and neighborhood change moves in VND (Variable Neighborhood Descent) under the GVNS framework. In the same vein, one could study modifications or specific combinations with more than one perturbation strategy during the perturbation phase in order to achieve even closer to optimal solutions, particularly on bigger asymmetric benchmarks.

Moreover, an idea for future work is to apply some parallelization techniques to accelerate the calculations of the algorithm in [1]. Finally, another possibility for future work could be a thorough study of the impact of perturbations on symmetric TSP and national TSP benchmarks and a computational analysis of the quality of the provided solutions driven from perturbation to perturbation.

References

1. Antoniadis, N., Sifaleras, A.: A hybrid CPU-GPU parallelization scheme of variable neighborhood search for inventory optimization problems. Electron. Notes Discrete Math. **58**, 47–54 (2017). 4th International Conference on Variable Neighborhood Search
2. Coffin, M., Saltzman, M.J.: Statistical analysis of computational tests of algorithms and heuristics. INFORMS J. Comput. **12**(1), 24–44 (2000). https://doi.org/10.1287/ijoc.12.1.24.11899
3. Curtin, K.M., Voicu, G., Rice, M.T., Stefanidis, A.: A comparative analysis of traveling salesman solutions from geographic information systems. Trans. GIS **18**(2), 286–301 (2014)
4. Defryn, C., Sörensen, K.: A fast two-level variable neighborhood search for the clustered vehicle routing problem. Comput. Oper. Res. **83**, 78–94 (2017)
5. Dey, S., Bhattacharyya, S., Maulik, U.: New quantum inspired meta-heuristic techniques for multi-level colour image thresholding. Appl. Soft Comput. **46**, 677–702 (2016)
6. Fang, W., Sun, J., Chen, H., Wu, X.: A decentralized quantum-inspired particle swarm optimization algorithm with cellular structured population. Inf. Sci. **330**, 19–48 (2016). https://doi.org/10.1016/j.ins.2015.09.055
7. Giannakis, K., Papalitsas, C., Kastampolidou, K., Singh, A., Andronikos, T.: Dominant strategies of quantum games on quantum periodic automata. Computation **3**(4), 586–599 (2015)
8. Hansen, P., Mladenovic, N., Todosijevic, R., Hanafi, S.: Variable neighborhood search: basics and variants. EURO J. Comput. Optim. **5**, 423–454 (2016)
9. Hansen, P., Mladenović, N., Todosijević, R., Hanafi, S.: Variable neighborhood search: basics and variants. EURO J. Comput. Optim. **5**(3), 423–454 (2017)
10. Huber, S., Geiger, M.J.: Order matters-a variable neighborhood search for the swap-body vehicle routing problem. Eur. J. Oper. Res. **263**, 419–445 (2017)
11. Jarboui, B., Derbel, H., Hanafi, S., Mladenovic, N.: Variable neighborhood search for location routing. Comput. Oper. Res. **40**(1), 47–57 (2013)
12. Liebig, T., Piatkowski, N., Bockermann, C., Morik, K.: Predictive trip planning - smart routing in smart cities. In: Proceedings of the Workshops of the EDBT/ICDT 2014 Joint Conference (EDBT/ICDT 2014), Athens, Greece, 28 March 2014, vol. 1133, pp. 331–338. CEUR-WS.org (2014)
13. Lu, T.C., Juang, J.C.: Quantum-inspired space search algorithm (QSSA) for global numerical optimization. Appl. Math. Comput. **218**(6), 2516–2532 (2011)
14. Mladenovic, N., Hansen, P.: Variable neighborhood search. Comput. Oper. Res. **24**(11), 1097–1100 (1997)
15. Mladenovic, N., Todosijevic, R., Urosevic, D.: An efficient GVNS for solving traveling salesman problem with time windows. Electron. Notes Discrete Math. **39**, 83–90 (2012)
16. Mladenovic, N., Todosijevic, R., Uroševic, D.: Less is more: basic variable neighborhood search for minimum differential dispersion problem. Inf. Sci. **326**, 160–171 (2016)
17. Papalitsas, Ch., Giannakis, K., Andronikos, Th., Theotokis, D., Sifaleras, A.: Initialization methods for the TSP with time windows using variable neighborhood search. In: IEEE Proceedings of the 6th International Conference on Information, Intelligence, Systems and Applications (IISA 2015), Corfu, Greece, 6–8 July (2015)

18. Papalitsas, C., Karakostas, P., Andronikos, T., Sioutas, S., Giannakis, K.: Combinatorial GVNS (general variable neighborhood search) optimization for dynamic garbage collection. Algorithms **11**(4), 38 (2018)

19. Papalitsas, C., Karakostas, P., Giannakis, K., Sifaleras, A., Andronikos, T.: Initialization methods for the TSP with time windows using qGVNS. In: 6th International Symposium on Operational Research, OR in the digital era - ICT challenges, Thessaloniki, Greece, June 2017

20. Papalitsas, C., Karakostas, P., Kastampolidou, K.: A quantum inspired GVNS: some preliminary results. In: Vlamos, P. (ed.) GeNeDis 2016. AEMB, vol. 988, pp. 281–289. Springer, Cham (2017). https://doi.org/10.1007/978-3-319-56246-9_23

21. Pavithr, R., Saran, G.: Quantum inspired social evolution (QSE) algorithm for 0–1 knapsack problem. Swarm Evol. Comput. **29**, 33–46 (2016)

22. Rego, C., Gamboa, D., Glover, F., Osterman, C.: Traveling salesman problem heuristics: leading methods, implementations and latest advances. Eur. J. Oper. Res. **211**(3), 427–441 (2011)

23. Sifaleras, A., Konstantaras, I.: General variable neighborhood search for the multi-product dynamic lot sizing problem in closed-loop supply chain. Electron. Notes Discrete Math. **47**, 69–76 (2015)

24. Sifaleras, A., Konstantaras, I., Mladenović, N.: Variable neighborhood search for the economic lot sizing problem with product returns and recovery. Int. J. Prod. Econ. **160**, 133–143 (2015)

25. Silva, R.F.D., Urrutia, S.: A general VNS heuristic for the traveling salesman problem with time windows. Discrete Optim. **7**(4), 203–211 (2010)

26. Sze, J.F., Salhi, S., Wassan, N.: A hybridisation of adaptive variable neighbourhood search and large neighbourhood search: application to the vehicle routing problem. Expert Syst. Appl. **65**, 383–397 (2016)

27. Tsiropoulou, E.E., Baras, J.S., Papavassiliou, S., Sinha, S.: RFID-based smart parking management system. Cyber-Phys. Syst. **3**(1–4), 22–41 (2017)

28. Van Haute, T., et al.: Performance analysis of multiple indoor positioning systems in a healthcare environment. Int. J. Health Geographics **15**(1), 7 (2016)

29. Voigt, B.F.: der handlungsreisende, wie er sein soll und was er zu thun hat, um aufträge zu erhalten und eines glücklichen erfolgs in seinen geschäften gewiss zu zu sein. Ilmenau. Neu aufgelegt durch Verlag Schramm, Kiel, Commis-Voageur (1981)

30. Woo, H., Lee, H.J., Kim, H.C., Kang, K.J., Seo, S.S.: Hospital wireless local area network-based tracking system. Healthc. Inform. Res. **17**(1), 18–23 (2011)

31. Wu, Q., Jiao, L., Li, Y., Deng, X.: A novel quantum-inspired immune clonal algorithm with the evolutionary game approach. Prog. Nat. Sci. **19**(10), 1341–1347 (2009)

32. Zheng, T., Yamashiro, M.: A novel hybrid quantum-inspired evolutionary algorithm for permutation flow-shop scheduling. J. Stat. Manag. Syst. **12**(6), 1165–1182 (2009)

A GVNS Algorithm to Solve VRP with Optional Visits

Manel Kammoun[1]([✉]), Houda Derbel[1], and Bassem Jarboui[2]

[1] MODILS, FSEGS, Route de l'aéroport km 4, 3018 Sfax, Tunisia
kamounemanel@gmail.com, derbelhouda@yahoo.fr
[2] Emirates College of Technology, Abu Dhabi, United Arab Emirates
bassem_jarboui@yahoo.fr

Abstract. In this paper we deal with a generalization of the multi-depot capacitated vehicle routing problem namely the multi-depot covering tour vehicle routing problem (MDCTVRP). This problem is considered more challenging since it deals with some situations where it is not possible to visit all the customers with the vehicles routes. In this problem, a customer can receive its demand directly by visiting it along the tour using a set of vehicles located at different depots or by covering it. A customer is considered as covered if it is located within an acceptable distance from at least one visited customer in the tour. The latter can satisfy its demand. We propose a general variable neighborhood search algorithm to solve the MDCTVRP. In this paper we use a variable neighborhood search (VNS) with a variable neighborhood descent (VND) method as a local search. Experiments were conducted on benchmark instances from the literature.

Keywords: Vehicle routing problem · Covering ·
Variable neighborhood search

1 Introduction

This paper investigates the solution of a multi-depot covering tour vehicle routing problem (MDCTVRP) by means of a general variable neighborhood search (GVNS). The Vehicle Routing Problem (VRP) has extensive variants studied in the literature. We focus on an extension of the classical multi-depot capacitated vehicle routing problem (MDVRP) in which covering option was added and each customer can be served by visiting or covering it. Obviously, our problem is more challenging than the MDVRP where all customers must be visited by the tours. The problem aims at minimizing the total cost which is composed by the following costs: The routing cost and the covering one. When the demand of customer was satisfied directly, the routing cost will be occurred by visiting the customer on the vehicles routes. Whereas, when the demand of customer was satisfied indirectly by 'covering' it, a covering cost will be associated. A customer is considered as covered if it is located within a given covering distance of at least one visited customer. Therefore, the distance travelled by the covered customers to reach their nearest destination on the route is proportional to the covering cost.

© Springer Nature Switzerland AG 2019
A. Sifaleras et al. (Eds.): ICVNS 2018, LNCS 11328, pp. 303–314, 2019.
https://doi.org/10.1007/978-3-030-15843-9_23

To the best of our knowledge, the MDCTVRP has been introduced for the first time by [1] where they developed two mixed integer programming formulations (flow-based and node-based formulation). They developed a hybrid meta-heuristic combining a greedy randomized adaptive search procedure (GRASP), iterated local search (ILS) and simulated annealing (SA). Since then, this problem has not been addressed in the literature. The MDCTVRP is a combination between two problems namely the MDVRP and the covering salesman problem (CSP). The covering problems receive recently more attention as for example covering tour problem (CTP) which attract so much attention than the classical routing problem. Recently, [2] solve the mutli-vehicle version of the CTP (m-CTP) exactly by a branch ant cut algorithm where they propose a new integer programming formulation based on a two-commodity flow model. They propose also a meta-heuristic based on evolutionary local search method (ELS) which provide good results and the solution is within 1.45% of optimality for the test instances. Few years later, GVNS algorithm was proposed by [6] to solve the same problem. The computational results show the effectiveness of the proposed method to solve small and large sets of instances in a reasonable time. The provided results is outperform those in [2]. More recently, [7] propose two meta-heuristics to solve a generalized variant of the m-CTP called the multi-vehicle multi-Covering Tour Problem (mm-CTP-p) where a customer must be covered several times to be completely served. In the MDCTVRP, each vertex must be along the tour or covered by already visited customers. Whereas, in the m-CTP, the set of customer vertices is divided into two groups where we need to visit a subset of vertices from the first group in order to cover all the set of customers from the second group.

This paper is organized as follows: Sect. 2 describes the MDCTVRP problem. Section 3 presents the proposed algorithm. Section 4 compares the performance of the proposed algorithm with an existing hybrid approach in the literature that combine GRASP, iterated local search and simulated annealing. Finally, Sect. 5 concludes the paper and presents some perspectives.

2   Problem Description

The MDCTVRP is defined by [1] as a direct graph $G = (N, A)$ where N represents the set of vertices, $N = N_c \cup N_D$. More precisely, N_c is the set of customers $N_C = 1, 2, ..., n_c$ and N_D is the set of depots, $N_D = 1, 2, ..., n_d$. Let A be the set of arcs $A = (i, j)/i, j \in N$, a routing cost c_{ij} was associated with traversing the $arc(i, j) \in A$ using the vehicle v and c'_{ij} is the allocation cost of customer i to the visited customer j for each $i, j \in N_C$. The main goal of the MDCTVRP is to minimize the total cost in such a way the entire demands of the customers were satisfied. In this problem it is not necessary that each customer is visited by a vehicle due to the introduction of the covering option. A customer is instead covered when it is located within an acceptable distance from at least one visited customer from which it can receive its demand. A fleet of homogenous vehicles is located at each depot. All the depots have a capacity H and each vehicle is also characterized by his own capacity Q. A solution representation of the studied problem is given in the figure below:

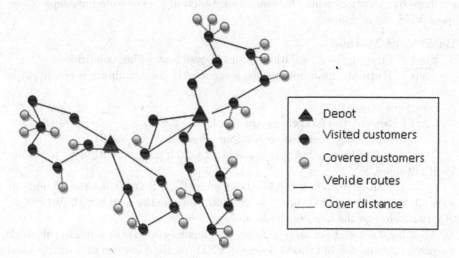

Fig. 1. A solution representation of the MDCTVRP.

An example of a feasible solution for the MDCTVRP is presented in Fig. 1. We have 2 depots, where two vehicles are located to the first depot and three vehicles are located to the second one. We have 47 customers; overall 25 of them are visited by the tours and used to cover the demand of the remaining unvisited customers.

3   General Variable Neighborhood Search Approach for the MDCTVRP

The variable neighborhood search (VNS) proves its effectiveness to solve various combinatorial optimization problems. The VNS is a meta-heuristic approach based on a systematic change of neighborhood structures within the local search algorithm ([3–5]). The basic VNS combines both deterministic and stochastic ingredients. Unlike many other meta-heuristics, the VNS is simple to understand and easy to implement since it requires few parameters.

The VNS is divided into two phases: The descent phase is used to find a local optimum by exploring different neighborhoods of the solution. The second phase avoids the search to be trapped in a local optimum. A perturbation is then used to escape from the corresponding valley. Despite its simplicity, VNS is able to produce a solution of high quality in a reasonable time.

The basic VNS method started by a perturbation phase where a neighbor of the current solution was selected, followed by running a local search to reach a local optimum, and then moved to the new solution if there has been an improvement. In the following, we give a formal description of the different steps of the basic VNS method. Let $N_k, k = 1...k_{max}$ be the set of neighborhoods and $N_k(S)$ be the neighbors of a solution S via a neighborhood structure N_k. We evaluate our solution using the evaluation function f. The different steps of the basic VNS are as follow:

Basic VNS Method

Step 1: Find an initial solution S and choose a stopping condition.

Step 2: **Repeat** the following sequence **Until** the termination condition is met:

    (i) set $k \longleftarrow 1$

    (ii) **Repeat** the following sequence **Until** $k = k_{max}$:

        (a) *Shaking*: Generate a random neighbor S' in $N_k(S)$.

        (b) *Local Search*: Apply some local search method within S' to obtain local optimum S''.

        (c) *Move or not*: if the local optimum S'' is better than the incumbent solution S ($f(S'') \le f(S)$) then $S \leftarrow S''$ and continue the search with $N_1(k \leftarrow 1)$ otherwise change the neighborhood and set $k \leftarrow k + 1$.

Most local search heuristics use only one neighborhood structure. In this work we use a variable neighborhood descent (VND) method instead of a simple local search to obtain a GVNS algorithm. The different steps of our GVNS algorithm are described as follows: First, we generate randomly an initial feasible solution. We start by inserting randomly the customer that cannot be covered within a predefined distance β. Then, we select the vertex that cover a maximum number of vertices and we insert it in a set, denoted by I, randomly. We repeat this process until a feasible solution is reached. Finally, we obtain the set of visited vertices I and I_1, the set of covered vertices in the solution where all customers were served. Inserting the customer that cannot be covered in the beginning aims to minimize the size of problem and in consequence to minimize computational time.

Let $N_k, k = 1...k_{max}$ be the set of neighborhoods used in the GVNS algorithm and $N_k(S)$ be the neighbors of a solution S via a neighborhood structure N_k. In our GVNS algorithm five neighborhood structures have been considered.

- Insertion: This operator is performed in the same route or between two different routes denoted respectively by the neighborhoods N_1 and N_2. For the first one, a neighbor of a solution is obtained by removing a customer and inserting it into a new position. However, the second neighborhood was performed between two different routes in which one customer is removed from its position and inserted into another route. The neighborhood N_2 changes not only the order of customers in the route but also the number of customers in the two routes. If the inserted customer has some vertices that are allocated to it, such allocation will be kept.

- Swap: The swap move attempts to swap the positions of each pair of vertices in the solution. It is performed by randomly selecting two customers, from the same route or from two different routes and exchanging them. This move represented respectively by the neighborhoods N_3 and N_4. In addition, if one or both of the swapped vertices have some customers allocated to them, such allocation will be kept.

- Remove: It consists on removing a set of vertices from the set I and denoted by N_5. this move is more detailed in the sequel.

The GVNS is composed by three main phases:

Shaking Phase: In this step, a restricted solution S' is obtained from the initial solution S_0 by removing randomly a given set of customers. In this procedure we use the neighborhood structure N_5 to allow diversification of the search. More precisely, we remove l vertices from $I \setminus N_{mc}$ randomly where $N_{mc} \subset N_c$ represents the set of customers that cannot be covered and so they always must be part of the solution. If the removed customer has some covered customers that could be covered by another visited ones, we assigned them to the best feasible covered position in S' else we removed each customer which cannot be covered from the solution.

Denote by R the set of customers that do not belong to the solution, $R = N_c \setminus I, I_1, N_{mc}$. We explore new solutions by inserting randomly each time the vertex from R that maximizes the number of covered vertices. Then, we insert each customer that can be covered by the inserted vertices in the best feasible position in S'. Note that the search continues until the solution is feasible.

Algorithm 1. Shake (S,l)

1: Let I be the set of visited vertices in an initial solution S;
2: Remove l vertices randomly from $I \setminus N_{mc}$;
3: **repeat**
4:    Select $i \in R$ where $R = N_c \setminus \{I, I_1, N_{mc}\}$ that maximizes the number of covered vertices in the solution;
5:    Insert i in I randomly;
6: **until** a feasible solution;

The Local Search Phase: In this phase different neighborhood structures were used and a mixed variable neighborhood descent (Mixed-VND) was considered. The mixed-VND algorithm is based on two components: a swap move and a variable neighborhood descent (VND) method.

As already mentioned I is the set of visited vertices and I_1 represents the set of covered vertices in S'. We attempt to determine a better feasible solution by applying a mixed-VND for each vertex in I. More precisely, we apply a swap move in S_0 to get S_1 as follow: We select a vertex i from I and a vertex j from J_i where J_i is the subset of covered vertices in which we can apply a swap move between i and j without a loss of feasibility of the solution $J_i \subset I_1$. Then we apply a variable neighborhood descent (VND) method to improve the solution S_1.

The VND methods consists on finding the best neighbor $S"$ of the solution S_1, $S" \in N_k(S_1)$. If the obtained solution $S"$ is better than S_1 then $S_1 \leftarrow S"$

and $k = 1$ otherwise the VND continues the search with the next neighborhood $N_{(k+1)}$. In the following we present our VND algorithm (see Algorithm 2).

Algorithm 2. VND (S)

1: Input : An initial solution S, Neighborhood structures $N_k, k = 1...k_{max}$;
2: $k = 1$;
3: **repeat**
4:      $S' \leftarrow$ the best neighbor of S;
5:      **if** $f(S') < f(S)$ **then**
6:          $S \leftarrow S'$;
7:          $k \leftarrow 1$;
8:      **else**
9:          $k \leftarrow k + 1$;
10:     **end if**
11: **until** $k > k_{max}$;

Move or Not: Let f be the evaluation function used in our algorithm. In this step, we evaluate the solution and we decide to make a move and continue the search with the first neighborhood structure if the solution was improved ($S_0 \leftarrow S', k = 1$) else we continue with the next neighborhood structure ($k = k + 1$).

The entire proposed GVNS algorithm is summarized in Algorithm 3.

4   Computational Experiments

The experiments are performed on laptop ASUS Intel Core i5-4200U and 2.3 Ghz processor and 6 GB memory and the proposed algorithm has been coded in C++ programming language. We performed extensive computational tests to evaluate the performance of our algorithm. Our algorithm is then tested on the benchmark small-size and Large-size instances where the three sizes of instances are 20, 25 and 30 vertices. Each instance is characterized by the number of customers ($|N_c|$) and the number of depots ($|N_D|$). For each depot a capacity (H) was assigned and a maximum number of vehicles ($|P_k|$) to each depot $k \in N_D$ was defined. Each vehicle has a capacity (Q).

Tables 1 and 2 reports our results and compare them with the hybrid meta-heuristic proposed by [1]. [1] test the effectiveness of their proposed meta-heuristic on a PC running at 2.93 GHz with 3.21 GB of RAM. They designed a set of instances where the small-size instances are composed by three categories and the Large-size instances are composed by fours categories. Each category contains eight different groups of instances. For each group of data, five random instances are generated. The instances are labeled $XYZT$, where "X" represents the category of the problem, and "Y", "Z" and "T" represent the scenarios corresponding to the number of depots, capacity of the vehicles and the coverage coefficient respectively (for further information see [1]). Column "**Best**",

Algorithm 3. GVNS for MDCTVRP

1: Input: initial solution S_0;
2: Initialize: max number of iteration $iter_{max} = 1000$ and the number of neighborhood structures $k_{max} = 5, iter = 1$;
3: Begin
4: **while** $(iter < iter_{max})$ **do**
5:   k= 1;
6:   **repeat**
7:     $S' = Shake(S_0, k_1)$;
8:     Let $I \in N_c in S'$ and $I_1 = N_c \ I, N_{mc}$;
9:     **repeat**
10:       Select $i \in I$;
11:       Let $J_i \subset I_1$;
12:       **for** $j \in J_i$ **do**
13:         Swap move between i and j to get S_1;
14:         $S'' = VND(S_1)$;
15:         **if** $f(S'') < f(S')$ **then**
16:           $S' \leftarrow S''$;
17:           Update I;
18:           Go to Line 10;
19:         **end if**
20:       **end for**
21:     **until** No possible improvement;
22:     **if** $f(S') < f(S_0)$ **then**
23:       $S_0 \leftarrow S'$;
24:       k= 1;
25:     **else**
26:       k= k + 1;
27:     **end if**
28:   **until** $k = k_{max}$;
29:   $iter = iter + 1$;
30: **end while**;
31: End;

"**Worst**" and "**Avg.**" show respectively, the best, the worst and the average solution obtained over the five run. Column "**Best.Time**" and "**Worst.time**" represent, respectively, the running time of the best and worst solution whereas, the column "**Time**" shows the total running time. Columns "**Bestgap**" and "**Avg.gap**" report, respectively, the best and the average gap solution with respect to the best solution found by the hybrid meta-heuristic (GRASP × ILS) proposed in the literature [1]).

Let BKS be the value of the best known solution and UB is the value obtained by our GVNS algorithm. The percentage deviation of our method, (Gap_{UB}) is computed as follows:

$$Gap_{UB} = 100 . \frac{BKS - UB}{UB} \tag{1}$$

Our results show that our method performs much better than the GRASP×ILS with respect to the best feasible solution. Particularly, the overall gap over 120 instances is 7.75%. But the GRASP × ILS performs well for the average solution for which the overall gap over 120 instances is 3.17%. However, our algorithm cannot provide good results for Large-size instances. More precisely, the overall gap over 160 instances with respect to the best feasible solution is 8.39% and 18.60% when the comparison is made in relation to the average results.

We observe in Fig. 2 that the GVNS algorithm presents the best feasible solution for almost all small-size instances but cannot provide good quality of solution for Large-size instances. Bisides, Fig. 3 shows the performance of the hybrid meta-heuristic algorithm comparing with our GVNS algorithm with respect to the average solution.

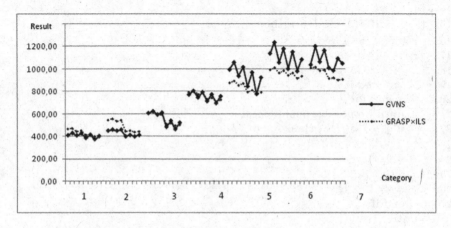

Fig. 2. The best feasible solution for the GVNS and GRASP×ILS.

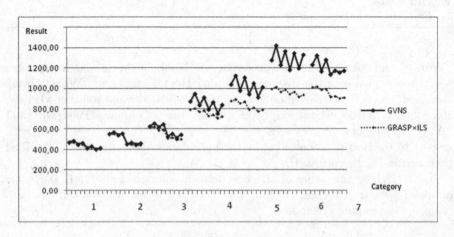

Fig. 3. The average solution for the GVNS and GRASP×ILS.

Table 1. Comparison of the GVNS meta-heuristic and GRASP × ILS methods on Small-size instances.

Category	Group	GVNS						GRASP × ILS			Best gap	Avg. gap
		Best	Best time	Worst	Worst time	Avg.	Avg. time	Best	Avg.	Avg. time		
Small	Input1000	409.30	0.03	527.71	2.19	472.71	0.74	468.34	468.34	3.52	14.42	−0.93
	Input1001	428.56	0.02	537.39	0.06	486.58	0.29	474.38	474.38	3.60	10.69	−2.51
	Input1010	407.70	0.16	503.09	0.02	451.38	0.98	445.88	445.88	3.44	9.36	−1.22
	Input1011	426.23	0.06	512.77	0.17	465.53	2.67	453.27	453.27	3.44	6.34	−2.63
	Input1100	385.07	0.00	455.85	0.02	416.72	0.05	414.08	414.08	3.44	7.53	−0.63
	Input1101	414.11	0.03	465.54	0.08	434.52	0.26	419.4	419.4	3.60	1.28	−3.48
	Input1110	373.74	0.43	443.41	0.36	404.49	0.32	401.36	401.36	3.21	7.39	−0.77
	Input1111	401.50	0.04	453.09	0.00	418.40	0.62	406.53	406.53	3.28	1.25	−2.84
	Input2000	449.54	0.22	627.65	0.35	555.27	0.24	546.98	546.98	4.22	21.67	−1.49
	Input2001	459.49	0.21	648.42	0.06	572.01	0.58	557.95	557.98	4.14	21.43	−2.45
	Input2010	449.54	0.01	616.97	0.08	542.34	0.49	536.68	537.39	4.22	19.38	−0.91
	Input2011	459.49	0.56	639.26	0.45	558.88	2.53	546.19	546.37	4.69	18.87	−2.24
	Input2100	398.95	1.62	457.88	1.18	456.39	0.86	447.1	447.1	3.99	12.07	−2.04
	Input2101	410.46	2.02	571.02	0.01	470.28	0.46	452.79	452.87	3.91	10.31	−3.70
	Input2110	398.20	3.25	541.33	0.02	450.53	0.80	440.07	440.07	4.14	10.51	−2.32
	Input2111	410.46	8.16	564.16	0.02	463.12	1.74	444.88	444.88	4.14	8.39	−3.94
	Input3000	606.29	0.13	681.64	0.66	628.26	0.47	615.46	615.8	5.08	1.51	−1.98
	Input3001	622.82	1.54	731.90	0.43	658.52	0.66	623.74	624.2	5.00	0.15	−5.21
	Input3010	593.64	1.03	664.03	1.03	620.59	0.68	588.77	589.08	4.77	−0.82	−5.08
	Input3011	611.74	1.03	715.38	0.24	651.30	0.54	596.48	596.54	4.77	−2.50	−8.41
	Input3100	483.88	0.02	566.00	1.39	529.43	0.70	512.37	512.52	4.14	5.89	−3.19
	Input3101	538.62	0.47	592.22	0.05	557.69	0.67	520.63	521.08	4.22	−3.34	−6.56
	Input3110	463.40	0.78	538.33	3.01	516.64	0.98	498.49	498.6	3.98	7.57	−3.49
	Input3111	522.19	0.02	573.47	0.87	548.63	0.21	504.5	504.51	4.22	−3.39	−8.04
Average		463.54	0.91	567.85	0.53	513.76	0.77	496.51	496.63	4.05	7.75	−3.17

Table 2. Comparison of the GVNS meta-heuristic and GRASP×ILS methods on large-size instances.

Category	Group	GVNS						GRASP × ILS				
		Best	Best time	Worst	Worst time	Avg.	Avg. time	Best	Avg.	Avg. time	Best gap	Avg. gap
Large	Input4000	771.70	2.26	1046.85	1.22	875.14	1.33	795.54	796.21	47.20	3.09	−9.02
	Input4001	806.57	5.26	1157.69	3.30	949.38	2.14	805.92	806.95	50.51	−0.08	−15.00
	Input4010	746.49	0.04	1022.18	1.15	838.53	2.65	775.66	775.66	47.43	3.91	−7.50
	Input4011	794.94	0.09	1094.73	0.24	913.79	0.17	787.64	787.64	49.95	−0.92	−13.81
	Input4100	713.87	6.67	969.16	0.04	796.82	4.07	733.26	733.72	39.53	2.72	−7.92
	Input4101	773.70	4.95	1044.15	0.83	866.27	2.11	745.55	745.62	45.61	−3.64	−13.93
	Input4110	696.04	2.74	919.27	0.04	756.00	1.45	711.55	711.94	42.24	2.23	−5.83
	Input4111	756.94	0.04	1010.85	0.47	841.41	0.44	727.84	728.17	46.40	−3.84	−13.46
	Input5000	996.95	0.04	1120.35	0.01	1041.02	0.48	879.49	880.01	58.60	−11.78	−15.47
	Input5001	1061.48	1.22	1184.41	0.01	1126.12	0.30	895.7	895.99	65.24	−15.62	−20.44
	Input5010	939.38	0.01	1030.35	0.86	979.83	0.47	857.12	857.27	58.51	−8.76	−12.51
	Input5011	1018.75	0.01	1195.17	0.92	1108.96	0.83	873.99	873.99	63.85	−14.21	−21.19
	Input5100	846.69	0.39	1016.96	1.20	946.00	0.42	796.05	796.88	50.55	−5.98	−15.76
	Input5101	971.59	2.93	1136.49	1.40	1051.94	1.69	812.73	813.7	57.60	−16.35	−22.65
	Input5110	777.66	3.51	991.04	1.20	915.65	2.03	776.19	777.99	50.47	−0.19	−15.03
	Input5111	926.24	0.09	1077.91	2.51	1014.72	1.42	794.32	794.78	57.92	−14.24	−21.67
	Input6000	1141.28	2.67	1434.20	0.54	1278.45	1.98	996.71	997.18	80.67	−12.67	−22.00
	Input6001	1240.46	0.08	1545.34	2.69	1423.58	1.83	1014.8	1016.1	95.32	−18.20	−28.62
	Input6010	1060.48	0.04	1411.75	1.83	1231.11	0.79	967.61	968.81	79.77	−8.76	−21.31
	Input6011	1183.71	6.01	1546.41	3.84	1367.56	2.96	987.88	989.6	88.48	−16.54	−27.64

(continued)

Table 2. (*continued*)

Category	Group	GVNS						GRASP × ILS				
		Best	Best time	Worst	Worst time	Avg.	Avg. time	Best	Avg.	Avg. time	Best gap	Avg. gap
	Input6100	1003.76	1.72	1312.36	3.78	1183.03	2.06	946.01	946.88	70.42	−5.75	−19.96
	Input6101	1154.64	5.12	1572.02	2.02	1347.67	2.32	967.12	967.23	80.98	−16.24	−28.23
	Input6110	982.54	5.62	1374.40	0.78	1199.63	1.75	919.35	920.24	70.95	−6.43	−23.29
	Input6111	1087.27	0.08	1474.89	3.99	1331.42	1.86	939.07	939.51	82.09	−13.63	−29.44
	Input7000	1040.32	0.70	1391.20	6.51	1233.43	4.12	1013.7	1013.9	188.72	−2.56	−17.80
	Input7001	1205.10	9.68	1438.02	14.36	1324.17	7.33	1019.1	1019.3	192.60	−15.44	−23.02
	Input7010	1066.37	8.63	1249.66	9.56	1169.52	7.74	988.18	989.22	185.36	−7.33	−15.42
	Input7011	1167.50	10.65	1353.56	6.25	1281.39	4.13	994.37	995.08	189.86	−14.83	−22.34
	Input7100	1011.43	0.32	1208.90	0.20	1138.85	1.34	918.64	918.91	173.97	−9.17	−19.31
	Input7101	988.39	0.18	1283.33	6.44	1175.85	3.10	924.49	924.61	180.90	−6.47	−21.37
	Input7110	1094.85	4.92	1240.76	0.38	1157.44	3.25	904.14	904.92	171.18	−17.42	−21.82
	Input7111	1051.66	9.27	1251.83	0.54	1174.22	2.21	910.1	912.05	178.19	−13.46	−22.33
	Average	971.21	3.00	1222.07	2.47	1094.97	2.21	880.62	881.25	91.91	−8.39	−18.60

5   Conclusions

In this work we have considered a MDCVRP which is a combination of multi-depot vehicle routing problem (MDVRP) and the covering salesman problem (CSP). This variant have attracted recently so much attention than the classical MDVRP problem and it is considered more challenging since it deal with some situations where it is not possible to visit all the customers with the vehicles routes. We have proposed a general variable neighborhood search approach to solve the problem. Experimental results show the efficiency of our approach to solve small-size instances but it is steel unable to solve efficiently Large-size instances.

References

1. Allahyari, S., Salari, M., Vigo, D.: A hybrid metaheuristic algorithm for the multi-depot covering tour vehicle routing problem. Eur. J. Oper. Res. **242**(3), 756–768 (2015). https://doi.org/10.1016/j.ejor.2014.10.048
2. Ha, M.H., Bostel, N., Langevin, A., Rousseau, L.: An exact algorithm and a meta-heuristic for the multi-vehicle covering tour problem with a constraint on the number of vertices. Eur. J. Oper. Res. **226**(2), 211–220 (2013). https://doi.org/10.1016/j.ejor.2012.11.012
3. Hansen, P., Mladenović, N.: An introduction to variable neighborhood search. In: Voß, S., Martello, S., Osman, I.H., Roucairol, C. (eds.) Meta-heuristics, pp. 433–458. Springer, Boston (1999). https://doi.org/10.1007/978-1-4615-5775-3_30
4. Hansen, P., Mladenović, N.: Variable neighborhood search: principles and applications. Eur. J. Oper. Res. **130**(3), 449–467 (2001). https://doi.org/10.1016/S0377-2217(00)00100-4
5. Hansen, P., Mladenović, N.: Variable neighborhood search: principles and applications. In: Glover, F., Kochenberger, G.A. (eds.) Handbook of Metaheuristics. International Series in Operations Research & Management Science, vol. 57, pp. 145–184. Springer, Boston (2003). https://doi.org/10.1007/0-306-48056-5_6
6. Kammoun, M., Derbel, H., Ratli, M., Jarboui, B.: An integration of mixed VND and VNS: the case of the multivehicle covering tour problem. ITOR **24**(3), 663–679 (2017). https://doi.org/10.1111/itor.12355
7. Pham, T.A., Ha, M.H., Nguyen, X.H.: Solving the multi-vehicle multi-covering tour problem. Comput. Oper. Res. **88**, 258–278 (2017)

Author Index

Printed in the United States
By Bookmasters